COST-BENEFIT ANALYSIS
AND
WATER POLLUTION
POLICY

CONTRIBUTORS

Abel, Fred H.
Environmental Protection Agency
Bishop, John
University of Wisconsin
Cicchetti, Charles
University of Wisconsin
Crocker, Thomas D.
University of California, Riverside
d'Arge, Ralph C.
University of California, Riverside
Fisher, Anthony C.
University of Maryland
Freeman, A. Myrick, III
Bowdoin College
Gutmanis, Ivars
National Planning Association
Hanke, Steve H.
The Johns Hopkins University
Haveman, Robert H.
University of Wisconsin
Kneese, Allen V.
Resources for the Future
Krutilla, John V.
Resources for the Future
Peskin, Henry M.
National Bureau of Economic Research
Portney, Paul R.
Resources for the Future
Rowen, Henry
Stanford University
Seskin, Eugene P.
National Bureau of Economic Research
Stockfisch, J. A.
The Rand Corporation
Tihansky, Dennis
Environmental Protection Agency
Weisbrod, Burton A.
University of Wisconsin

COST-BENEFIT ANALYSIS
AND
WATER POLLUTION
POLICY

Edited by
Henry M. Peskin and Eugene P. Seskin

Papers prepared for a symposium sponsored by the
United States Environmental Protection Agency, and
conducted in Annapolis, Maryland, September five,
six, and seven, 1973.

THE URBAN INSTITUTE ● WASHINGTON, D.C.

The preparation of the papers included in this volume was funded by the United States Environmental Protection Agency. Publication is sponsored jointly by EPA and The Urban Institute. Opinions expressed are those of the authors and do not necessarily reflect the opinions of the Environmental Protection Agency or The Urban Institute.

Published as a public service in 1975.

LC 74-82074

ISBN 87766-119-7

U.I. 169-5006-1

A/75/2M

Refer to URI 75000 when ordering

List price: $12.00

Available from:
The Urban Institute
2100 M Street, N.W.
Washington, D.C. 20037

CONTENTS

PART ONE: BENEFIT MEASURES

v

PART TWO: COST METHODOLOGY
AND MEASUREMENT

PART THREE: THE FUTURE AND UNCERTAINTY

PART FOUR: POLICY ISSUES
AND INSTITUTIONS

FOREWORD

SOCIETY HAS MADE CLEAR ITS DESIRE to improve the quality of the environment. This decision is made with full knowledge that the social benefits of improved environmental quality are obtained only at some sacrifice in the quantity or quality of other services which society provides. With many of the actions aimed at environmental improvement, it is readily apparent that the social benefits greatly exceed the social costs. With many other actions that have been proposed, however, the social desirability is much less apparent. There is a consequent need for methods by which the costs and benefits of a broad range of environmental proposals can be consistently and systematically determined.

Recognizing this, Congress, in the Federal Water Pollution Control Act Amendments of 1972, directed the Environmental Protection Agency to submit a report on the most effective practicable tools and techniques for cost-benefit analysis applicable to the abatement of water pollution. To that end, the Environmental Protection Agency invited experts on environmental cost-benefit analysis to prepare papers on the major problem areas. Most of the papers are theoretical, but comprehensive reviews of empirical studies are also included. These papers comprise the four primary divisions of the present book, for which the editors have prepared an extensive introduction in which they present a comprehensive overview of environmental cost-benefit analysis.

Many issues are discussed in these papers, including the basic theory underlying cost-benefit analysis, empirical problems, institutional problems, and the role of cost-benefit analysis in policy making. Although not all are resolved, this book will hopefully contribute to their resolution by stimulating further research on the improvement of existing techniques and methods.

RUSSELL E. TRAIN
Administrator
Environmental Protection Agency

PREFACE

IN SEPTEMBER 1973 THE ENVIRONMENTAL PROTECTION AGENCY sponsored a symposium designed to determine and advance the state of the art of cost-benefit analysis with special reference to water pollution policy. At that time a number of draft papers were presented and discussed by the conference participants. Taking account of the conference proceedings, the papers were revised and, in some cases, extensively edited. The resulting papers along with excerpts from the symposium discussions comprise the major portion of this volume. In addition, the editors have provided a comprehensive introduction and overview.

While the editors have relied heavily on the conference papers for this introductory material, other relevant subject matter has also been included. Following this introduction, the thirteen symposium papers and excerpts from the pertinent discussions are grouped under four categories, or parts. Part One concerns both theoretical and applied issues relating to the measurement of benefits. Part Two contains similar material pertaining to the assessment of costs. Part Three deals with aspects of uncertainty which are relevant to cost-benefit analysis. Finally, Part Four relates the specific technique of cost-benefit analysis to the broader area of public policy.

ACKNOWLEDGEMENTS

IN A WORK OF THIS NATURE a number of people must be cited as important contributors to the end product.

The Environmental Protection Agency not only sponsored the symposium, but also provided valuable assistance in the actual conference preparations. In addition, several members of the agency's staff, most notably Fred Abel (Conference Cochairman and EPA Coordinator) made noteworthy suggestions on the "Introduction and Overview," which served in one version as a report to Congress.

Robert Haveman also read and supplied beneficial comments on the "Introduction and Overview." The editors feel that his remarks made an especially important contribution to the final manuscript.

At the symposium itself, Ralph Turvey served as a "guest participant," adding his keenly perceptive observations to the conference proceedings.

Finally, our thanks go to Ernest Strauss for his excellent editorial contributions and to Elaine Liang for her administrative skill as Conference Secretary and her clerical skill in preparing the final manuscript.

HENRY M. PESKIN
EUGENE P. SESKIN

COST-BENEFIT ANALYSIS AND WATER POLLUTION POLICY

Introduction and Overview

HENRY M. PESKIN
and
EUGENE P. SESKIN

*Basic Concepts Used in
Cost-Benefit Analysis*

COST-BENEFIT ANALYSIS HAS BECOME an increasingly familiar term in both public and private circles, and one might expect it to have a precise and generally accepted meaning. Instead, the basic concepts used in cost-benefit analysis are the center of much confusion, especially among many policy makers. There is no general agreement on definitions, nor is there concensus on what the scope of analysis is or should be. The intent of this section, therefore, is to define three fundamental terms as they are used in this book—in particular, to examine the concepts of *efficiency, benefit,* and *cost.*

Cost-benefit analysis is a formal procedure for comparing the costs and benefits of alternative policies.[1] It differs from more informal comparisons of costs and benefits in two principal ways. First, the terms *cost* and *benefit* are defined more narrowly than in general English usage. Second, the formal procedure and basis of comparison rely on specialized techniques and principles, most of which are derived from economic theory.[2] These special definitions and procedures will be discussed in more detail below.

A less formal comparison of costs and benefits that usually includes distributional and political effects is sometimes termed *policy evaluation* to distinguish it from the formal cost-benefit analysis.[3] While the relation-

---

[1]We wish to stress the word *policies* because traditionally cost-benefit analysis has been most often applied to large public investment projects. It is equally relevant in assessing the consequences of alternative public policies. For example, a particular water pollution problem might be handled either by a system of taxes or by direct regulation. Neither of these options might involve significant public investments, but a cost-benefit comparison might be extremely useful in making the final choice between the two alternatives.

[2]There are many assumptions underlying the application of this theory. See the paper by *Haveman and Weisbrod* and the paper by *Freeman* for a discussion of most of the assumptions. NOTE: Throughout the "Introduction and Overview," papers included in this volume are signaled by the use of italics for the names of the respective authors.

[3]Actually, policy evaluation analysis could also become rather formal. Such a formal approach is suggested by Dorfman and Jacoby [5].

1

ship between these two concepts will be discussed in more detail later,[4] it is generally accepted that cost-benefit analysis is a component of policy evaluation.

A related activity, *cost-effectiveness analysis,* will not be discussed in this introduction and overview. Cost-effectiveness analysis estimates the costs of alternative methods to achieve a *given* policy objective. Since the objective is taken as given, it is not necessary to compare it with other policy objectives. Thus, a crucial aspect of cost-benefit analysis, the need to measure the value of the benefits in commensurate dollar units, is avoided.

The Concept of Efficiency

Cost-benefit analysis is concerned with analyzing public decisions on the use of the economy's resources. As such, its purpose is to aid in the fundamental economic task of allocating scarce resources to alternative uses. Thus, its primary goal is one of achieving *allocative efficiency,* where

> Allocative efficiency as an economic goal reflects the fact that it is sometimes possible to reallocate resources—perhaps increasing or decreasing the amount of resources used for water pollution control—in ways which will bring about an increase in the net value of output produced by those resources. [*Haveman and Weisbrod, p. 38*]

In other words, an improvement in economic efficiency is attainable in the economy, if it is possible to increase the value of the economy's output for any given amount of resource input.

Analysts often recognize that policy makers must evaluate alternative projects in terms of other goals, e.g., equity and political.[5] However, these other goals are irrelevant to the principal objectives of a formal cost-benefit analysis. It should be noted that the analyst undertaking a formal cost-benefit analysis often generates useful information relating to these broader policy issues in the course of his work.

The Concept of Benefits

One step in evaluating a project's efficiency goals is to enumerate the relevant benefits which would stem from the project's implementation. By *relevant* we mean only those benefits which represent an improvement in the allocation of society's resources.

Examples of benefits stemming from a project which enhances water quality might include: improvements in human health; reductions

---

[4]See pp. 26-32.

[5]The goal of equity must be defined with respect to some concept of "fairness." Often the term refers to distributional equity; e.g., whether the distribution of income resulting from a project is more or less "fair" than it was prior to the project. Since fairness is a social as well as a moral issue, equity and political goals may overlap.

in the water treatment costs for industrial water users; increases in sport and commercial fishery yields (for given levels of labor and capital); improvements in water recreation opportunities; reductions in household costs which are associated with water hardness; as well as increases in those aesthetic values of water which are based on appearance, taste, and odor.[6]

It is useful to distinguish between *private* or *individual benefits* which each individual can choose to enjoy in varying amounts, and *collective benefits* which affect a number of individuals simultaneously and for which each individual has little choice over the amounts he consumes. For example, the irrigation water a number of farms receive, *in principle,* can be marketed on an individual basis. However, for many goods and services associated with high water quality (such as the aesthetic enjoyment provided by an unpolluted river), there is no way to parcel out the benefits individually. Therefore, their allocation cannot be handled in the marketplace.[7]

The distinction between private and collective benefits becomes crucial when the analyst attempts to measure them in terms of the value placed on them by the affected individuals. For private benefits, this can usually be accomplished by looking at the market prices for the flow of goods and services from the project. However, since collective benefits are seldom reflected in the marketplace, the analyst must infer the amount the individuals would be willing to pay. In essence, the analyst must devise a hypothetical experiment in which the affected parties are asked what they would be willing to pay for the benefits that are thought to be forthcoming from a proposed project.[8]

Thus, for purposes of a cost-benefit comparison, the *benefits* of a project are defined by how much people would be willing to pay for project outputs.[9] The actual empirical measure of *willingness to pay* is very much dependent on the distribution of income and wealth among the beneficiaries.[10] In performing a cost-benefit analysis, the distribution of income and wealth must be taken as given. Otherwise the measure of benefits will not be unique. Thus, a formal cost-benefit analysis is restricted to considerations of attaining allocative efficiency.

[6]For more details on man's uses of water, see *Freeman, pp. 70-73.*

[7]In such cases, we usually refer to the goods in question as *public goods.* In practice, however, many seemingly public goods, such as clean air, are somewhat "private" in character, since an individual can control his consumption by his choice of domicile. Perhaps quasi-public would be a more accurate term. For more on this distinction, see Morhing and Boyd [18].

[8]It should be pointed out that whether people actually make any payments does not change the value of the social benefits of the project.

[9]Note that this definition of benefits is far more restrictive than the usual definition of the word. For example, a benefit is defined by Webster as "something that promotes well-being."

[10]This dependency is further discussed below, pp. 28-29.

The Concept of Costs

In ordinary usage, *cost* usually means the monetary expenditure required to purchase a good or service, e.g., the money outlay for pollution control. To economists, however, "*cost* means 'opportunities foregone' "[*Kneese, p.176*]. Thus, this cost concept is termed *opportunity cost* and refers to the value foregone of employing a resource in one activity rather than in its next best alternative use. For example, when a water pollution control program is undertaken, resources (such as the labor and materials needed to construct a sewage disposal plant) must be diverted from other uses. Consequently, production and consumption may be reduced elsewhere in the economy. The value of these foregone opportunities is the opportunity cost of the pollution control program.

The distinction between simple monetary outlays and true opportunity costs must be kept in mind by the cost-benefit analyst when evaluating projects. For reasons to be discussed below, market costs (prices) may diverge substantially from opportunity costs, and the direction of such differences is not always apparent without careful analysis.

Issues in
the Assessment of Benefits

A PUBLIC POLICY INEVITABLY RAISES certain issues which are of major concern to those who may be affected by that policy. It is, therefore, essential for the analyst to understand what the issues and their consequences are so that he can conduct an effective cost-benefit analysis and communicate its results. The purpose of this section is to clarify and contrast several of the most important issues associated with the assessment of benefits.

Allocative vs. Distributional Effects

The dichotomy between efficiency criteria and equity criteria carries over to the distinctions made between types of benefits resulting from a project. One classification involves *allocative (e.g., real)* versus *distributional (e.g., pecuniary)* effects. Allocative benefits refer to those effects of a project which are characterized by increases in consumer satisfaction or decreases in the amount (cost) of resources required to produce goods and services. For example, an improvement in the water quality of a lake may be reflected by the amount people are willing to pay for the additional enjoyment of fishing, swimming, and boating.

Distributional effects, on the other hand, refer to changes in some people's well-being at the expense of the well-being of others. For example, improvement in the water quality of the lake and an accompanying increase in fishing might bid up the price of fishing tackle. The sellers of tackle would experience increased profits, but it would represent a dollar-for-dollar loss to the purchasers of the tackle. Hence, the effect would amount to an income transfer from the purchasers to the sellers of fishing tackle.

It is generally accepted that in a cost-benefit analysis, the allocative benefits should be valued whenever possible, while the strictly distributional effects should be omitted from any valuation.[11] The cost-benefit analyst should not assume the role of judging the dollar gains of one group versus the dollar losses to another group.

It should be mentioned that there have been some recent attempts to integrate distributional considerations into the formal cost-benefit framework.[12] However, there is by no means a general concensus on this matter.

Primary vs. Secondary Effects

The distinction between allocative and distributional effects is closely related to the classification of *primary* versus *secondary benefits*.[13] Increases in well-being resulting from a project are usually taken as the primary (or direct) benefits, while *demand-inducing effects* (usually in the form of monetary payments) and other effects generated by (or stemming from) the direct outputs are considered secondary (or indirect) effects.

In the water control example above, the willingness to pay for the improved water quality of the lake would comprise the primary benefit, while the increased profit accruing to the tackle sellers would be a demand-induced secondary benefit. If the improved condition of the lake also resulted in increased lodging and eating facilities, the net profits from these facilities would constitute "stemming" secondary benefits.

While, for reasons already discussed, the distributional issues surrounding changes in the tackle market cannot be formally evaluated in a cost-benefit analysis, it is less clear how the analyst should handle the other activities stemming from the project.

The answer depends on the employment conditions in the economy. If there is relatively full employment, any resources used for hotel and restaurant construction around the lake would be diverted from other productive activities. Hence, overall national benefits associated with the employment shift would be essentially unchanged.

---

[11]We, again, emphasize that a listing of such effects would be useful to decision makers for purposes of policy evaluation.

[12]See, for example, the paper by Hochman and Rodgers [11] and the subsequent comment by Mishan [17] for an interesting debate on the merits of this effort.

[13]It is also closely related to the classification of *internal* versus *external effects*. This distinction depends on project boundaries which will be discussed further below.

Thus, these secondary effects should not enter the picture.[14] However, if significant unemployment exists (or, if there are important obstacles to resource mobility) then secondary effects among projects may differ substantially in terms of foregone opportunities, and attempts should be made to account for this. Thus, if the increased numbers of hotels and restaurants make use of inputs (waiters, cooks, etc.) that would have been otherwise unemployed, the analyst, should, if possible, enter the value of the newly employed resources into his calculations.[15]

Tangible vs. Intangible Effects

Thus far, the valuation of the benefits resulting from a project has been discussed in terms of the affected parties' willingness to pay. Implicit in this notion is the fact that such benefits can be measured in dollar terms, i.e., that they are *tangible benefits*. On the other hand, benefits which cannot be valued in monetary terms are categorized as *intangible benefits*.

According to *Haveman and Weisbrod*, classifying benefits as intangibles does not mean that it is impossible to conceptualize a monetary value for them. Instead, it implies that available data and empirical methods have not permitted such imputation. This view is also shared by *Bishop and Cicchetti*. However, there is some debate as to whether unique monetary values for all project effects can be conceptualized. While techniques have been suggested for assigning monetary values to such intangibles as the aesthetic quality of a clean lake, there is some concern that these assigned values can be misleadingly arbitrary.

In any event, failure to monetize certain project effects does not mean that they should be or will be excluded in decision making. While the cost-benefit analyst should not assume the role of placing arbitrary monetary values on such intangibles,[16] he should describe such unmeasured effects as quantitatively as possible. Such quantitative information is a significant by-product of a cost-benefit analysis, since there will always exist intangibles which a decision maker must consider.

Partial vs. General Analysis

Empirical studies of benefit estimation often make two basic assumptions: "The first is that all prices other than for the good in question are held constant. . . . The second assumption is that the benefits

---

[14]However, if a case can be made that the shifted resources are somehow more productive in their new activities, this gain in productivity should be counted as a benefit. In practice, however, it is difficult to prove that such a gain has occurred.

[15]A method for adjusting cost-benefit ratios to account for unemployment is discussed by Haveman [8].

[16]As Mishan [16] points out, no one, including the policy maker, should place arbitrary monetary values on intangibles and then compare these to tangible values. To do so, subverts the logical principles of cost-benefit analysis.

do not change the real income of the beneficiary" [*Freeman, pp. 74-75*]. These two assumptions substantially simplify the task of assessing the affected parties' willingness to pay (and also simplify the job of estimating project costs). In short, it is usually more convenient and practical to choose a *partial* method, as opposed to a *general* approach that allows for changes in prices and real incomes.

The primary difference in these two techniques is that when partial analysis is employed, the analyst concentrates on the close interdependence between a few factors, while in a general framework, the analyst attempts to account for the fact that everything depends on everything else. In a practical sense there is a parallelism between this distinction and the previously discussed distinction between primary and secondary effects: partial analysis generally neglects secondary effects.

These distinctions are, in large part, a matter of degree. The improvement in water quality constitutes the basic change in the example above. Its consequences on the markets for other related goods and services comprise effects which may or may not be included in a cost-benefit analysis. The decision is usually made on the basis of whether the project is expected to have large impacts on particular markets. Thus, one might argue that the secondary impact on the tackle market can be safely ignored in evaluating the project's costs and benefits because these impacts would, most likely, be negligible. Of course, whether this is true is an empirical matter and ultimately depends on the project being analyzed.

Implementation of Benefit Assessments

THE MOST CRITICAL PROBLEM IN ASSESSING BENEFITS is the lack of information and the scarcity of data—possibly because in some cases researchers do not know what to look for or where to find it. At the same time, many techniques for benefit assessment do exist but, since they entail certain limiting assumptions, they cannot be applied indiscriminately without endangering the validity of the analysis. The first part of this section discusses the kinds of information needed by the analyst; the second part clarifies the conditions under which a variety of approaches to benefit assessment might be used.

Informational Requirements in Benefit Estimation[17]

A critical prerequisite to any cost-benefit analysis is the availability of relevant data. In the water pollution area, this is a matter of particular

[17]Many of these issues are relevant to the cost side of the picture; we have somewhat arbitrarily chosen to place them in this section.

concern. There are various types of data needed for an analysis of the benefits (and costs) associated with changes in environmental characteristics. Indeed, the analyst faces a multifaceted problem.

The Relationship between Physical Attributes (Pollution Levels) and Environmental Quality

The primary goals of a water pollution control program are to achieve certain levels of "environmental quality." Thus, the first type of necessary information involves the relationship between physical attributes of the environment (e.g., pollution levels) and characteristics defining environmental quality (e.g., the number of lost recreation days, the damage to human health, or the negative effects on manufacturing).

Some of these relationships are more behavioral than others (e.g., the association between pollution and recreation) and are especially difficult to determine. Others tend to be more physical (e.g., the association between pollution and a production process), and information can be gathered by carefully controlled experimentation and technical analysis. However, even in the case of the more physical relationships, existing knowledge is quite limited. This point is emphasized in *Freeman's* discussion of the health effects of water pollution [*Freeman, pp. 95-97*].

Another point which should be noted is that, underlying the determination of these relationships, is the assumption that both the physical attributes and the measures of environmental quality have been quantified. This, in itself, may be a formidable task as evidenced by the techniques discussed below (see pp. 10-13).

Transfer Functions

The usual objective of a water pollution control program is to attain ambient concentrations of pollutants which are consistent with a desired level of environmental quality. In practice, however, despite standards which are based upon ambient concentrations, most control programs must be designed in terms of monitoring pollutant discharges. Thus, the second type of information needed concerns the quantitative link between residual discharges and ambient concentrations of pollutants, a purely technical issue.

The formal mathematical relationships associating residual discharges with ambient water pollution levels are called *transfer functions*.[18] Basically, these are models which describe the fluid-flow characteristics of the receiving waters and the physical, chemical, and biological reactions taking place in these waters. Although the models are often both difficult to formulate and to operate (requiring in many cases the use of

---

[18]For a detailed discussion of transfer functions, see Appendix I of *Kneese's* paper.

large-scale computers), some transfer-function concept is essential if effective controls are to be evaluated.

Distinguishing between Natural and Man-Made Sources of Pollution

A topic related to the understanding of transfer functions is raised in the paper by *Tihansky*. As he notes, it is often impossible (with current technology) to distinguish man-made from naturally-occurring pollution. Thus, a third type of essential information pertains to distinguishing between sources of pollution. If one cannot determine the degree to which a given pollutant is occurring naturally, it becomes impossible to assess the real economic impact of any proposed changes in the level of emissions.

For example, consider expenditures on water softeners. While these devices remove dissolved solids, it is difficult to gauge what portion of the solids come from natural vs. man-made sources. Consequently, if a project which is being analyzed only reduces the emissions of dissolved solids from man-made sources, it may not be possible to estimate accurately any associated decrease in consumer spending for water softeners.

Experimentation in the Water Pollution Area

As we have noted above, there is a dearth of information concerning the physical effects of water pollution. For example, the presence of coliform bacteria in water has been used in the past by public health officials as an index of water quality. However, there is no evidence of a *direct* relationship between coliform bacteria and dangerous pathogens.[19] If the costs of coliform control are to be compared to possible benefits from improved health, this relationship must be determined.

Controlled laboratory experimentation can be expected to fill such data gaps. However, the fact that harmful physical effects may result only after prolonged exposure to pollution may make experimental analysis quite costly. In addition, certain researchers feel that accepted experimental techniques are questionable in investigating some of these relationships.[20] For example, the attempt to simulate human response to water pollution using laboratory animals may not be persuasive, since casual observation suggests that water "unfit" for human consumption often has no ill effect on animals used in laboratory research.

[19]Nevertheless, the existence of fecal coliforms may indicate the existence of pathogenic organisms of a fecal origin. For a defense of the use of coliforms as an index of water quality, see Geldreich [6, 7].

[20]On this point, see the discussion by Lave and Seskin [13] on the usefulness of laboratory experimentation in investigating the association between air pollution and health.

Suggested Approaches to Benefit Estimation[21]

When a Project Affects a Production Input

As a general rule, if a project lowers production costs, there are two types of beneficiaries. First, benefits accrue to consumers of the final product to the extent that the decreased costs lead to a reduction in output price. Second, under certain circumstances, benefits accrue to factors of production in the form of higher returns.

To evaluate these effects generally requires a complex analysis of the supply and demand conditions facing the affected firms. However, in certain specialized circumstances, the analysis can be greatly simplified. If the affected firm is small relative to the market and it is reasonable to assume that output prices, costs of labor and materials, and the stock of industry capital are unaffected, then any increase in the firm's income can be used as a measure of the project benefits.

According to *Freeman*, this practice has been applied in cost-benefit analyses of irrigation projects when the major benefits accrued to agriculture. Providing the affected farms' outputs constitute only a small fraction of the total supply, the application of this approach is partially justified. However, the assumption of fixed capital, even in agriculture, is hard to accept. Given sufficient time, most industries can adjust all factors of production to meet new output demands. Thus, this approach might serve only to estimate short-term benefits, since the increase in income to any given firm may be completely eroded in time.

When a Project Affects a Complement to Another Good

If a project affects a public good which can be considered a complement[22] to a private good in consumption, the public benefits can be measured from information on the demand for the private good. For example, water quality can be considered a complementary public good to a beach area. Hence, in theory, the benefits from improvements in water quality can be measured by estimating the effects on the demand for the recreation site.

One approach which has been used to estimate the demand for recreational areas is the Clawson-Knetsch travel cost method.[23] Briefly, this technique estimates the effect which an incremental increase in the

---

[21]Much of this discussion is based on *Freeman's* paper.

[22]Loosely defined, two goods are complements (or complementary) if the demand for one affects the desire for the other in the same direction. Shoes and shoelaces are examples.

[23]For discussions of the Clawson-Knetsch method, see Berry [4] and *Freeman, pp. 87-91.*

cost of travel will have on the number of visits to a recreational site, and then uses the estimate to derive the demand curve for the site.[24]

This procedure has been criticized for a number of reasons, one being its assumption that there are no substitute sites. In many urban areas this assumption is simply not valid; hence, the technique becomes inapplicable. Another difficulty arises with respect to *nonuser benefits*.[25] These are benefits which individuals derive from the existence of the site (or improvement in it), despite the fact that they may not actually visit it. In the Clawson-Knetsch method, such benefits necessarily go unmeasured. A more fundamental criticism has been raised regarding the valuation of leisure time by using travel costs. The basic argument is that most activities involve the "consumption" of time and money in fixed proportions; hence, the value of time cannot be isolated. Thus, if one wants to value the time people consume in getting to fishing grounds, alternative routes involving different time and money-cost combinations must be examined. The value of time obtained from studying commuter behavior is irrelevant to the valuation of time spent on a fishing trip; i.e., the minute which the commuter might save in going to work on Wednesday does not give him an extra minute for fishing on Sunday.

When a Project Affects a Characteristic of Land

Environmental characteristics such as air and water quality may affect the productivity of land either as an input to commercial production or as a direct input to private consumption (e.g., for residential use). In such cases, land-value information is sometimes used to measure benefits due to changes in environmental traits.[26]

Where land is used primarily as a factor of production, this is a reasonable procedure if there are freely competitive land markets insuring that differences in productivity are reflected in the structure of land rents. However, where the principal use of land is as a consumer good, several difficulties may arise. Most of these stem from the fact that productivity is not uniquely defined in this case. A parcel of land usually has many attributes (only some which are environmental in nature). Hence, it is difficult to associate differences in land prices with any particular attribute.

Freeman believes that a land-value approach to benefit assessment is "of limited practical significance" [*Freeman, p. 82*]. He argues that using

---

[24]Two underlying assumptions are that the primary purpose of the trip is to visit the site and that there are no close substitutes for the site within the area being studied.

[25]See the discussion of option value on pp. 23-24.

[26]A practical problem surrounds the actual measurement of land value. Since transactions are relatively infrequent and often inaccurately recorded, sale price is often unsatisfactory. Other possibilities (e.g., independent appraisals, owner self-reporting, and tax assessments) are subject to error and potential biases.

changes in land rents as a measure of benefits requires that everyone share the same tastes for the attributes (i.e., they all have identical willingness-to-pay functions). This assumption, along with the assumption of competitive land markets is not very realistic. However, in some cases, land-rent differentials may offer the only indication of certain types of benefits derived from control programs.

When a Project Reduces the Level of Defensive Outlays

When the environment is degraded by pollution, it is not uncommon for households and industries to make certain expenditures in an effort to offset the adverse effects of the pollution. Painting a house more frequently than would be desired in the absence of pollution is an example of this type of *defensive outlay*. If a project results in a reduction of such defensive expenditures, the decrease can serve as a measure of project benefits.[27]

However, the analyst should be aware of a practical difficulty with this approach. Identifying those expenditures which are truly "defensive" is seldom an easy task. For example, to determine the extent to which home air conditioning represents a defense against polluted air versus a defense against humidity and high temperatures, would require knowledge about the motives behind each purchase of an air conditioner. Unfortunately, this type of information is rarely available to the analyst.[28]

When a Project Improvement Represents a Public Good and the Previous Techniques Are Inappropriate

If a project results in an improvement in environmental quality which can be considered a public good, "the possibility of simply asking people about their willingness to pay should be investigated" [*Freeman, p. 93*]. The major problem with an interview technique is inducing individuals to reveal their true willingness to pay. Generally, if the interviewees feel that they may be taxed in proportion to their expressed willingness to pay, they may be tempted to underestimate their willingness to pay. On the other hand, if they feel that their tax share will inevitably be lower than the benefits they will derive from the project, they may overstate their willingness to pay in order to promote affirmative action.[29]

---

[27]See *Freeman, p. 91.*

[28]In addition, *Freeman* points out that in some cases there is displeasure associated with pollution which cannot be fully alleviated by defensive expenditures. In such instances, reductions in defensive outlays will result in an underestimate of the "true" benefits of pollution control.

[29]This difficulty in ascertaining the true willingness to pay for public projects is known as the "free-rider problem."

Various strategies have been proposed for minimizing biased responses.[30] Most of these involve carefully designed questionnaires that employ different types of taxing schemes. However, an underlying problem with using any interview technique to value potential improvements in water quality is that people often find it difficult to visualize the personal implications of hypothetical changes in an environmental parameter.

Issues in
the Assessment of Costs

IN A MATTER AS COMPLEX AS ASSESSING the true opportunity costs of water pollution control, the need for understanding some of the underlying analytical problems cannot be overemphasized. The two major issues examined in the following pages are: (1) whether the appropriate assessment of costs can be accomplished through a "partial" analysis or whether it requires a systems approach, and (2) is it appropriate to estimate costs on a short-run basis, or is the long-run cost estimate essential.

Partial vs. System Costs

One method of assessing the costs of a water pollution control program would be to cost all the *separate* control activities of each "actor" (industrial establishments, households, and governmental enterprises) that contribute to the pollution reduction. These separate control costs could then be summed to determine total program cost. Neglecting for the moment the practical merits of this approach, the following considerations argue for a more simultaneous assessment of costs—a *systems* approach rather than the *partial* approach usually employed for benefit analysis.

Technical Interdependency

In principle, if there are several possible control strategies for each relevant actor in the system and if the choices among strategies are independent of the choices of other actors, a partial analysis raises no special problems. Each actor can be analyzed separately and in each case the least-cost strategy can be determined. However, in practice, the choice of a control strategy is often dependent upon the choices made by others in the system. For example, whether a manufacturing plant dilutes its wastes or impounds them depends on the availability of unpolluted influent. However, the accessibility of clean influent depends, in

---

[30]For a discussion of some of these, see *Freeman, pp. 93-95.*

turn, on the waste disposal practices of possible polluters upstream. Such technical interdependencies necessitate a systems approach.

Cost Interdependencies

A frequent assumption made in partial analyses is that the prices of inputs to producers remain unaffected by whatever control decisions are made. In many circumstances, this may be a reasonable assumption. For example, if the major source of water pollution in a particular region is a single firm, the prices of pollution control inputs under a wide choice of abatement strategies would probably be fairly insensitive to the particular alternative chosen. This follows from the fact that a single firm (even moderately large) would not be apt to affect substantially the supply prices of those materials and equipment purchased for the cleanup (provided that these inputs were not very specialized).

However, if many firms are required to control their water pollution emissions, it is no longer reasonable to assume that these prices will remain fixed. Depending upon supply conditions, the prices of control equipment may rise, fall, or remain constant. Consequently, an apparent least-cost control strategy under a partial analysis may no longer constitute the least-cost strategy when changes in input prices are accounted for. Thus, costing major pollution control programs again suggests the desirability of a systems approach.

Demand Effects

As has been pointed out by *Kneese*, often the "best" (least-cost) control strategy for a firm involves some change in the firm's final products. This change may involve an alteration in the quality of a product (durability, styling, packaging), a reduction in the quantity of output produced, a modification in product mix, etc.

In order to determine the full opportunity cost associated with a given control strategy, some measure of the costs incurred by users (or final consumers) because of the firm's altered output should be added to any increases in manufacturing costs. If, for example, factory A in response to a pollution control program limits the production of an input to factory B, B's costs could be expected to increase. To the extent that B's increased costs are not offset by increased profits to A, the costs should be considered part of the opportunity cost associated with the control strategy.[31]

Unfortunately, the opportunity costs associated with product changes that result from control programs cannot be analyzed using a partial approach. The ability of the users of the products to substitute

---

[31]Another example would be the reduction in consumer satisfaction if the quality of a consumer good fell in response to a control strategy. This loss in consumer satisfaction should similarly be included as part of the opportunity cost.

other goods from other suppliers, again, necessitates a systems approach.

Intermedia Effects

It is quite possible that a particular strategy designed to clean up one facet of the environment results in polluting another. Often, these intermedia effects are not visible in the sector for which pollution control has been initiated.[32] For example, the effect which a shift to low-sulfur fuels has on strip mining is difficult to detect. The linkages between fuel-input substitution and strip mining are indirect and depend upon complex relationships determining the most economical fuel substitutes. Clearly a systems analysis is needed to deal with these complexities.[33]

Initial vs. Fully Adjusted Unit Costs

Input prices for a control project may reflect adjustments, depending upon whether or not the supply of the factors of production have fully adjusted to the demands of the project. For example, a proposed pollution control strategy may require specialized labor skills that are currently in short supply. Accordingly, the initial labor costs might be considerable, since a high wage rate would be required to employ the necessary workers. However, the very fact that the wage rate would be high attracts new labor into this skill category. Subsequently, the increase in labor supply would result in a fall of the wage rate. Thus, the fully-adjusted labor costs should be lower than the initial labor costs.

Which estimate—the current input price or the generally lower fully-adjusted price—is the appropriate one for cost-benefit calculations? As Kneese and Bower [12] point out, only the use of fully-adjusted input prices is consistent with the goal of assessing the true opportunity cost of the project. Since a cost-benefit analysis should attempt to ascertain the total "opportunities foregone" if a project is realized, all resource adjustments engendered by the project should be captured in the cost estimates used. Thus, the analyst should endeavor to employ only his best estimate of fully-adjusted input prices and unit costs for cost-benefit comparisons.[34]

Of course, the attempt to estimate opportunity cost with complete precision may not be successful since forecasting ultimate adjustments is quite difficult. However, even if one is forced by practical considerations

---

[32]The reader is referred to the discussion of spillovers on pp. 27-28.

[33]Of course, in many cases a partial analysis is sufficient to deal with spillovers *within* a given sector. For example, it is fairly easy to detect the spillover effect when a firm chooses to burn a residual that was previously dumped into a river.

[34]Kneese and Bower [12] also point out, however, that current costs are relevant for managerial decisions involved in the day-to-day operation of the project, whether they correspond to markets in fully-adjusted (long-run) equilibrium or not.

to rely on initial costs, a qualitative appraisal of the fully-adjusted prices should be included in the analysis to give additional perspective to the resulting cost-benefit comparison.

Estimating
Control Costs

THE PURPOSE OF THIS SECTION is to highlight some of the major empirical problems faced in estimating the costs of reducing water pollution. Should pollution control focus on one pollutant at a time or on groups of several pollutants? How does one analyze costs when pollution control is an integral part of the production process? In what way should control costs be measured if pollutants represent commercially viable products? How can the costs of control be disentangled and allocated if they are associated with joint reductions of more than one pollutant? What implications does regional pollution control have on highly interdependent areas? The following discussion centers on these and related issues and deals with the pros and cons of various approaches used to determine the costs of pollution abatement.

General Considerations in the
Estimation of Costs

Pollutant-Specific Analysis

While one can speak loosely of the cost of reducing "pollution," cost-benefit analysis usually requires estimates of the cost of reducing specific pollutants. This involves the nontrivial task of making cost estimates pollutant-specific. A number of difficulties arise with regard to this procedure. First, to cost out the complete list of all affected pollutants, defined according to their physical and chemical properties, can be quite laborious, even when one is considering a single pollution source, such as a factory. In practice, such a detailed list is usually far too long for a realistic costing effort.

Second, even if a detailed list of individual pollutants and the associated costs of reduction could be generated, it is not clear that the information would be comparable to data on the benefits of reducing the pollutants. If the benefits of reducing C were known, (where C is some simple aggregation of individual pollutants A and B, e.g., $C = A + B$) it would be relatively easy to compare the costs of reducing units of C (equal to the cost of reducing A plus the cost of reducing B) with the benefits.[35] However, if C was not a simple aggregation of the individual

---

[35] This statement assumes that A and B can be reduced independently of each other. This is not always possible. See the discussion below concerning the problem of joint reduction of more than one pollutant.

pollutants the comparison would be more difficult. For example, *A* might represent wood fiber, *B* might represent soluble organic waste, and *C* might represent BOD$_5$ (biochemical oxygen demand over 5 days). Then, knowledge of the cost of reducing the components of BOD would provide little information to assess the costs of reducing BOD itself.[36]

Thus, to assure comparable cost and benefit data, it is not necessarily optimal to generate pollutant-specific information. Instead, prior to undertaking the analysis, the analyst should develop compatible classifications for the pollutants being controlled and the pollutants causing "damages."

"Jointness" of Production

Although it is widely recognized that pollution emissions often increase with the expansion of production activities, the implications of this technological relationship are not always taken into account in estimating the cost of pollution reduction. If pollution is a joint product with normal production outputs, decreases in pollution will be accompanied by decreases in production outputs unless some other alteration in the manufacturing process takes place, e.g., the introduction of new technology. Too often, cost estimates of pollution control have been predicated on the assumption that the least-cost strategy will reduce pollution but will somehow leave all other aspects of production unaltered.

For example, the most simple "end-of-pipe" strategy, such as the installation of a filter or grate to trap solids before they enter a river, can significantly alter the engineering characteristics of the production process by altering pressures and flow rates. Hence, in principle, the opportunity cost of the strategy should account for the direct cost of the "end-of-pipe" pollution control equipment,[37] and/or any costs associated with production changes (which could include either the costs of substituting more expensive inputs, or a decline in profits due to a drop in the sales of certain products).[38]

Pollutants as Secondary Products

In addition to the above costing problems, there is another related point, namely, that today's pollutants may be tomorrow's commercially

---

[36]While it is true that both wood fiber and organic waste have BOD$_5$ equivalents, these equivalents must be determined empirically. The combination of two separate BOD equivalents is not merely the sum of those two measures.

[37]This is the basis for the most commonly provided cost estimates.

[38]It should be noted that these production changes could also lower certain costs to the producer. For example, pollution control equipment may require other, more modern, capital equipment. This, in turn, may lead to increased production and revenue which would offset the control costs.

sold, secondary products.[39] Presumably, many potential secondary products are presently being dumped as residuals because collection and marketing costs would exceed the revenue from sales. However, even under these circumstances, it is possible that a producer required to reduce his pollutant emissions would sell the residuals at a loss if this represented the most economical (i.e., the least-cost) control strategy. Unfortunately, it is difficult to ascertain the relevant opportunity cost. If it is estimated by asking the producer, he may report only collection and marketing costs, since it is unlikely that he will regard the new source of revenue as a negative cost.[40] Thus, the analyst must be wary of overestimating the opportunity costs associated with strategies involving secondary products.

Joint Reduction of Pollutants

It is quite likely that for technological reasons, a control method to reduce a particular pollutant will also result in the reduction of other pollutants, generally by differing percentages. This can lead to difficulties in making cost-benefit comparisons.

Suppose, for example, in controlling BOD, suspended solids are also reduced. More specifically, assume that it costs ten dollars to reduce a combination of one pound of BOD and two pounds of suspended solids. Furthermore, assume the policy maker is interested only in a program designed to decrease BOD (implicitly assuming that no benefits emanate from a reduction of suspended solids) and he knows that reducing each pound of BOD generates a benefit valued at eight dollars.[41] Given these assumptions, there may be a temptation on the part of the policy maker to prorate the ten dollar control cost between the two pollutants. For example, the costs might be assessed according to their weight ratio; i.e., the "cost" of reducing one pound of BOD would be assigned a value of $3.33, and the result would be a favorable cost-benefit ratio. However, this method of estimating costs should be avoided, since it takes the full ten dollars to reduce each pound of BOD. If the correct cost figure is used, the BOD reduction program will not pass the cost-benefit test.

However, should the analyst be interested in a program that requires the removal of some other combination of BOD and suspended solids, he faces a rather severe analytical problem. If, by assumption, he knows only the cost of removing these pollutants in the fixed ratio of one

[39]In large processing industries, it is also possible that the residuals may be profitably used in the actual production process; e.g., a combustible gas which was previously emitted may be recycled and used as a substitute for other fuel inputs.

[40]Other problems associated with survey methods to determine control costs are discussed below.

[41]Note, in this simple example, both costs and benefits are assumed to remain constant with respect to changes in pollution level.

pound of BOD to two pounds of suspended solids, there is no way he can use this information to deduce the cost of another combination, such as two pounds of BOD to two pounds of suspended solids. Of course, he may estimate the cost by using the $3.33 per-pound prorated figures, but this procedure has no analytical basis. As long as technology results in the joint fixed-ratio reduction, there are no shortcuts for determining costs of individual reductions. Cost functions that relate costs to a wide range of different ratios of pollution reduction must be developed.

Generalizing from Specific Cases

Often a pollution control program may be regional or national in scope and, consequently, may apply to a large number of actual or potential polluters. In such cases, it is unrealistic to analyze the costs of control for each and every polluter. Therefore, two basic approaches have been adopted to handle this problem: (1) generalization—or extrapolation—from specific studies of individual polluters; and, (2) surveys of individual polluters. We will first mention problems of generalization and will follow with a discussion of surveys.

One of the disturbing aspects of extrapolations made in the pollution area, regardless of the methods used,[42] is that the usual aggregate is the "industry," where an industry is defined according to the Standard Industrial Classification (SIC).[43] There is, however, no inherent reason why the SIC, which is primarily based upon product classification, is the relevant aggregate for cost-benefit comparisons.

For example, although one might be interested in contrasting the costs and benefits of "cleaning up" the primary metal industry, it is more likely that one would be concerned with comparing the costs and benefits of reducing particular pollutants such as BOD emissions, regardless of industrial source. In this latter instance, cost estimates based upon a grouping of BOD-generating processes would be more useful than cost estimates based upon an SIC-defined aggregate. Thus, extensive efforts to produce cost estimates by standard definitions of industries may be a case of misplaced emphasis.

Another difficulty with generalizations (whatever the aggregation principle chosen) is that they usually require an assumption of homogeneity. That is, they assume that each member of the group experiences "similar" costs in reducing pollution discharges. Implicit in this assumption, is that each member of the group uses approximately the same control technology.[44] Unfortunately, emphasis on technological similarities may also be misplaced.

[42]Some of these techniques are described in detail by *Hanke and Gutmanis.*

[43]EPA has supported extensive studies based upon this industrial classification.

[44]When the group is clearly dissimilar, an attempt can be made to construct more homogeneous subgroups. Thus, EPA has created industrial subgroups which are characterized as using similar processing technologies.

For example, two pulping firms could be identical with respect to
their processes and products, but only one of the firms might be located
near a potential buyer of waste liquor. The opportunity costs of reducing water pollution caused by the waste liquor would then differ substantially between the two firms. Thus, both technological similarities and
economic dissimilarities should be accounted for in generalizations from
specific cases.

Suggested Approaches to Cost Estimation

The Survey Approach

Seemingly the most direct way to ascertain the costs associated with
a pollution reduction program is to rely on the polluter for providing
the estimates. While there has been little use of survey techniques to
determine prospective costs, surveys have been widely used to discover
past expenditures on pollution control. These past outlays have formed
the basis for cost estimates of planned reduction programs.[45]

Although surveys can be valuable, reliance on them requires the
analyst to assume: (a) the respondent is knowledgeable, and (b) the
respondent is honest. The first assumption—that the respondent can, in
fact, provide meaningful cost estimates—is rather stringent. It implies
that the respondent can estimate his uncontrolled residual levels, define
what are and what are not residuals, recognize control measures other
than those applying "end-of-pipe" technologies, allocate his costs properly, etc. To the extent a respondent lacks such knowledge, it may be
best if a survey is not attempted, since there is a danger that the responses will be guesses (albeit "best" guesses) and that they will be interpreted by the cost analyst as "facts."

The second assumption—that responses will be honest—may also be
a stringent one. This depends on whether the polluter has powerful
incentives to distort the facts. If he feels that his cost estimates will
become public, he may overestimate both past and proposed expenditures in order to make a case for financial support. On the other hand, if
he feels that he may become subject to new regulations, he may misrep-
resent his costs by overstating the success he has had controlling pollution, i.e., he may understate his current discharges. This could lead the
analyst to underestimate future control costs.

One final aspect of the questionnaire technique is that its implementation does not require the same degree of technical knowledge that the
engineering approaches (discussed below) require. As long as the sur-
veyor is willing to accept the responses as reasonably accurate, he need

[45]Estimates based on survey techniques of the industry cost to meet current (as of
January 1, 1973) pollution control standards are presented in McGraw Hill [15].

not be especially knowledgeable as to the technical and financial justifications of the cost estimates provided. This fact—that surveys can be undertaken by generalists rather than specialists—must be considered an advantage of the technique.

The Engineering Approach

The engineering approach relies heavily on technical analysis for control cost estimates rather than on information provided by the polluter.[46] Knowledge of the type and quantity of outputs as well as the underlying technical processes enables the analyst to estimate: (a) the uncontrolled levels of residuals; and (b) the expenditures relating to the best (least-cost) methods of achieving various target levels of control. As was the case for the survey approach, the validity of these estimates depends on several crucial assumptions.

First, the analyst must assume that the polluter uses a "typical" production technology so that by looking at outputs, he is able to estimate the corresponding level of uncontrolled emissions. Second, the analyst must determine what pollution-control technology is in current use so that he is able to calculate existing pollution levels. Third, the analyst must forecast the method that the polluter will choose to attain future control levels. Fourth, an accurate cost estimate of the selected technology is necessary.

A further difficulty with the engineering approach is that the technical expertise to conduct the analyses is not widespread. In addition, those analyses which have been undertaken have required considerable time.[47] Unfortunately, the need for skilled technical analysts is shared by the combined survey-engineering approach described below.

The Combined Survey-Engineering Approach

The basic strategy of the combined survey-engineering approach is to use surveys in gathering information for which the respondent is knowledgeable and for which he has no incentive to be misleading. Essentially, surveys are used to obtain information on product outputs[48] and on explicit pollution control strategies. All other costing information is acquired by engineering analysis.

Accurate reporting of output and control information can be expected. The likelihood of the polluter to underreport existing control expenditures is counterbalanced by a natural desire to publicly demonstrate his clean-up activities. Overestimation can still present problems. However, engineering analyses can uncover gross overestimates of re-

---

[46]Of course, a minimal amount of survey information (e.g., on output levels) is still required.

[47]Some of these analyses are referred to by *Kneese.*

[48]Information on product inputs is also useful, but not critical, for the approach.

ported expenditures for control equipment and for supporting labor and material inputs, if they are inconsistent with census reports on current and capital expenditures. In fact, experience has shown that a combined engineering-survey approach can serve as a check on the accuracy of census data by bringing inconsistencies to the attention of the respondent.[49] Thus, the quality of both the information generated by engineering analysis and the data gathered in surveys can be improved by using the combined approach.

The Future and Uncertainty

IT IS APPARENT THAT DECISIONS ON ANY POLICY must be made in the presence of uncertainty surrounding the future streams of costs and benefits. This uncertainty arises because of the lack of knowledge concerning such factors as the rate of growth in the economy, changes in consumers' tastes, and unanticipated price variations. The question is: how should aspects of an uncertain future enter a cost-benefit analysis? This can be analyzed by separate consideration of three related issues: the social discount rate, option value and irreversibility, and the role of decision theory.

The Social Discount Rate

Present benefits are usually considered more "valuable" than future benefits, since they mean increased consumption now rather than later.[50] Thus, not only the magnitude of benefits, but also the timing of their occurrence is relevant to cost-benefit analysis. While there is little dispute over this point, there has been substantial controversy over what discount rate should be used to convert future costs and benefits into *present values*. Choice of a discount rate will not only influence the type of projects which will be undertaken; it will also affect resource allocation between the public and private sectors of the economy.

Many economists share the view that "the correct discount rate is the opportunity cost [of capital] in terms of the potential rate of return in alternative uses on the resources that would be utilized by the project" (Baumol, [2] p. 491). Even for those who accept this concept, it is a severe practical problem for the analyst to determine that rate of return, especially in view of the fact that opportunity costs are likely to vary

---

[49]See the pilot study by Peskin [20] which used a combined approach to analyze the Norwegian paper pulp industry.

[50]The argument is symmetric in that future costs are usually preferred to present costs, since they also mean increased present consumption (or reinvestment).

throughout the economy.[51] In principle the correct discount rate is a weighted average of the various opportunity-cost rates for the sectors from which a project withdraws resources.[52] However, these opportunity-cost rates are not directly observable and, hence, must be inferred from generally poor data series on corporate profits.

Some have maintained, however, that this principle will not provide adequate investment for public projects. This view is based on the belief that capital put to public use often has a higher social return than the same capital put to private use because of beneficial spillover effects. Because of this "higher social return," they consequently argue that the cost of capital *should* be lower for governmental than for private projects and that the lower discount rate is justified.[53] While this argument may be true for certain public projects (which may, indeed, include water pollution control projects), it would be wrong to assume this to be the case for all public projects. Moreover, it should be kept in mind that determining whether the spillover argument is valid, is often the purpose of a cost-benefit analysis. To assume its validity by using a low discount rate in the analysis can make the projection of a favorable cost-benefit ratio self-fulfilling.

Baumol has argued convincingly that the government's use of an artificially low discount rate is not the proper way of achieving the goal of increased welfare for future generations. Most analysts who share this view argue, further, that using an artificially low discount rate is not compatible with the goal of good cost-benefit analyses.

Option Value and Irreversibility

There is some belief that the "measured" willingness to pay for the benefits of a project underestimates the "true" benefits of the project in certain situations where the future demand for project outputs is uncertain. Suppose, for example, that the principal output of a water-clean-up project is the creation of more recreational facilities. Assume also that if the project is not undertaken, it will be impossible for technical reasons to provide these facilities at a later date, i.e., the decision not to clean up is *irreversible*.

Under these circumstances, it has been argued that the benefits of the project can be logically separated into two parts. First, there are the present and future benefits to those who are certain to demand the recreational facilities. Second, there are the benefits to those who are not sure they will use the recreational facilities, but who will support the

---

[51]Reasons for these variations include taxation policies, the existence of risk, and the presence of externalities. See *Fisher and Krutilla, pp. 280-83.*

[52]The correct weights are somewhat controversial. For a discussion of alternative weighting schemes and their means of implementation, see Haveman [9].

[53]See Marglin [14] on this point.

project in any event because they feel it is important to keep their (and future generations') options open.[54] This second component of the benefits is termed *option value.*

Controversy has surrounded this concept, ranging from the question of whether usual empirical measures of willingness to pay already capture this component, to the question of whether option value even exists in situations where risks of adverse outcomes are widely shared.[55] It should be kept in mind that, as a practical matter, the controversy arises only when a project involves the possibility of irreversible consequences or of consequences which are only reversible at high cost. *Haveman and Weisbrod* suggest that the decision to build a sewage treatment facility is not of this character, although a plan to clean up Lake Erie may be.

Fisher and Krutilla also share the view that option value may be important.[56] In addition, they have extended the analysis of irreversibility and provided a more precise formulation of how its existence can affect the cost-benefit assessments of the project—even under the assumption that future demands and outcomes are known with certainty. Briefly, they argue that it may be wise to forego the investment in a project—such as a dam—even if a conventional comparison of discounted costs and benefits looks favorable. This recommendation is based upon the fact that their analysis explicitly accounts for the opportunity cost of the foregone benefits emanating from the preserved state of nature which would be irreplaceably destroyed by the dam.

The Role of Decision Theory

The formal analysis of the relationship between uncertainty and decisions is known as *decision theory*. Certain aspects of this theory can provide guidance to the analyst concerned with treating uncertainty in his cost-benefit assessments.

Two types of uncertainty are sometimes distinguished; statistical uncertainty (e.g., the uncertainty about the outcome of an event with known probability), and uncertainty about the future (e.g., the uncertainty of the state of the economy two decades hence). "Statistical uncertainty stems from the chance elements which occur in the real world" (*Benefit-Cost Analysis Guide,* Fifth Draft, [3] p. 31). On the other hand, uncertainty about the future stems from such factors as technological

---

[54]As noted before, these are sometimes termed *nonuser benefits.*

[55]For a discussion of this controversy as well as examples of when option value is likely to be important, see *Haveman and Weisbrod, pp. 59-61.*

[56]They also prove that option value exists even when risk is widely shared, providing the project output has the characteristics of a public good. Also, they argue that a type of option value exists even if individuals are neutral towards risk—contrary to the usual assumptions of adversity to risk required for option value.

developments (e.g., inventions), future political changes (e.g., elections), and unforeseen natural occurrences (e.g., earthquakes).

One common way of incorporating uncertainty in analyses is to apply the concept of *expected value*. The expected value of a decision is found by multiplying the value associated with each possible outcome by the probability of the outcome's occurrence and then summing the products across the possible outcomes. An illustration is shown below:

Possible Outcomes of Project	Probability of Outcome	Expected Payoff
$ −50,000	.5	$ −25,000
500,000	.1	50,000
200,000	.4	80,000
Expected Value of Choosing Project		$ 105,000

Faced with a number of projects, a decision maker would then be expected to choose the alternative with the highest expected outcome.[57]

An inherent difficulty in using this approach is the assignment of reasonable probabilities in the first place. Especially with uncertainty about the future, there is no objective way to determine such probabilities. In other cases, even when the probabilities of individual events can be estimated, the sequences of events generated by the project may be so complicated and interrelated that computation of the probabilities associated with ultimate outcomes is extremely difficult. Under such circumstances, *Monte Carlo simulation* may be useful. This technique involves using a model of the project, with which repeated calculations are made to simulate the possible outcomes from a chain of uncertain events. These events, in turn, are assumed to be generated by a known probability distribution. The probability distribution of the outcomes is estimated by inspection of the Monte Carlo simulation.

Another technique often employed to incorporate uncertainty in cost-benefit analysis is *sensitivity analysis*. Sensitivity analysis examines the effects on the cost-benefit ratio of changing key parameters. It provides the analyst with a measure of how crucial certain assumptions are. A typical example of this approach would involve computing the present value of the costs and benefits using a range of discount rates,[58] a prac-

[57]This decision rule is not inherently optimal, but it is optimal under certain reasonable assumptions concerning the decision maker's attitude toward risk. Essentially these assumptions are embodied in the von Neumann-Morgenstern utility postulates. For a nontechnical explanation of these postulates, see Baumol [1], pp. 512-28.

[58]This is not to be confused with arbitrarily increasing the discount rate to allow for uncertainty. "This practice is of dubious validity" *(Benefit-Cost Analysis Guide.* Fifth Draft, [3] p. 33). It assumes that the uncertainty associated with the stream of benefits increases with the passage of time; i.e., future benefits will be smaller than predicted. Furthermore, to the extent that discount rates reflect a weighted average of private rates of return, uncertainty is embodied, since these private rates incorporate risks.

tice recommended by *Fisher and Krutilla*. This would then tell the decision maker how insensitive the conclusions of the analysis were with respect to the parameter in question.

Cost-Benefit Analysis and Institutional Considerations

COST-BENEFIT ANALYSES ARE CONDUCTED within a framework of institutions; consequently a number of important institutional factors which the analyst should take into consideration are discussed in the first part of this section. These issues involve not only policy costs, such as expenditures for enforcement and administration, but also adjustments which must be made between the geographic scope of the project and the political jurisdiction. In addition to these issues, this part of the section examines the way in which the distribution of property and wealth can affect the assessment of costs and benefits. The second part of this section deals with institutional structures and raises the question whether cost-benefit analysis should be conducted by the sponsoring agency or by outside analysts. The concluding part of this section discusses the relationship between cost-benefit analysis and policy evaluation.

Institutional Factors in Cost-Benefit Analysis

Policy Costs

Policy costs represent an important, although often neglected, element of project opportunity costs.[59] *D'Arge* has enumerated three types of these costs: *information, enforcement,* and *administrative. Information costs* are those expenditures associated with obtaining knowledge on the various aspects of a water quality project; *enforcement costs* refer to expenditures made by an agency to insure that parties comply with emission standards; and *administrative costs* are simply the associated overhead costs of the agency.

There is a close tie between the institutional arrangements of water control agencies and policy costs. In discussing institutional structure, *d'Arge* distinguishes between the use of *direct* and *indirect controls.* "Direct controls are those which are applied at or to the source of the water

---

[59]An important point to note is that since the public agency making a regulation often bears the associated policy costs, it may try to minimize these costs and, in so doing, choose a policy which is suboptimal for society as a whole.

quality problem and which threaten sufficient penalties to make avoidance extremely costly" [*d'Arge, p. 219*]. For example, the closing of a factory emitting wastes over a prespecified level would represent a direct control. "Indirect controls are defined here as management strategies with at least two links of expected causation between problem source and application of control" [*d'Arge, p. 219*]. For example, taxation of waste discharges from the factory would typify an indirect control.

In general, the more indirect the control, the greater the information costs necessary for its satisfactory implementation are likely to be. This stems from the fact that indirect controls usually contain many technological and behavioral links; hence, if they are to be effective, they will require considerable knowledge of the probable responses by the affected parties. On the other hand, the more direct the control, the higher the enforcement costs are likely to be. This follows, since rigid controls, which leave little room for private decision making, may provide incentives for the polluters to circumvent the regulations. Thus, there is likely to be a trade-off between indirect controls, with their uncertain behavioral assumptions and high information costs, and direct controls, which eliminate some uncertainty, but have high enforcement costs.

Project, Jurisdictional, and Analytical Boundaries

Ideally, the kind and size of area used for environmental planning should be small enough for manageable planning, but large enough to encompass the entire area for which significant economic, political, and social interactions exist. [*Abel, p. 336*]

Unfortunately, in defining project boundaries, these ideal conditions are likely to be characteristic of differing areas of control. Consequently, in selecting the project boundary for a cost-benefit analysis, a number of criteria must be considered.

One aspect to consider is the political realities involved. There is little reason to believe that a geographical region which contains a water pollution problem will coincide with any existing political entity. As *Abel* has pointed out, this type of difficulty is always present if one takes an all-encompassing view of a control program. For example, "The air shed does not correspond with the water basin, and neither corresponds with areas of high economic and social interdependence" [*Abel, pp. 336-37*].[60]

Another consideration which should be made in defining project boundaries is the possible creation of *external effects*[61] or *spillovers*. These effects are defined as those impacts of a control program which are not

[60]This example also highlights the importance of the interrelationships among environmental media.

[61]"External effects" should not be confused with "externalities," a technical term used by economists to describe the effects of consumption and production activities that are not reflected in normal markets. Most water pollution is, by this definition, an externality.

contained within the project bounds. A frequently cited example in the water development area is the spillover effect associated with the level of recreational activity in an improved area vis-à-vis the level in nearby unimproved areas. Other things being equal, in designing pollution control programs and their associated project boundaries, these external effects should be minimized.

Once project boundaries are defined, there remains the problem of establishing the boundaries of the cost-benefit analysis. In principle, there is no reason for the project boundary and the analytical boundary to coincide. If fact, the theory of cost-benefit analysis requires one to examine the relevant cost and benefit effects throughout the economy, even if the project in question pertains to a small geographical area.

However, as a practical matter, obtaining the data and other information necessary to assess the cost and benefit impacts of a project throughout the economy would probably require resources well beyond those available to local and state planners. Thus, operationally, one can expect analytical boundaries to approximate project or jurisdictional boundaries. As a consequence, costs and benefits falling outside the jurisdictional concerns of the policy maker may be assigned zero values. To the extent that the "true" values are substantially greater than zero, what may appear optimal from a local or regional point of view, may be suboptimal from the national point of view. A solution to this problem might be a sharing of analytical and data resources between local, state, and regional planners.

Distributional Aspects

We argued earlier (p. 3) that the distributional impacts of a proposed project were beyond the scope of cost-benefit analysis. Nevertheless, certain kinds of distributional considerations may affect an assessment of the willingness to pay for some project outputs. Before going into this, some discussion is needed on the issue of property rights.

Stockfisch makes the useful distinction between *property* and *wealth*. Property is defined in terms of activities (e.g., the capability of performing specific services) and resources (e.g., endowments of physical assets), while wealth refers to the relative earnings which accrue to these activities and resources as determined through market interactions. *Property rights* relate to the rules and laws governing the disposition of property. Furthermore, the *distribution of property* refers to the allocation of service capabilities and physical endowments, as between households and other entities, while the relative incomes received by these services and endowments constitute a *distribution of wealth*.[62]

The principles regulating the distribution of property rights and the resulting distribution of income and wealth, can play a key role in

---

[62]See *Stockfisch, pp. 316-22.*

determining consumer preferences and can thereby influence the public projects undertaken by society. We can illustrate this point with a hypothetical example. Suppose there existed a small lake which for many years was used solely as a receptacle for the discharges from a single firm. Assume, further, that over time, demand for the lake as a recreational facility occurred, and the surrounding community consequently decided that the pollution should be controlled.

Two polar actions could be taken. The first would not require the firm to bear the clean-up costs, reasoning that the firm was not responsible for the new demand for recreation. Such a position, in effect, would assign the property rights of the lake (for purposes of dumping residuals) to the firm. If, on the other hand, the firm was required to bear the clean-up costs, in effect, the property rights would be assigned to the recreational users. The point is that the measured willingness to pay will differ in the two situations.[63] In the latter case it could be greater, reflecting the fact that recreational demand is positively related to the wealth of the users, which has increased as a result of the property rights assigned to them.

On the other side of the picture, laws can also be enacted to compensate those who are affected by the costs and the benefits of public programs, thereby altering the existing distributions of property and wealth.[64] Although this latter point concerns a normative issue which is beyond the scope of cost-benefit analysis, it may be quite relevant to the policy maker.

In evaluating alterations in resource allocation, it seems clear that if everyone is made better off because of a change, then the change can be deemed a good thing. Unfortunately, in real world situations (especially those dealing with public programs), this seldom occurs; i.e., at least one person or group (often taxpayers in general) is made worse off. Consequently, other criteria must be used to judge the desirability of alternative resource allocations.

A number of these criteria are based upon *compensation principles*. Basically these suggest that a reallocation can be considered favorable if the persons who have gained from the change can compensate those who have lost from the change so as to leave the latter as well off as before the change. In practice, difficulties may arise with regard to assessing the "correct" payments and enforcing the actual transfers. Often this is the responsibility of the courts. What is important for the cost-benefit analyst, is that if the courts can surmount these operational difficulties, the determination of a just compensation can provide a useful measure of willingness to pay.

[63]In practice, however, this difference may not be great enough to substanitally alter the cost-benefit analysis.

[64]These issues are, of course, closely related to the existing political process.

Institutional Structures for Conducting
Cost-Benefit Analysis

Throughout this introduction we have employed the rather imprecise term *policy maker* to describe the recipient of a cost-benefit analysis. Use of this word, however, neglects to point out that policy is made and executed by many different bureaucracies—federal, state, and local. Thus, it would be perhaps more accurate to speak of particular bureaucratic institutions and their respective policy makers, each interested in specific projects, and accordingly, in specific cost-benefit analyses.

It has been common practice for a bureaucracy with the responsibility of making and implementing policy decisions to undertake in-house cost-benefit analyses of proposed programs. Arguments in favor of this procedure include: (1) the operating agency has the best access to relevant data;[65] (2) the agency administrator can be apprised of results more quickly and easily; and (3) the agency is more sensitive to political considerations which may be inseparable from other aspects of the analysis.

Nevertheless, there is also an argument for having a completely independent organization undertake cost-benefit analyses of proposed programs. This is based on the hypothesis that individual bureaucrats are motivated by a desire to increase the magnitude of both their budgets and their staffs.[66] If this hypothesis is correct, an operating agency will look for additional programs to control. Consequently, whenever there is a prospective project to examine—either by a cost-benefit analysis or a more general policy evaluation—there will be an inherent bias on the part of the agency to arrive at a result favorable to the implementation of the proposed project. Since in most cost-benefit analyses, there is considerable opportunity to make self-serving assumptions—for example, in using unrealistically low discount rates—it is fairly easy to doctor the analyses.

The problem of conducting a reputable cost-benefit analysis in a bureaucratic context has been pointed out, but it has not been subjected to careful analysis. This is, in part, because relatively few analysts have had experience conducting cost-benefit analyses within operating bureaucracies; and, hence, few analysts may be sensitive to the problems involved. This is an area which appears to be too important to continue to ignore.

Policy Evaluation vs. Cost-Benefit Analysis

In the first section of this chapter a distinction was made between policy evaluation and cost-benefit analysis. We used the term *policy evaluation* to describe the broad activity of comparing all the favorable

---

[65]The agency may have the only access to the data if it is classified or confidential.
[66]This view is formally treated in Niskanen [19].

and unfavorable aspects of a proposed public program. In contrast, we reserved the term *cost-benefit anlaysis* to designate a rather formal method of comparing the efficiency gains and losses of such a proposed project. Furthermore, the comparison of costs and benefits was discussed in terms of achieving the goal of "efficiency," where the words *costs, benefits,* and *efficiency* were given special meanings. Thus, a complete policy evaluation of a project would not be limited to a cost-benefit analysis but, in addition, would include considerations of its distributional effects, its political feasibility, its legality, etc.

While the above dichotomy is not universally accepted, there seems to be fairly wide agreement that policy evaluation involves something more than cost-benefit analysis. The disagreement arises over the question, *how much more?* Specifically, some analysts appear to believe the areas covered in cost-benefit analysis should not differ greatly from those considered in policy evaluation. If, at present, the differences between the two are great, they argue (sometimes implicitly) that these differences should be narrowed by expanding the scope of cost-benefit analysis.

The belief that cost-benefit analysis should be more encompassing may stem from a special interpretation of the rather limited efficiency objective. For example, Hill has written:

> The cost-benefit method was designed not only to choose the course of action that maximized "economic efficiency," but it was also assumed that in the process of doing so economic welfare was maximized. . . . (Hill, [10] p. 20).

In continuing, he argues that welfare maximization requires the assumption that the existing income distribution is "in some sense, 'best.' " At the same time he criticizes cost-benefit analysis by stating that this assumption is "questionable" (Hill, [10] p. 20).[67]

There is nothing inherently wrong with this type of criticism, provided one accepts the proposition that the attainment of economic efficiency implies the maximization of economic welfare. If that is the case, Hill's criticism is valid and cost-benefit analysis is in need of some alteration. However, there are others who subscribe to the view that attaining the efficiency objective has only the following modest welfare implication: to the extent that a situation is inefficient, there exists the *potential* of making at least one person better off without making anyone else

[67]Hill also argues that two other assumptions are necessary for welfare maximization: (1) that the project leaves the income distribution unchanged; and (2) that prices equal social opportunity costs in the economy except in the sector requiring the project. However, Hill has confused the concepts of economic efficiency and welfare. These two assumptions are required for economic efficiency, a fact that Hill fails to appreciate. He incorrectly states that even where these two assumptions do not hold, cost-benefit analysis "identifies the most efficient course of action" (Hill, [10] p. 20).

worse off.[68] The surplus required to accomplish this might accrue from the implementation of a new project or from alteration of an existing program to make it more efficient. It should be noted, though, that there is no assurance such a surplus would be distributed without making anyone worse off; hence, the above potential might never be realized.

As Mishan points out, in order to make an unambiguous statement concerning economic efficiency, the restrictive assumptions and special definitions of cost-benefit analysis must be accepted (see page 3, above). If they are dropped in order to expand the scope of cost-benefit analysis and make it more relevant for a complete consideration of welfare, it becomes impossible to make unambiguous statements about economic efficiency. Thus, these assumptions do not weaken cost-benefit analysis. Rather, the assumptions are fundamental to a clear analysis of economic efficiency.

It might be argued, however, that there is a more valid criticism of cost-benefit analysis: namely, that although it allows one to make a scientifically precise statement regarding the economic efficiency of a project, efficiency may be of minor importance to policy evaluation.[69]

This critical comment on cost-benefit analysis may be a result of confounding the *outcome* of cost-benefit analysis with the *analysis* itself. Perhaps it is true that determining whether a project is efficient or not is of little importance to the ultimate policy decision. However, the analytical process behind this determination may be useful in itself. In his effort to compare costs and benefits, the analyst must clarify project objectives, identify affected parties, assemble relevant technical data, enumerate tangible and intangible program effects, etc.—information that is critical for the broader purpose of policy evaluation.

Thus, one might conclude that, with respect to cost-benefit analysis, while the specific ends do not always justify the elaborate means, the means may have significant justification in themselves.

REFERENCES

1. Baumol, W. J. *Economic Theory and Operations Analysis.* 2nd ed. Englewood Cliffs, New Jersey: Prentice-Hall, 1965.
2. ———. "On the Discount Rate for Public Projects." In *The Analysis and Evaluation of Public Expenditures: The PPB System—Vol. 1.* Washington: Government Printing Office, 1969, pp. 489-503.
3. *Benefit-Cost Analysis Guide.* Fifth Draft. Planning Branch, Treasury Board Secretariat, Canada (Xerox).

[68]See Mishan [16] on this point.
[69]This view is expounded by *Rowen.*

4. Berry, D. "Environmental Protection and Collective Action. The Case of Urban Open Space." Regional Science Research Institute Discussion Paper Series, No. 61, January 1973.
5. Dorfman, R., and Jacoby, H. D. "A Model of Public Decisions Illustrated by a Water Pollution Policy Problem." In *The Analysis and Evaluation of Public Expenditures: The PPB System–Vol. 1*. Washington: Government Printing Office, 1969, pp. 226-74.
6. Geldreich E. E., et al. "The Faecal Coli-aerogenes Flora of Soils from Various Geographical Areas. *Journal of Applied Bacteriology* 25 (1962): 87.
7. ———. "Type Distribution of Coliform Bacteria in the Feces of Warm-blooded Animals" *Journal Water Pollution Control Federation* 34 (1962): 295.
8. Haveman, R. H. "Evaluating Public Expenditures Under Conditions of Unemployment." In *The Analysis and Evaluation of Public Expenditures: The PPB System—Vol. 1,* Washington: Government Printing Office, 1969, pp. 547-61.
9. ———. "The Opportunity Cost of Displaced Private Spending and the Social Discount Rate." *Water Resources Research,* October 1969, pp. 947-57.
10. Hill, M. "A Goals-Achievement Matrix for Evaluating Alternative Plans." *Journal of the American Institute of Planners*, January 1968, pp. 19-29.
11. Hochman, H. M., and Rodgers, J. D. "Pareto-Optimal Redistribution." *American Economic Review*, September 1969, pp. 542-57.
12. Kneese, A. V., and Bower, B. T. *Managing Water Quality: Economics, Technology, Institutions*. Baltimore: Johns Hopkins University Press, 1968.
13. Lave, L. B., and Seskin, E. P. *Air Pollution and Human Health*. Baltimore: Johns Hopkins University Press, forthcoming.
14. Marglin, S. A. "The Social Rate of Discount and the Optimal Rate of Investment." *Quarterly Journal of Economics*, February 1963, pp. 95-112.
15. McGraw-Hill Publications Company. "Sixth Annual McGraw-Hill Survey of Pollution Control Expenditures." Key findings by McGraw-Hill Publications Company's Economics Department. New York: May 18, 1973.
16. Mishan, E. J. *Economics for Social Decisions: Elements of Cost-Benefit Analysis*. New York: Praeger, 1972.
17. ———. "The Futility of Pareto-Efficient Distributions." *American Economic Review,* December 1972, pp. 971-76.
18. Mohring, H., and Boyd, J. H. "Analyzing 'Externalities': 'Direct Interaction' vs. 'Asset Utilization' Frameworks." *Economica*, November 1971, pp. 347-61.
19. Niskanen, W. A., Jr., *Bureaucracy and Representative Government*. Chicago: Aldine Press, 1971.
20. Peskin, H. M. *National Accounting and the Environment*. Oslo: Central Bureau of Statistics, 1972.

Part One

BENEFIT MEASURES

The Concept of Benefits in Cost-Benefit Analysis: With Emphasis on Water Pollution Control Activities

ROBERT H. HAVEMAN
and
BURTON A. WEISBROD

A CLEAR UNDERSTANDING of the meaning of *benefit* is the fundamental requirement for undertaking any sound cost-benefit study of public activities. In this paper, this concept is explored in some detail—with emphasis on the meaning of the benefits of water pollution control activities. In the first section, the theoretical basis of the concept of benefits is explored. The propositions of the "new welfare econòmics" which give the concept meaning are described and analyzed, and the willingness-to-pay criterion is developed. In particular, the relationships between benefits and the economic concepts of efficiency and equity are examined. The second section focuses on efficiency and tries to determine which types of benefits are or are not properly included in a correctly defined concept of efficiency benefits. The discussion focuses on the distinctions between primary and secondary benefits, real and pecuniary benefits, external and internal benefits, and tangible and intangible benefits. Finally, two special issues in the concept of benefits are explored: the issue of option price and option value, and the issue of donor benefits and equity effects.

The Concept of
Benefits and Its
Theoretic Base

IT IS USEFUL TO THINK of program "benefits" as the extent to which the program produces desirable results. What is or is not "desirable" depends, in turn, on the goals or objectives of the program. This is to say that the first step in the process of project evaluation should be a statement of goals. The second step should be an attempt to state these goals in operationally measurable form. For each goal an operational measure of the degree to which the program achieves the goal is indispensable for program evaluation. Later in this paper we will return to the relationship between these two steps in the evaluation process—namely, the step in which goals are stated in idealized, conceptual terms, and the step in which the conceptual goals are restated in operationally measurable forms. The third step in the evaluation process is the development of a set of weights that reflect judgements about the comparative importance of progress toward each of the goals—the goal trade-offs.

At this point in the analysis, we need to consider the general nature of the goals—and hence the benefits—of public expenditure programs in general and of water pollution control activities in particular. While there may be numerous goals, they can be grouped into two principal categories: (1) those related to economic *allocative efficiency,* and (2) those related to *distributional equity.* The efficiency goals involve recognition of the alternative uses to which limited resources can be put, and also involve recognition of the existing pattern of demand for the various uses of resources. That is, given the distribution of income in the society and given the preferences people have for various goods and services, there will be particular patterns of "effective demand," although—and this is critical—such demand may not be reflected in the private marketplace because of organizational and related problems. By effective demand is meant not only the desire for some commodity but also the willingness to pay for it. Demand is, thus, dependent on both preferences and on the distribution of income and wealth, since both determine willingness and ability to pay.

Allocative efficiency as an economic goal reflects the fact that it is sometimes possible to reallocate resources—perhaps increasing or decreasing the amount of resources used for water pollution control—in ways which will bring about an increase in the net value of output produced by those resources. For such reallocations, the increase in the value of the output of the good whose production is expanded must be greater than the decrease in the value of the output of the good whose production is being decreased. To determine allocative efficiency one must ignore considerations of which particular people are made better off or worse off as resource allocation alternatives are considered. The issue of how alternative resource allocations affect the well-being of particular people is captured by the distributional—or equity—goals.

Concern with the goal of allocative efficiency is the preoccupation —indeed the sole concern—of the "new welfare economics," and, at least until recently, of cost-benefit analysis. For most of the 200-year history of economics, the distinction between allocative efficiency and distributional equity as social goals was not clearly made—or if made, was not regarded as a very important distinction. It is primarily over the last several decades when economics has become more analytical that economists have attempted to evaluate the efficiency effects of policy decisions in abstraction from equity effects. In this process, equity became generally regarded as a matter of judgment—and economists have presumed that their judgments as to what is equitable are no more valuable than the equity judgments of any other citizen. As the "old" welfare economics gives way to the "new" in the search for a relatively value free economics, the emphasis of analysts has been focused on allocative efficiency. An implicit assumption of this efficiency focus is that a "more efficient" allocation of resources is desirable, quite apart from the distributional consequences of that improvement in resource allocation efficiency. Although the goal of efficiency is itself a normative judgment, economists regarded it as a rather noncontroversial, widely held goal. Moreover, economists were able to show that if an efficient reallocation of resources could be made, it would be possible to distribute the net gain so that no one would find the reallocation inequitable. Hence, equity concerns could be eliminated from the analysis. We turn now to an elaboration of the new welfare economics and its implicit definition of program benefits in terms of allocative efficiency. Later we will turn to consideration of distributional equity goals.

The "New Welfare Economics"—Allocative Efficiency and the Concept of Benefits

Cost-benefit analysis has been concerned primarily with efficiency in the allocation of resources. As such, it flows directly from the body of neoclassical economic theory (or resource allocation theory). But whereas the main body of neoclassical theory deals with the nature of *private* decisions by consuming and producing units, cost-benefit analysis forcuses on *public* decisions. Rather than predicting such decisions, cost-benefit analysis, by focusing on efficiency, seeks to set forth the requirements for any resource-using criteria to be efficient—that is, to be a decision which leads to an "improvement" in the allocation of economic resources. The basic question to which most cost-benefit analysis has sought an answer then is: "Will an alteration in the current pattern of resource allocation improve efficiency in the use of national resources (thereby increasing social well-being) or will it not?"

Insofar as cost-benefit analysis is directed at allocative efficiency, it can be viewed as an attempt to replicate, for the public sector, the decisions that would be made if private markets worked satisfactorily. This is to say that the efficiency orientation of cost-benefit analysis can be viewed as an attempt to develop a public sector analogue for private market decision making. In private markets, there is no explicit attention given to equity goals; private markets are expected to respond to actual demand patterns and resource constraints, where demand is partly a reflection of the distribution of income. Economists do not generally regard private markets as operating unsatisfactorily simply because they fail to produce goods for people who cannot afford to pay for them. Similarly, the focus of cost-benefit analysis on allocative efficiency implies that public sector decision making *should* be preoccupied with considerations of demand (not desire) and resource costs, relegating to a secondary position concerns about either the acceptabilty of the income distribution that contributed to the demand pattern or the impact of the program's benefits and costs on the income distribution. By "impact" we mean the manner in which the program's benefits and costs are distributed among people —that is, who benefits and who pays.

Before proceeding to a discussion of formal welfare criteria, it is useful to distinguisn between two kinds of economic analysis: normative analysis, by which is meant prescriptive analysis—statements regarding how things *should* be done, how resources *should* be allocated—and positive, predictive analysis, in which the objective is to state what *will* happen as a result of some particular decision. Cost-benefit analysis has, in fact, been used in both ways. While its principal use, and the one analyzed in this paper, has involved normative economics—attempts to guide government decision makers regarding resource allocation decisions—it has also come to be used for predictive purposes. Here the objective has been to describe and ultimately to predict how government decision makers will respond to particular kinds of information regarding the benefits and the costs—the advantages and the disadvantages —of alternative uses of resources. In this paper we concentrate on the normative aspects of cost-benefit analysis—those aspects which hold out some hope for guiding public decision makers and, in particular, those decision makers having responsibility for water pollution control activities.

Pareto Optimality and Other Welfare Criteria

Basic in the literature of theoretical welfare economics is the question of the criteria by which to judge the desirability of proposed alterations in the allocation of resources. Perhaps the best-known social welfare criterion is that proposed by Vilfredo Pareto: any change in the social state is "desirable" if at least one person judges himself to be better off because of the change while no one else is made worse off by the change.

While the Pareto criterion seems unexceptional as a basis for making judgments on changes in social states—induced by, say, public policy decisions—it fails to cover a wide range of those public policy decisions most common in the real world. They are decisions in which some persons are inevitably made worse off while, at the same time, others are benefited. Indeed, if the Pareto criterion is accepted, no normative judgments at all can be made regarding the desirability of a change in a given social state even if some people are benefited greatly while but one person is made worse off. It is difficult to contemplate real world decisions that would fail to make at least one person worse off. Pollution control policies would appear to be classic cases of the simultaneous imposition of costs and bestowing of benefits due to a change in the allocation of resources.

Because of the limitation of the Pareto criterion, efforts were made to develop a normative welfare criterion with more general applicability to actual economic decisions which inevitably help some people and harm others. In short, recognizing that all resource-allocation decisions, whatever their intended effects, alter the distribution of real income, economists have attempted to develop propositions regarding what constitutes an increase in economic welfare which would have wider applicability than the Pareto criterion yet would be generally accepted as value judgments. Two more general criteria are now widely cited in the literature underlying cost-benefit analysis and the concept of benefits. The first of these criteria, proposed by both Nicholas Kaldor and J. R. Hicks,[1] states that a change in the allocation of resources should be regarded as increasing welfare if either the Pareto criterion is met *or* if the persons who have gained by the resource reallocation *could* compensate those who have been harmed by it so as to leave the latter at least as well off as without the reallocation.[2] The second criterion, proposed by I.M.D. Little,[3] accepts a compensation notion similar to the Kaldor-Hicks criterion,[4] but, in addition, requires the decision maker to

[1]N. Kaldor, "Welfare Propositions of Economics and Interpersonal Comparisons of Utility," *Economic Journal* 49 (1939), and J. R. Hicks, "The Foundations of Welfare Economics," *Economic Journal* 49 (1939).

[2]E. J. Mishan has referred to this criterion as a *potential* Pareto improvement criterion; a "diluted" version of the original. See E. J. Mishan, *Cost-Benefit Analysis* (New York: Praeger Publishers, 1971), p. 316. This criterion, it should be noted, assumes the existence of costless income transfers and, again, should be qualified to the effect that no superior reallocation is precluded by the one being considered.

[3]I.M.D. Little, *A Critique of Welfare Economics*, 2nd ed. (Oxford: Clarendon Press, 1957), p. 109.

[4]The possibility of a paradox of inconsistent evaluation resulting from the effect of the resource reallocation on the structure of relative prices has been noted by T. Scitovsky, "A Note on Welfare Propositions in Economics," *Review of Economic Studies* (1942). The probability that a resource reallocation of small proportions will significantly affect the relative price structure in a large economy is not great. Nevertheless, the possibility of the paradox should be noted.

judge that the change would cause a "good redistribution of wealth." Thus, while the Kaldor-Hicks criterion is concerned only with the economic efficiency effects of a resource reallocation attempting to abstract from redistributional effects, the Little criterion inserts equity considerations explicitly into the discussion.[5]

The main thrust of these efforts has been, and is, to expand the set of situations and policies about which statements may be made as to whether economic welfare has been increased, decreased, or left unchanged. The objective has been to find agreement on how to define the net effects of a resource-using public expenditure. The outcome of this is general acceptance of the proposition that a project with positive net effects yields benefits in excess of costs, in the aggregate sense that the project *could* make everyone better off if appropriate compensation payments were made, although in actuality some persons would be made worse off.

The Economics of Compensation

This net benefit concept has not been without its critics. For one thing, it has been argued—in our judgment, properly—that a resource reallocation does not unambiguously increase economic welfare unless the compensation is actually paid, so everyone is really better off, or unless an explicit judgment is made that real economic welfare *should* be redistributed—that is, that those people who are made worse off deserve to be made worse off. In short, an explicit decision regarding the desirability of the income redistributional effects is required before a meaningful statement can be made about the effect of net benefits on real welfare.

A second criticism recognizes that redistributions of gains and losses are not costless. Thus, if compensation payments were actually to be made, the informational costs of determining beneficiaries and cost-bearers and the amounts gained or lost, the administrative costs of levying the taxes and making the transfer payments, and the work-incentive (and other efficiency) effects on taxpayers and transfer recipients might well be large enough to bring into question the desirability of making the compensation and, hence the desirability of the project itself.

The relevance of the compensation question has often been recognized, although less frequently in actual public policy decisions than in the research literature. Seldom recognized in even the research literature, however, is the question of how the compensation, if it is to be paid, should be financed. The answer again hinges on an ethical judgment

[5]In the discussion of the concept of benefits in this paper, the efficiency criterion (that is, the Kaldor-Hicks criterion or the first part of the Little criterion) will receive primary attention. In the third section of this paper, however, the problem of equity effects will be discussed and efforts to integrate efficiency and equity effects will be described.

about the desired income distribution, and it is not necessarily desirable that the payments be made by those who would benefit from the project. To see this, consider a project that would benefit a group of poor people, and would bring some harm to a group of people who are more well-to-do. Improvements in air quality in the central city through required treatment of industry stack emissions may be an example. In this case an ethical judgment might be made both to award compensation to those harmed and to finance the compensation by levies on persons higher in the income distribution—say, through an increase in the general personal income tax—rather than through levies imposed on those benefited.

It should be emphasized that, in a political economy context, a decision to make, or not to make, compensation payments may have great influence on whether or not a program is actually approved. Consider a program whose aggregate net benefits are positive, so that it is possible, at least in principle, for program beneficiaries to fully compensate all persons who suffered losses as a result of the program, leaving a positive residual. Assume, however, that the compensation would not actually be paid. Then, in the situation where people have more equal representation in the political system than they do in the economic system (where their "votes" depend on their income and wealth), a majority of persons, if they would be hurt by the project, could delay or ultimately even defeat a decision to undertake it. This is possible notwithstanding our assumption that the overall benefits exceed the overall disbenefits, so that the majority who would be hurt could be fully compensated. Thus, in such a political context, the failure to make compensation arrangements can lead to the demise of an allocatively efficient project.

A similar result might occur even if the uncompensated persons were only a minority of voters—for the political system is more complex than a simple one-man-one-vote characterization would suggest. If the basic majoritarian system is modified, as it actually is, to provide protection for minorities, then even a relatively small minority of persons, if they would be made worse off by some project, might be able to defeat or greatly delay the project through petitions, hearings, litigation, etc. Conversely, the political system might be one in which a minority of individuals could induce the public sector to undertake an inefficient project which yielded them a stream of benefits. This could occur even in a one-man-one-vote majoritarian situation if organizational and informational costs differed between large and small groups. Thus, a small homogeneous group, such as irrigators along a watercourse, would find it easier to organize political pressure than would a large number of losers (taxpayers) each of whom incurred a very small loss because of the project. If the compensation of those who lost due to the project was not required of the gainers (because of, say, the absence of a user charge or cost-sharing arrangement), beneficiaries would have substantial incentive to lobby and to use their influence to have the project undertaken. Were full compensation required, however, the incentive to seek the inefficient project would be eliminated; were only partial compensation required, some incentive to seek the project would remain.

Under still other political arrangements, additional patterns resulting from the payment or nonpayment of compensation could result. For example, a requirement that compensation payments had to be made could result either in failing to undertake an allocatively efficient project or in undertaking an inefficient project. Consider the case in which a given project is allocatively efficient, but assume that the total costs as seen by project decision makers are synonymous with the budgetary cost to their agency. Thus, any increase in the amount of compensation payments is seen as an increase in the "cost" of the program. Of course, the point is that if the compensation were not paid, certain people would suffer a real loss; the compensation payment arrangement is to determine not the total amount of the project benefits but only precisely who would receive the benefits and who, if anyone, would be left worse off—that is, with negative benefits. Without compensation, some will reap benefits and others, losses. With compensation, the potential losers will be helped, and taxpayers, who would provide the compensation funds, would bear the losses. In short, even though the existence of compensation payments does not change the real cost (or benefit) of the program, but changes only the distributional impact, if project decision makers view costs in simple budgetary terms, economically efficient programs could fail to gain approval. As a consequence, inefficient projects without the losses requiring compensation could be approved.

Recognizing the effects of the payment or nonpayment of compensation, we can usefully consider the magnitude and pattern of compensation that would result in undertaking efficient projects and in rejecting those which were not efficient. Such compensation might be termed "efficient compensation."

In addition to the desirability and consequences of making compensation payments, the mechanisms to be developed for making compensation should also be discussed. Early in this paper we talked about the relationship between program goals at the conceptual level and program goals in their operationally defined form. We pointed out that there are problems of developing operational measures of program goals and benefits that are fully congruent with the conceptual goals of the program. This distinction is useful once again in connection with the desirability of developing compensation arrangements. Despite the lack of perfect congruity between conceptual goals and operational benefit measures, the development of such operational measures of benefits is indispensable; and this is also true with regard to compensation payments. That is, at the conceptual level it is clear what we wish to do —determine precisely who it is that benefits and who it is that suffers losses as a result of a given project, determine the amounts of the benefits and losses suffered by each person, and then make payments and levy taxes accordingly. So much for the compensation.

At the operational level, however, these ideals cannot be fully achieved. It would seem feasible, though, to develop operational rules for determining the amount of compensation payments and the source of funds for those payments in ways that would reasonably approximate our compensation ideals. For example, in the case of projects to clean polluted lakes, it may be possible to develop rules of thumb relating benefits (perhaps as measured by increases in land values) to distance from the lake. The task of determining the amount of compensation to be paid to various classes of persons is not a simple one; neither is it an impossible one. And the task is worth pursuing not only for the obvious equity reasons, but also for efficiency reasons. In our judgment, as pointed out above, the failure to make compensation payments significantly influences the organization of political opposition or support and may lead to the defeat of allocatively efficient projects or the approval of others which are not efficient.

What Are "Social" Benefits?

When a project is evaluated, the analyst should bear in mind that his immediate target is to determine if the value of the output in the economy (the national income) *with* the proposed project is greater than the value of the economy's output *without* the project. It should be emphasized that this evaluation is prior to determination of whether it should be undertaken and whether, if it is, those who are harmed should be compensated, and whether, if they should be compensated, the compensation should be paid by those who would be benefited or, instead, by others in the society. This conceptual goal is not easy to attain, and, as a practical matter, analysts often resort to comparing the value of the economy's output *before and after* a project is undertaken. The danger in this comparison is that events unrelated to the project may change over time, influencing the apparent, though not the true, effects of the project.[6]

The analyst should also bear in mind that the accounting framework for the analysis must be a comprehensive one in which all of the adverse and beneficial effects of the project are tabulated, regardless of who is harmed or helped by them. The impacts of a resource alloca-

---

[6]For an excellent discussion of the "with-without" principle, see M. Regan and E. Weitzell, "Economic Evaluation of Soil and Water Conservation Measures and Programs," *Journal of Farm Economics* (November 1947).

tion project may erroneously appear to be localized. With a comprehensive accounting framework, the cost-benefit evaluation of the project will not be restricted to the localized impacts but will estimate its full effect on social well-being. To measure only the real effects of a project on a locality, region, firm, or industry when there are additional, more pervasive effects would lead to choices which are inefficient from a national point of view.[7] Recognition of this possibility is especially important for decision makers who apply cost-benefit analysis within limited jurisdictions: given the incentives faced in decision making processes with narrow objectives, suboptimization (from a national point of view) may be a serious problem.

One of the implications of our discussion is that the total benefits to society from some public action bear no particular relationship to the flow of revenues to the government, or to the specific government agency that undertook and financed the project. Social benefits can be thought of as being reflected by the willingness of people to pay for the flow of goods or services from the project rather than do without them. (This "ex post" willingness to pay is termed *compensating variation*.[8]) Whether people actually make those payments or whether they pay less, retaining a "consumers' surplus," does not change the value of social benefits; it only changes the distribution of those benefits.

Note, however, that this, the most prevalent concept of social benefits, makes the benefits of a project depend on the initial income and wealth distribution in the society. This is so because willingness to pay depends in part on the income or wealth one has to begin with. In general, a particular public project will produce a different amount of social benefits for each different pattern of consumers' demands and, thus, for each different initial income distribution in the society. Hence, there is no unique value of social benefits that can be measured independently of that distribution. As a consequence, judgments about the desirability or acceptability of alternative income-distributional patterns are conceptually inseparable from estimates of social benefits. This is true as long as we retain the view that "social" welfare is simply the welfare of individual citizens—not the welfare of an abstraction termed

[7]This, of course, assumes that the effects of a resource reallocation are contained within national boundaries or that extranational effects are deemed irrelevant.

[8]For an explication of the tie between the equivalent and compensating variations and the demand curve, see E. J. Mishan, op. cit., Note B. It should be noted that if the income effect of the impact in question is positive—the normal case—the area under the compensated demand curve is an overestimate of the compensating variation but an underestimate of the equivalent variation. When the income effect is zero, the three concepts coincide. As Mishan has stated, however:"Goods having zero income effect are hard to come by, but for a great many purposes the income effect involved is small enough for economists to make use of the area under the demand curve as a close approximation of the relevant benefit or loss" (p. 338).

the "state" that is separate from its citizens. Later (pp. 61-64), we will return to the inseparability of efficiency and equity when we discuss donor benefits, a concept which suggests that equity concerns may themselves be part of an efficiency criterion.

In Figure 1 the willingness-to-pay concept of benefits is illustrated. One person's demand curve for a specific project, the output of which is in units of X, is given by D. If the beneficiary receives OX_1 units of the output, his willingness to pay—a compensating variation—is equal to the shaded area under the demand curve.[9] If the beneficiary were consuming OX_2 of the output prior to the resource reallocation, the benefit attributable to the reallocation would be shown by the cross-hatched area at the right. The gross benefits of a project, then, must include a willingness to pay for the outputs of the project by all those who are recipients of them.

Figure 1

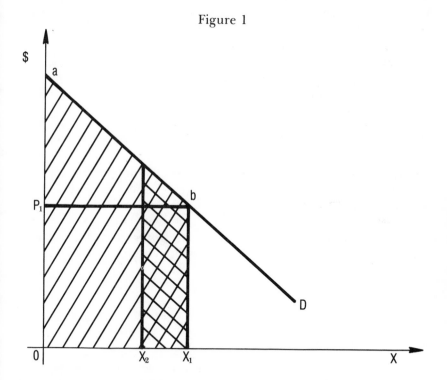

It should be noted that in addition to generating its own outputs, a public project may alter the pattern of demands for other goods and services. These altered demands may generate increases or decreases in

[9]If the beneficiary is forced to pay a price of, say, P_1 for the output, his compensating variation or willingness to pay can be decomposed into two parts—the amount which he actually does pay (OP_1bX_1) and his consumers' surplus (P_1ab).

consumers' or producers' surplus. The beneficiaries of these changes are willing to pay to gain or to avoid the change, and this willingness to pay must also be included in the estimate of benefits along with the value of the direct outputs of the project.[10]

In the discussion of the benefit concept to this point, we have appeared to assume that the public project generated outputs of a private-goods type—a private good being one which can be consumed by only one person, or at most, by one person at a time. Public projects, however, may yield benefits in a form that allows a number of persons to benefit simultaneously. When such "public" outputs are yielded by a project, the total willingness to pay of all beneficiaries is the maximum amount that *all* beneficiaries would be willing to pay for the outputs.[11] This theoretic welfare economics basis of cost-benefit analysis has been pointed up again recently by A. C. Harberger, who urges acceptance by cost-benefit analysts of "three basic postulates" which, if accepted, would permit stronger conclusions to be made about the desirability of a government project.[12] These postulates are:

1. The competitive demand price for a beneficial effect of a resource reallocation measures its value to the beneficiary.
2. The competitive supply price for an adverse effect of a resource reallocation measures its value to the cost bearer.
3. The net benefits of a resource reallocation are the sum of the beneficial and adverse effects valued by competitive supply and demand prices.

The measurement problem—that is, of determining how close an observed price is to a "competitive demand price" and a "competitive supply price"—implied by statements 1 and 2 is severe. In a real world economy that is full of monopoly and monopsony elements, taxes, subsidies, externalities, price controls, and other governmental regulations, one might accept propositions 1 and 2 without accepting the judgment

---

[10]For a discussion of the concept of rent related to an economic change and the need for this surplus to factor owners to be incorporated into the measurement of economic benefits, see E. J. Mishan, op. cit., Note B.

[11]In the case of private outputs, individual demand curves of beneficiaries are summed horizontally to get a group demand curve, the area under it representing the willingness to pay. For collective goods, individual demand curves are summed vertically and the area under the curve for the output yielded by the change (and made available to all beneficiaries simultaneously) is the willingness to pay. A distinction with relevance for benefit estimation is that between public outputs which are final outputs and those which are intermediate goods. See R. Musgrave, "Cost-Benefit Analysis and the Theory of Public Finance," *Journal of Economic Literature* 8 (September 1969).

[12]See A. C. Harberger, "Three Basic Postulates for Applied Welfare Economics: An Interpretive Essay," *Journal of Economic Literature* 9 (Sept. 1971), p. 785.

that *existing* prices satisfactorily measure either of the effects stated. If one rejects observed prices, however, "shadow" prices which more accurately reflect social values must be formed. The task of developing appropriate shadow prices is a difficult and treacherous one.[13] This difficulty is reinforced by the fact that these postulates require that the value of consumers' and producers' surplus be included in the measurement of net benefits.[14]

Proposition 3 requires, if it is to be of any normative relevance, that the income distribution which generated the net benefits is ethically satisfactory. Even under perfectly competitive market conditions, if the income distribution is judged unsatisfactory, then one might accept propositions 1 and 2 but yet reject the proposition—seemingly (though not logically) suggested by 3—that a project *should* be undertaken if its net benefits are positive. Our point is not to suggest rejection of Harberger's postulates but to underscore once again the essentiality of recognizing the policy relevance of the distributional and the efficiency consequences of public programs.

Many issues remain to be resolved as cost-benefit analysis continues to be refined and standardized. Some of the most important of these issues involving the concept of benefits will be examined in the section which follows. The chief objective of this section, however, has been to emphasize that cost-benefit analysis rests not on a set of arbitrary, *ad hoc* rules and procedures, but on a consistent conceptual foundation. This foundation, constructed out of the theories of allocative efficiency and income-distributional equity, is not only the basis of cost-benefit analysis but of all economics.

---

[13]See R. McKean, "The Use of Shadow Prices," in S. Chase, ed., *Problems in Public Expenditure Analysis* (Washington, D. C.: Brookings Institution, 1968); and J. Margolis, "Shadow Prices for Incorrect or Non-Existent Market Values," in R. Haveman and J. Margolis, eds., *Public Expenditures and Policy Analysis* (Chicago: Markham, 1971).

[14]*A technical note:* Harberger defines the change in welfare (net benefits) from any resource allocation (project) as the product of the change in an output of a good or service induced by the project and the competitive supply or demand *price* of the output. Where distortions are present in the production of the output induced by the project, the competitive supply or demand price must be modified by the deviation of marginal social benefit from competitive demand price or marginal social cost from competitive supply price. Where pre-existing distortions are present in activities indirectly affected by the resource reallocation, a further adjustment must be made, this time to the outputs which are not directly affected by the project. These indirect or secondary effects will be discussed later.

Harberger's postulates, it should be noted, do not explicitly address the question of the valuation of pure public goods generated by a resource reallocation. Nor do they capture all welfare effects of an exogenous resource reallocation when the reallocation alters the availability of resources, the technological options, or the trading conditions in external markets faced by an economy.

Evaluating Efficiency Benefits—
What to Include and What Not to Include

T HE LITERATURE OF COST-BENEFIT ANALYSIS contains references to numerous concepts of benefits which, it is argued, should or should not be included in evaluating net efficiency benefits under various sorts of economic circumstances. In this section, several types of benefits and their place in a theoretically sound concept of benefits will be discussed and evaluated. The concept of efficiency benefits developed in the preceding section forms the principal frame of reference for this discussion, but the relevance of distributional-equity effects (benefits) will also be noted. In addition, we will consider the possibility of accurately measuring the value of some of these types of benefits.[15] Since a given project effect may be characterized in a number of ways, the following benefit concepts are not mutually exclusive.

Primary (Direct) and Secondary
(Indirect) Benefits

The distinction between primary and secondary benefits, and the relevance of each, has long plagued the discussion and practice of benefit evaluation. Typically, primary benefits are taken to be the direct outputs or effects of—or, the increases in well-being resulting from—a resource reallocation project, such as the construction of a dam for water quality improvement, while secondary benefits are taken to be those indirectly generated by the resource reallocation either through the stimulative effects of the direct outputs or through the demand-inducing effects of the expenditures (inputs) required by the project.[16] For example, an investment in water quality improvement for a lake may directly increase the recreational use of the lake. The willingness to pay for the additional enjoyment of fishing, swimming, and boating is a measure of the primary benefit of the improvement. Now, because of the increased recreational use of the lake, lodging and eating facilities may be constructed around the lake, and these represent secondary effects of the improvement. The issue is: Should the profits (or any

---

[15]For general discussion of all the concepts examined below, see A. R. Prest and R. Turvey, "Cost-Benefit Analysis: A Survey," *Economic Journal* 75 (December 1965); and B. Weisbrod, "Concepts of Costs and Benefits," in S. Chase, ed., *Problems in Public Expenditure Analysis* (Washington: Brookings Institution, 1968).

[16]Two types of indirect benefits have sometimes been distinguished. These are "induced" benefits, caused by the demands induced by monetary payments generated by the resource reallocation—and "stemming" benefits, caused by the additional activities generated by the primary outputs of the resource reallocation. For a discussion of the economics of secondary benefits, see O. Eckstein, *Water Resource Development* (Cambridge: Harvard University Press, 1958); and J. Margolis, "Secondary Benefits, External Economies, and the Justification of Public Investment," *Review of Economics and Statistics* (August 1957).

other sum) associated with these secondary effects be added to the willingness to pay of the users of the lake in determining the total benefits of the project?

The answer to this question depends on the employment conditions in the economy. With relatively full employment, there is, in general, no basis for claiming the existence of any secondary benefits. This is not to say that there would be no hotels or restaurants built around the lake if it were improved, for surely there would be. The point is that in a full-employment economy, where there are essentially no idle resources, such resource-using construction requires that some other activities be foregone or reduced. Thus, under full-employment conditions—or more generally under conditions in which unemployment is constant and equal across activities—there is no presumption that the secondary activities add anything to aggregate national income or welfare. One expects that, on balance, the secondary benefits of a project will be cancelled by a reduction in benefits from the productive activities which were displaced. Hence, with normal circumstances prevailing in the economy, there is no basis for including indirect effects in the benefit concept.[17] If, however, there is significant unemployment in the economy—nationally or regionally—or if there are important obstacles to resource mobility, there may be a disparity between the indirect benefits generated by the project under consideration and the indirect benefits that would be foregone elsewhere if the project were undertaken. If the first of these benefits exceeds those that are foregone, an addition to primary benefits is warranted (or, what is equivalent, a reduction in costs); if the latter benefits exceed the former, a subtraction from primary benefits is required.

Let us apply the foregoing to our example, where a water pollution control investment stimulates additional recreational use of the lake and, as an indirect effect, generates increased hotel and restaurant activity nearby. If the inputs brought into use by the increased hotel and restaurant activity (waiters, cooks, etc.) would have otherwise been unemployed, the real social costs of their use is close to zero rather than the payments they actually receive. Hence, an addition to primary net benefits approximately equal to the payments to them is warranted. In effect, because their productivity is a net addition to national output, the income which they receive—and which is regarded as a cost by their employer—is, in fact, a social benefit.

Note that in a fully employed economy the *primary* benefits, too, are achieved at the cost of displacing some other project and its benefits.

[17]For further discussion of this conclusion, see A. R. Prest and R. Turvey, op. cit.; B. Weisbrod, "Concepts of Costs and Benefits," op. cit.; and M. Brewer, R. Haveman, C. Howe, J. Knetsch, and J. Krutilla, "Federal Natural Resources Development: Basic Issues in Benefit and Cost Measurement," National Resources Policy Center, George Washington University, May 1969.

Such "opportunity costs" must indeed be considered by the cost-benefit analyst. They enter in through the inclusion of an interest rate (cost of capital) and through the estimated cost of purchasing the resources required for the project—resources that must be bid away from competing uses. The fundamental cost-benefit problem, of course, is to determine which is greater—the benefits that would be produced by the project being evaluated, or the benefits that would be produced if those resources were used in some other way. The nature of that "other" way is a key issue. If some of the employed resources would otherwise be idle, it is clear that the proposed project will require less displacement of other activities and their benefits than would be required if these resources had been fully used. As a result, if all other things are held constant, a project which employs idle resources will yield greater net benefits than one which requires that resources be diverted from some alternative activity.

Although it is generally acknowledged that a project's use of unemployed resources must effect the cost-benefit analysis of the project, various ways of accounting for the use of otherwise idle outputs have been suggested. One government report[18] proposes that the market costs of otherwise unemployed resources be included as a benefit of the project. Others have argued that the deviation between the real cost and the market cost of the resources used be subtracted from project costs. The adjusted costs are, in effect, shadow prices designed to reflect the real social costs of the project.[19] In calculating net benefits, it does not matter whether the difference between the real social cost of resources used and their market cost is added to project benefits or subtracted from project costs. It should be noted that the ranking of projects by cost-benefit ratio is affected by the procedure adopted. However, the decision makers should not focus on the cost-benefit *ratio* in choosing between projects. Rather, the correct criterion is the maximization of the difference between benefits and costs.

In this same context, it should be noted that net efficiency benefits can be generated by a project if its outputs bring into use additional resources which would otherwise have been unemployed. This will occur if the income paid to unemployed inputs used by the project is spent in a way which leads to the employment of still other resources which would

[18]President's Water Resources Council, *Policies, Standards, and Procedures in the Formulation, Evaluation, and Review of Plans for Use and Development of Water and Related Land Resources,* Senate Document 97 (1962).

[19]For a more complete analysis of the effects of unemployed resources on public investment evaluation, see R. Haveman and J. Krutilla, *Unemployment, Excess Capacity, and the Evaluation of Public Investments* (Baltimore: Johns Hopkins Press, 1968).

otherwise have been unemployed. In either of these cases, what we have referred to as secondary effects become real efficiency effects.[20]

Real and Pecuniary Benefits

Although the distinction between real and pecuniary benefits is related to the previous discussion of primary and secondary benefits, it is a meaningful dichotomy of benefit categories in its own right. Real benefits consist of an increase in consumer satisfactions or a decrease in the amounts (costs) of resources required to produce goods and services. For example, an increase in the willingness of users of a lake to pay for an improvement in water quality reflects a real improvement in the consumers' satisfactions, and it must be counted as a benefit. Or, if a new municipal waste treatment plant reduces the expenses a downstream vegetable processor incurred in purifying river water prior to use, this reduction in the real amount of resources required by the vegetable processor is also a real benefit and must be included in the benefit estimate of the treatment plant.

A pecuniary benefit, on the other hand, represents a change in some people's well-being at the expense of the well-being of others; that is, it represents a redistribution of income. It should be termed a pecuniary "effect," not benefit, for while such an effect is a benefit to some people, it is a disbenefit to others. Pecuniary effects stem from price changes induced by a resource reallocation project. For example, an improvement in river quality might result in a bidding up of the price of fishing tackle. The sellers of tackle would benefit by an increase in profit, but their gain would be a dollar-for-dollar loss to the purchasers of fishing tackle. The pecuniary effect represents an income transfer from the purchasers of fishing tackle to the sellers of tackle. As a consequence, strictly pecuniary effects should be omitted in any estimate of the efficiency benefits of the proposed project.[21] For the economy as a whole there is no net welfare gain, unless some separate social judgment is made on whether a given dollar magnitude of gains to the particular group of gainers is or is not more "important" than the equal dollar magnitude of losses to the particular group of losers. Here, again, we see how efficiency and distributional equity are entwined.

---

[20] For a discussion of these effects and their treatment in current federal water resources project evaluation, see U.S. Water Resources Council, "Principles and Standards for Planning Water and Related Land Resources," *Federal Register* 38 (Sept. 10, 1973).

[21] For a discussion of a pecuniary benefit that is required by legislation to be included in an efficiency cost-benefit analysis see R. Haveman, *The Economic Performance of Public Investments* (Baltimore: Johns Hopkins Press, 1972), chap. 3 and app. B.

External and Internal Benefits

We have argued that it is important, for the purpose of project evaluation, to recognize that some benefits from a project expand total consumption (or production) potential for the entire society, while other benefits cancel out when viewed from a larger societal perspective. We turn now to the question of how the "society" is to be defined. Since decision makers—whether government officials or individual citizens in their daily lives—are expected to be concerned about the effect of their actions on some people (the included or "internal" group) but not on other people (the excluded or "external" group), we must understand the consequences of different ways of defining "internal" and "external."

In distinguishing external from internal effects, it is necessary to establish a definition that will include some individuals while excluding others. Those included and affected favorably by a resource reallocation project realize internal benefits; any benefits accruing to those excluded are, by definition, external benefits.

The definitional boundaries drawn to include or exclude people can be of many types. If a national accounting stance is taken in benefit evaluation, a boundary is drawn between persons within a country and those outside. In cost-benefit analysis the boundary is often drawn between the recipients of the direct outputs of a project and those who are incidentally affected by it. This distinction approximates that between primary (direct) and secondary (indirect) benefits discussed above. A third possible boundary can be drawn between effects which are within the domain of the agency undertaking a resource reallocation and those which are external to that agency's domain. A fourth boundary could lie between the "target group," for whom the benefits are intended, and the nontarget beneficiaries.[22]

Several points should be made with respect to any of these distinctions or boundary lines. First, for a comprehensive national benefit evaluation framework, the only meaningful boundary is that between persons inside and outside the economy within which the proposed resource reallocation project is to be undertaken. All welfare effects on persons in that economy should be included, whether the effects are of an efficiency or of a distributional equity variety. Welfare benefits or disbenefits to persons external to this group are appropriately excluded (in the absence of any special concern with them).

Second, given a national perspective, the boundary line between direct output recipients and the recipients of indirect effects is artificial and undesirable. Effects on all members of the society should be consid-

[22]"Target efficiency"—"vertical" and "horizontal"—is defined and discussed in B. Weisbrod, "Collective Action and the Distribution of Income: A Conceptual Approach," in R. Haveman and J. Margolis, op. cit.

ered although, as we have already noted, those project benefits which are termed "pecuniary" cancel out when the corresponding disbenefits to other persons in the society are considered. Welfare effects "to whomsoever they accrue" are all relevant for an evaluation of project benefits, but it is important to recognize that some effects serve only to redistribute economic welfare (the pecuniary effects) while others expand aggregate economic welfare opportunities (the real effects).

Note that if a subnational perspective were taken—regional, for example—adverse effects (disbenefits) accruing to persons outside the region might be disregarded by a planner-analyst. As a result, effects that were pecuniary for a country as a whole would be erroneously regarded as real, simply because the consequences outside the region were disregarded.

To see the problems resulting from planning decisions that take a subnational perspective, consider a locality that is required to install a major waste treatment facility to meet a federally-imposed water quality standard. In the absence of an explicit policy, the full cost of the facility would have to be borne by residents in the municipality, while the bulk of the benefits might accrue to downstream residents and river users. It is important for achieving both allocative efficiency and distributional equity—the two coincide here—that a larger perspective be taken by planners so as to encompass the downstream users. Governmental reorganization, or budgetary transfers—taxes and subsidies—are possible devices for "internalizing" the external effects.

A similar example would involve the construction of a dam to regulate stream flows for water quality purposes by a water pollution control agency. If the dam also generated flood control services (external to the domain of the water pollution control agency), an issue of interagency compensation might arise in which the external-internal benefit distinction would be crucial to an appropriate compensation or governmental reorganization scheme.

Tangible and Intangible Benefits[23]

In the above discussion of the concept of efficiency benefits, the welfare effects of a resource reallocation were measured in terms of willingness to pay—a notion which employs dollars as the common unit of account. If all of the welfare effects of, say, a public water pollution control project could be measured in dollar terms, we could say that all of the benefits are tangible. Intangible benefits, then, are those which cannot be valued in monetary terms—at least not at present. To designate effects as intangible, however, does not imply that it is impossible to

---

[23]For an excellent discussion of the importance of intangibles in benefit-cost analysis, see R. McKean, *Efficiency in Government through Systems Analysis* (New York: John Wiley and Sons, 1958), pp. 58-64.

conceptualize a monetary value for them. Rather, it implies that with the available data or empirical methods it is not feasible to place a money value on them. Indeed, the use of the term *intangible* to describe unmeasurable benefits and costs is, we believe, unfortunate. First, if the term means "unmeasurable," then it would surely be better to use that term. More important, however, is the fact that what is, and what is not measurable is far from clear, although this is not the place for a philosophic discussion of measurability in economics, let alone in more general terms. Suffice it to say that in the historical process of developing measures, variables that were not measured at one point later came to be measured, and that in the course of time constantly improved measures were developed. This is surely also the case for the benefits and costs relevant for economic analysis. Rather than talk about intangible or unmeasurable effects of programs, we should be talking about how to develop improved measures of those things which we are already measuring, and how to develop some measures of those program effects that we are currently not measuring.

The most important point to note is that the failure to measure quantitatively certain program effects—whether they are benefits or costs—does not in any way simplify or change the nature of a rational decision making process. Indeed, even though such program effects are not measured explicitly, there is no alternative but to take them into account in some way in the decision process. To the extent that the effect does not explicitly enter into the cost-benefit calculation, a decision regarding the project or program establishes an implicit value for the effect. Assume a proposed project yielding a reduction in water pollution with all benefits and costs measured except for the improved aesthetic value of the cleaner waterway. If the net measured benefits were negative—say, minus $100—and if the decision makers, after considering all effects, judged the project to be worthwhile, they would be placing an implicit value of at least $100 on the improved visual attractiveness of the watercourse. Such an implicit valuation is far less subject to scrutiny than a more explicit measurement, but it is nevertheless a benefit measure. In actual decision making processes, this sort of implicit valuation of program effects which have not been explicitly measured is common. Again, the question is not whether those "intangible" effects *should* be measured, and it is not whether they *can* be measured. Rather, the question is *how* should they be measured to be of most assistance to decision makers.

In the case of resource reallocations which have the improvement of water quality as their primary output, the problem of unmeasured effects is a serious one. While it is reasonable to expect that physical units for describing improvements in water quality can be devised, one can hardly expect the full range of welfare impacts from the improvement to be confidently stated in money values. To be sure, substantial progress has been made in devising empirical methods for assigning values to the willingness to pay for additional recreation services (swimming, fishing, and boating) from a water quality improvement.[24] However, some people simply want to be near a clean watercourse and enjoy its sights and sounds. It is a real benefit, although, for the time being, no data or empirical methodology can translate this enjoyment into monetary terms. Moreover, many individuals who do not now have plans to "use" the improved stream may well be willing to pay some amount to preserve the option of doing so in the future.[25] Again, data are not likely to become available soon to translate this real, though "intangible," welfare effect into monetary terms. Finally, without question, water quality improvements will influence patterns of regional development and growth. And again, while such redistributional or equity impacts[26] generally cannot be valued with much confidence, it is clear that some individuals do place a value on them.

Given the importance of unmeasured benefits, it is essential that they be handled appropriately in cost-benefit analysis. One reasonable way of dealing with these effects is to describe their nature in as explicit terms as possible, and where feasible, describe them in nonmonetary, though quantitative, units. Such indicators at least provide information on the importance of the impact, even though its monetary value is still beyond measurement. A second way of proceeding is to establish a reasonable maximum and minimum bound for the value of the unmeasured effect and then to display the effect of both bounds on the final calculation of net benefits. This sensitivity-analysis approach has the advantage of bringing the intangible effect into an explicit cost-benefit framework. Finally, as indicated above, the cost-benefit analysis can be presented in such a way as to indicate how valuable the intangibles would have to be if the resource reallocation is to be either economically worthwhile or preferable to some alternative reallocation.[27]

[24]See C. J. Cicchetti, *Forecasting Recreation in the United States* (Lexington, Mass.: Heath-Lexington Books, 1973); M. Clawson, "Methods of Measuring the Demand for and Value of Outdoor Recreation," Resources for the Future Reprint No. 10, 1959; and J. L. Knetsch, "Outdoor Recreation Demands and Benefits," *Land Economics* 39 (1963).

[25]The question of option demand and its relationship to the concept of efficiency benefits is discussed in the next section.

[26]For a discussion of the treatment of regional development effects in a national economic efficiency context, see A. M. Freeman and R. Haveman, "Benefit-Cost Analysis and Multiple Objectives: Current Issues in Water Resources Planning," *Water Resources Research* 6 (1970).

[27]These "intangible" effects have been cited as the basis of the debate over multiple objectives in the cost-benefit literature. See A. M. Freeman, "Project Design and Evaluation with Multiple Objectives," in R. Haveman and J. Margolis, op. cit. The issue of the legitimacy of allowing the decison maker to place his own subjective value on displayed intangibles in reaching a decision is unresolved. McKean, for example, argues that the display of

58

Some Special Benefit
Concepts and Issues

I N EARLIER SECTIONS, the fundamental principles underlying the concept of benefits were presented and applied to water pollution control. Emphasis was placed on the concepts of social economic efficiency and of equity, and on their relationships in project evaluation. In discussing the benefit concept from an efficiency point of view, the willingness to pay of beneficiaries of the output of a project was seen to be the theoretically correct benefit concept, if the distribution of income that generates the willingness to pay is regarded as acceptable. While the willingness-to-pay concept appears to be a straightforward one, its application in empirical analyses often raises subtle issues of interpretation and measurement. In this section, two such issues are discussed and recent literature dealing with them is summarized.

The first of these issues is that of *option value*: Does the area under the demand curve for the output of a project represent the full social value of the output when future flows of that output are threatened with irreversible (or very costly) destruction and there are no close substitutes for the output? In the second issue—that of *donor benefits*—the question is whether or not the equity effects of a project are simply disguised efficiency or willingness-to-pay effects. It has been argued that if the cost-bearers of a project themselves receive utility from supporting an activity which yields benefits to others, income redistribution may well be justified on an efficiency basis. In discussing each of these issues, the nature of the benefit is described and its relationship to the concepts of efficiency and equity benefits as we have earlier discussed them is analyzed. When appropriate, the means of dealing with these benefit concepts in actual analyses will be suggested.

---

intangibles should be given a subjective value by the decision maker in reaching a judgment on the efficiency of an undertaking. See McKean, op. cit., esp. pp. 133, 206-8, and 240-42. Mishan, on the other hand, argues: "The determination of a project, or of any part or effect of that project, by the political process is either (economically) arbitrary, or else, if it arises from any other consistent body of principles, is in conflict with the allocative criterion on which a cost-benefit calculation proceeds." See Mishan, op. cit., pp. 322-24. Both Mishan and McKean warn against the placing of an arbitrary value on intangible benefits by the economist in the absence of sound data or empirical methodology.

For examples of the presentation of monetary and nonmonetary benefits of specific projects, see J. Rothenberg, "Urban Renewal Programs," and B. Weisbrod, "Preventing High School Dropouts," in R. Dorfman, ed., *Measuring Benefits of Government Investments* (Washington, D.C.: Brookings Institution, 1965). A discussion of the importance (value) that would have to be placed on income-redistributional benefits to make a project worthwhile is presented in the next section.

Option Price and Option Value

During the past decade, a controversy has existed regarding the comprehensiveness of the "consumers' surplus" measure of economic welfare. This controversy stemmed from the suggestion that, for some types of resource reallocations and some individuals, consumers' surplus captures only a part of the full welfare impact of an economic change.[28] The portion of the full welfare effect not captured was labeled *option value*. This extra value was judged to exist when the individual was uncertain regarding his future demand for an output (because of uncertainty regarding his future tastes and income), when that output had no close substitutes, and when the future availability of the output was threatened and could not be replaced.[29] While most goods or services do not appear to qualify for the attribution of option value benefits, the amenity services offered by a unique natural site *would* appear to yield such benefits. Were, say, the Grand Canyon threatened with destruction, the loss of these option value benefits would be over and above the loss of consumers' surplus as conventionally understood and would have to be counted as a cost of the canyon's elimination.

Subsequent to this initial suggestion, there has been substanital debate regarding the meaning of "option value" and the circumstances under which it is relevant to the evaluation of public investment proposals. One position is that option value is already included in an appropriately defined measure of consumers' surplus and that it was an error to suggest that it was a benefit over and above the standard willingness to pay.[30] A contrary point of view suggests that option value must exist for unique irreplaceable assets because of the costs associated with uncertain future demands when individuals are risk averse.[31]

A more recent exploration of this issue, using a game theory type of approach, analyzes the form of the relationship between consumers' surplus, risk aversion, and uncertainty in both demand and supply.[32] This analysis leads to a third position which is related to the second: if an individual is risk averse and has an uncertain demand for the future

[28]B. Weisbrod, "Collective-Consumption Services of Individual-Consumption Goods," *Quarterly Journal of Economics* 78 (Aug. 1964).

[29]It is with respect to these services that the implications of uncertain future demands are of particular relevance. It should be noted, however, that option value is relevant to the proposed elimination of future services in which replacement is simply "costly."

[30]M. F. Long, "Collective-Consumption Services of Individual-Consumption Goods: Comment," *Quarterly Journal of Economics* 81 (May 1967).

[31]C.M. Lindsay, "Option Demand and Consumer's Surplus," *Quarterly Journal of Economics* 83 (May 1969).

[32]C.J. Cicchetti and A.M. Freeman, "Option Demand and Consumer's Surplus: Further Comment," *Quarterly Journal of Economics* 85 (Aug. 1971).

services yielded by an asset, the expected value of his consumer's surplus fails to capture all of the value of the asset. Some positive increment representing option value must be added to consumers' surplus in evaluating a resource reallocation involving that asset.

In a still more recent development, a state-preference model of uncertainty has been employed in the analysis of this issue and a conclusion, somewhat different from the others, has been drawn.[33] This result is based upon a theorem which states that an individual is risk averse when faced with some set of conditional incomes and a fixed price system only if the marginal utilities of the incomes are equal. From this theorem, it is demonstrated that, if there are no markets for contingent claims, option value may be positive, negative, or zero. The sign depends on the relative degree of risk present in the world with-and-without the particular resource reallocation project. If the project would make conditions riskier, then the option value would be negative, and vice versa. If there were perfect (insurance) markets for contingent claims—a theoretical case without a real world counterpart—option value would come to equal zero, even though individuals are risk averse. The contingent claims markets would, in effect, transform uncertainty to certainty. There would be no remaining option value because a market —and a price—for each option would exist. From this analysis, it is asserted that in the real world, where tastes are uncertain and contingent claims markets are imperfect, the expected value of consumers' surpluses is the best available approximation to the sum of their option prices in estimating benefits.

In comparing the analysis behind this fourth position with the analysis behind the third, it appears that the conclusion regarding the correct sign of consumers' surplus depends upon a judgment regarding the conformity of actual circumstances with the characteristics of the two analytical models employed—the state preference perspective (underlying the fourth position) and the game theoretic perspective (underlying the third position).

The relevance of this issue to cost-benefit evaluations in the water pollution area is clear: proposed water pollution control investments may have those characteristics which entail a positive option value: they may generate outputs (higher water quality) for which the future demands (for recreation and aesthetic uses) of risk averse individuals are uncertain, and those outputs may be replaceable only at high cost if current action is not taken (e.g., water quality investment in Lake Erie).[34]

[33]R. Schmalensee, "Option Demand and Consumer's Surplus: Valuing Price Changes Under Uncertainty," *American Economic Review* 62 (Dec. 1972).

[34]It should be noted, however, that the cost of recoupment in the case of water pollution control investments is typically not as serious as in the standard preservation-development case. This is true because of the automatic self-renewing quality of most watercourses.

Examples of this sort of investment may include water impoundments for low flow augmentation which provide future recreation services both at the site and downstream, and water quality improvements to retard deterioration of a water body which, when deteriorated beyond a threshold level, becomes irretrievable. For other efforts to control water pollution—such as standard municipal or industrial waste treatment investment—option value would appear to be of little importance. In any case, it should be noted that option value may be attached to both the outputs of a proposed investment and the outputs of the alternative use of the resources (these latter outputs become the cost of the proposed reallocation). Net option value of the reallocation will be positive only if the option value attached to the proposed activity exceeds the option value attached to the alternative.

Donor Benefits and Efficient Redistribution

In analyzing the welfare economics basis of the concept of benefits at the outset (pp. 37 ff.), we emphasized the relationship and the distinction between equity effects and efficiency effects. Until recent years, the separation of efficiency from the income redistributive effects of a resource reallocation was a standard analytical distinction. Efficiency was viewed as something which altered the size of the total income pie —about which economists were equipped to speak—while equity pertained to a strictly normative division of the pie among individuals in the economy.

Although the distinction has been a standard one, it is essentially a pedagogical tool. Alterations in income distribution can have real welfare effects, as do any output effects—private or public. The difficulty then is the practical one of developing concepts for defining and ultimately measuring the benefits of these effects.

Efforts to integrate equity and efficiency effects have proceeded along two distinct lines in the welfare economics literature. The first line of approach begins with a Bergsonian social welfare function in which changes in income to different individuals receive different weights. Somehow, policy makers perceive these weights and attach them to increments of benefits and costs to different individuals in evaluating the social welfare effects of resource reallocation projects. To the extent that these weights are unequal among individuals, such a social welfare function in effect asserts that increments of income have different social welfare effects depending on who receives the income.[35]

[35] For a discussion of this approach, see O. Eckstein, "A Survey of the Theory of Public Expenditure Criteria," in *Public Finances: Needs, Sources, and Utilization* (Princeton: Princeton University Press, 1961).

If this approach is used to incorporate the equity effects of a resource reallocation into the measurement of welfare, some means of
ascertaining the appropriate set of weights is essential. It is evident that
development of such weights is indispensable if equity effects are to be
counted as benefits and disbenefits, but the basis for deciding on the
weights is not clear.[36]

A number of attempts have been made to facilitate the empirical
estimation of these welfare weights. In one of them, an explicit set of
weights was derived from the increase in the marginal effective personal
income tax rates—apparently using an instance of congressional valuations of income increments to different people.[37] These weights were
then used in an experiment in the evaluation of the social welfare effects
(efficiency plus equity) of federal water resource investments. Another
attempt used data on the composition of public expenditures on water
resources to solve for weights indicating the Congress' appraisal of the
marginal utility of income to various classes of people.[38] A further attempt was directed less toward inferring weights than it was toward
establishing a procedure which would enable Congress or the executive
branch to fix a set of weights for evaluating the benefits and costs of
resource reallocation projects. For example, a three-stage process (from
the agencies, to the executive, to the Congress) has been proposed as a
mechanism to secure a *consistent* set of publically determined welfare
weights.[39] A final attempt involved the establishment of an equity objective as an explicit constraint, with projects being designed to maximize
the value of nonequity efficiency effects while meeting this constraint.[40]
Each of these attempts to ascertain explicit or implicit welfare weights
has serious limitations and each has been subjected to substantial evaluation and criticism.[41]

The second line of approach to the integration of efficiency and
equity has been to incorporate equity considerations in a Pareto-

[36]See I.M.D. Little, op. cit., pp. 121-22; R. Mack, "Comments" in Samuel Chase, ed.,
op. cit., pp. 213-22; and R. Haveman, "Benefit-Cost Analysis: Its Relevance to Public
Investment Decisions," *Quarterly Journal of Economics* 80 (Nov. 1967).

[37]R. Haveman, *Water Resource Investment and the Public Interest* (Nashville: Vanderbilt
University Press, 1965). See also K. Mera, "Experimental Determination of Relative Marginal Utilities," *Quarterly Journal of Economics* 83 (Aug. 1969).

[38]B. Weisbrod, "Income Redistribution Effects and Benefit-Cost Analysis," in S.
Chase, ed., op. cit.

[39]A. Maass, "Benefit-Cost Analysis: Its Relevancy to Public Investment Decisions,"
Quarterly Journal of Economics 78 (May 1966). See also M. McQuire and H. Garn, "The
Integration of Equity and Efficiency Criteria in Public Project Selection," *The Economic
Journal* 79 (Dec. 1969).

[40]S. Marglin, *Public Investment Criteria* (Cambridge: MIT Press, 1967).

[41]See A. M. Freeman, op. cit. ; R. Haveman, "Comment," in S. Chase, ed., op. cit.; R.
Mack, ibid.; P. O. Steiner, "The Public Sector and Public Interest," in R. Haveman and J.
Margolis, eds., op. cit.; R. Musgrave, op. cit.; and A. Harberger, op. cit.

efficiency welfare model and then to continue to rely on this efficiency-based welfare criterion in evaluating the benefits of resource reallocations. In one discussion of this approach, a "shoe-horning" of equity concerns into an efficiency model has been accomplished through dependence on the alleged existence of utility interdependence.[42] In effect, this approach asserts that donors receive benefits by transferring some of their own net worth to (less fortunate?) others. To the extent that such benefits exist, they can be incorporated into the strict Pareto-efficiency criterion by evaluating the willingness of donors to pay for the pleasures of such transfers.

While this approach has value in extending the neoclassical welfare model, it has come under severe criticism. The reasons offered for rejecting this approach involve both the ethical basis of the efficiency criterion, and the extremely limited potential for actually estimating donor benefits. A prominent critique[43] of this approach has emphasized the following points:

1. The interdependent utility criterion has zero operational value. To make it operational, the willingness of each donor to pay for a proposed transfer would have to be known explicitly. Knowledge of average donor benefits for a *group* of people will not suffice since the Pareto criterion requires *no* dissent from a proposed reallocation.[44]

2. The concept is deficient in that it cannot ultimately resolve the question of the appropriate distribution of income. This is so because it provides no basis for choosing among an infinite number of potential Pareto-efficient distributions (assuming all informational problems can be solved) which can be obtained from any initial distribution of real income. The problem of distribution remains an ethical problem.

3. The criterion contains no guarantee that the implied "desirable" redistributions will not violate accepted ethical standards. Such standards include the inadmissability of tax exemptions for the nonbenevolent rich and the forbidding of individual charity because of the excitement of envy.

While research and debate continue on the theoretic basis for integrating efficiency and equity benefits, two approaches seem both reasonable and widely accepted. The first involves the complete display of the distributional consequences of a proposed resource reallocation as a supplement to the efficiency cost-benefit calculation, thereby enabling decision makers to understand the full range of impacts of the decisions

---

[42]H. M. Hochman and J. D. Rodgers, "Pareto Optimal Redistribuiton," *American Economic Review* 59 (Sept. 1969).

[43]E. J. Mishan, "The Futility of Pareto-Efficient Distributions," *American Economic Review* 62 (Dec. 1972).

[44]Mishan describes the tax system required to insure Pareto-efficiency as one which "would require some of the very rich to contribute next to nothing to the poor and some of the middle income groups to contribute handsomely."

to be made.[45] The second is for the analyst to display a series of welfare-weighted cost-benefit calculations using alternative weighting functions. In this way, all observers—both plain citizens and policy makers—can perceive the distributive implications of alternative value judgments. In this way, as one analyst has put it: "The economist can perform experiments in policy evaluation using specific objective functions, treating the results as free of absolute normative significance."[46]

Conclusion

W HAT FORMS ARE TAKEN by the benefits of public projects? How should each form of benefit be measured? To what extent can these benefits be evaluated in terms that make them additive and then comparable to estimates of project costs? These questions are difficult to answer, but each must be answered—as best the decision maker can—if wise expenditure decisions are to be made. With special emphasis on environmental projects, this paper has concentrated on the first question, for until it is resolved, measurement cannot proceed sensibly.

The paper has examined many conceptual issues that are important to understand before any measurement efforts are initiated. Our most fundamental point, however, is the distinction between the "real social" effects (benefits or costs) of a project—those effects that expand the total production and consumption potential for the entire society—and "pecuniary private" effects—those that do not expand the total potential but, rather, expand it for some persons and contract it for others. This distinction is critical for a number of reasons.

First, it should remind the decision maker to be cautious lest he inadvertently take an incomplete view and count positive pecuniary-private effects without counting the corresponding disbenefits. It should also remind the decision maker that although some project effects are of a pecuniary sort—serving essentially to redistribute economic well-being, benefiting some persons but hurting others—this does not imply that such effects should be disregarded. Rather, the decision maker, having ascertained the relationship of total social gains and costs, should ask: Who are the gainers? Who are the losers? and Is this redistribution a favorable or unfavorable one in terms of the society's objective? These are not simple questions to answer, but in the long run wiser resource allocation decisions in the public sector will be made if the importance of redistributional considerations in project selection is recognized explicitly.

[45]McKean, op. cit., p. 133. "It is deemed appropriate here to have the cost-benefit measurements shed light on efficiency in this limited [national income maximization] sense, and to have further exhibits shed light on redistributional effects." See also Harberger, op. cit.; and Mishan, *Cost-Benefit Analysis,* op. cit., chaps. 47 and 52.

[46]O. Eckstein, op. cit., p. 447.

Second, if the decision maker recognizes the distinction between real and pecuniary (or redistributional) effects he may avoid the danger of focusing too narrowly on project benefits for some particular region, some particular industry, or some particular subgroup in the population. Too narrow a focus can blind him to unfavorable effects in other regions or in other groups. In short, effects that are simply redistributional for the society as a whole may appear to be real benefits in a perspective which includes only a part of the society.

Third, understanding the difference between real and pecuniary effects lessens the likelihood that certain effects will be mistakenly counted more than once. For example, an improvement in the quality of water in a lake is clearly a real benefit. Assume that a value was estimated for such an improvement. It would be double-counting if the increased value of the land surrounding the lake were also added, for the increased value would be a reflection of the enhanced value of the water as a real benefit—adding to the total consumption or production opportunities of the society.

As the last example illustrates, the distinction between real and pecuniary effects is more clear-cut at the conceptual level than it is in practice. However, with the economic welfare of a society depending not only on its total output of goods, services, and leisure but also on how they are distributed among the population, public decisions should reflect an understanding of the often subtle difference between these terms.

DISCUSSION

DISCUSSANT: *John V. Krutilla*

Despite the general clarity of this paper, a few issues still seem unclear.

In the first place, where compensation is infeasible, without making interpersonal comparisons we cannot be certain that the project serves efficiency, defined as a net welfare gain. But, assuming there are many such projects, each trying to win some surpluses which conventional markets fail to achieve, we have relied on the belief that with random incidence of such gains over sufficiently long periods, almost everyone shares in the gains of productivity and in real income benefits.

The problem, of course, comes up when the question of desirability of compensation is broached. If the distribution of income does not enjoy social sanction despite the explicit transfer programs of the government, a case is made that compensation of the injured by the beneficiaries is undesirable, provided the beneficiaries are in some sense the deserving and the injured in any event are the undeserving. There are times when I have trouble with this approach. The problems are of the following nature.

One, if the distribution of income is not being taken care of by some explicit transfer programs, we must acknowledge that the constellation of relative prices has no normative significance. Hence, calculating benefits and costs with nonnormative prices and costs does not, with any particular assurance, provide guidance for the improvement of welfare.

Two, when we consider that public intervention in resource development is generally undertaken because of market failure and is largely due to technological externalities, we must confront the fact that we end up primarily with the production and marketing of intermediate private goods, which is not the case when explicit income transfers are involved. It is much more difficult to identify the deserving when the primary beneficiaries are commercial establishments than when they are consumers of final goods. The redistributional implications of non-compensation—working through the production and marketing of intermediate private goods or services—are often perverse.

DISCUSSANT: *Henry M. Peskin*

The paper suggests that there is a suboptimization to the extent that a state or local decision maker applies cost-benefit analysis to a proposed project and considers only his state or local concerns. The question is: how serious do you feel this suboptimization is? I am afraid that if the decision maker has to take a national view of every project in his domain, he will be put under a tremendous burden.

REPLY: *Robert H. Haveman*

My judgment is that you can evaluate the seriousness of the problem of suboptimization, or partial accounting, only by looking at particular cases. Consider the case of navigation evaluation in the water resources area. Presently, we have, mandated through legislation, a benefit estimation concept called "savings to shippers." That is an example of an accounting framework that, in many projects, includes but a small portion of the total group of people affected by the project, namely those who actually make shipments. The use of that partial—as opposed to national—accounting framework leads to an enormous overestimation of benefits and the approval of a large number of projects at enormous cost that are economically inefficient and wasteful of resources. An example on the other side would be one where all or nearly all of the costs and benefits accrue to individuals within a local or regional jurisdiction. In this case, accounting for the impacts on only these people will not lead to a serious misrepresentation of cost or benefit estimates. Local waste treatment plants may be examples.

A Survey of the Techniques for Measuring the Benefits of Water Quality Improvement

A. MYRICK FREEMAN III

MEASURING THE BENEFITS of water pollution control means somehow determining the monetary value of the results of efforts to improve the quality of water in some water body. Basically, the methodologies for estimating benefits are techniques for inferring values of economic goods where market processes fail to reveal these values directly. Thus, benefit estimation often involves a kind of detective work—piecing together the clues about the values that individuals place on these goods as they respond to other economic signals. This paper is primarily about the detective's tool kit. That is, it describes and evaluates the available techniques for obtaining monetary measures of the benefits of water pollution control.

The discussion of these techniques is both theoretical and empirical. From theoretical welfare economics we derive the appropriate definitions of benefits and develop the economic rationale against which we can determine the conditions and assumptions under which the various techniques are valid.

The choice of an appropriate technique for measuring benefits also requires an understanding, based on such empirical data as exist, of the uses which man makes of waterways and the ways in which these uses are affected by changes in water quality—in other words what benefits are being sought, and how are they obtained? To use a simple example, we need to know how much a day of recreational activity is worth. We also need to know how many additional days of recreation are produced by a given reduction in water pollution.

There are three essential conditions which any technique of benefit estimation must meet if it is to fulfill the needs of water resource planners. First, since the objective is the comparison of values, the technique must lead to monetary measures of value or benefits. For example, a technique which predicts only the increase in recreation user-days may be useful, but it does not fully meet the needs of water resource planners for benefit measures which are commensurable with their monetary estimates of pollution control costs.

The second condition is that the technique be firmly based on economic theory and the concepts of willingness to pay and of consumer and producer surplus. The third condition is that some measure of water quality explicitly enter the analysis as a variable. For example, analysis of the demand for water-based recreation is not sufficient unless the analysis shows how that demand is affected by changes in pollution and the quality of the water bodies being analyzed.

First in order, this paper reviews the concept of benefits, describes the several kinds of benefits which can result from improvements in water quality, and outlines a framework for discussing the problem of benefit estimation. It then provides a classification of benefits, based on the underlying economic theory, in terms of the ways in which changes in water quality enter into economic processes to become or produce economic goods with values attached to them. It next describes and evaluates the available techniques for estimating the kinds of benefits found in each of the classifications previously discussed. The concluding section summarizes the discussion and presents recommendations.

A Framework for Benefit Estimation

Benefits

The net economic efficiency concept of benefits, which follows from theoretical welfare economics, is the one adopted in this paper. The benefit associated with a particular public action can be viewed as being the sum of the willingness to pay of all individuals affected directly or indirectly by that action. If we picture an income-compensated demand curve being defined for that action, benefits are equal to the area under the curve.*

Benefit estimation involves attempts, either directly or indirectly, to determine the shape of the income-compensated demand curve. If water pollution control were a marketed good, this would be a fairly straightforward econometric problem. But in most cases water pollution control is a public good since only one level of water quality can be provided at a time, and individuals are not free to vary independently the level of water quality they consume. Information on the shape of the resulting demand curve is not directly revealed by the actions of individuals. The possibilities for inferring some of the properties of this

*An "income-compensated demand curve" shows the relationship between the quantity of a good demanded and changes in its market price, with the purchasing power of the demander held fixed by making adjustments in his money income. Note that as the price of a good changes, it alters the real purchasing power of a given amount of money income. See fn. 8, p. 46. [Editor's note.]

demand curve indirectly will be considered in subsequent sections. The rest of this section is devoted to a discussion of the process by which the benefits are produced.

The Production of Benefits

The process by which the benefits of water pollution control are produced has three distinct stages. First, a reduction in the quantities of waste products being discharged into a water body leads to an improvement in various parameters or measures of water quality. Second, these changes in the characteristics of the water body lead to changes in the way individuals make use of the water body—that is, changes in the stream of services extracted from the water body. Finally, these changes in use, or service, have their counterpart in changes in aggregate willingness to pay for the uses of the water body—that is, changes in consumers' and producers' surpluses.

The first stage is almost totally noneconomic in nature since it involves a variety of physical and biological processes and relationships. The third stage is wholly within the realm of economics since it involves demand and production theory and economic values. The second stage represents the interface between the noneconomic and economic stages of the production of benefits. Understanding this stage is essential if empirical estimates of benefits are to be made. Yet this seems to be the least well understood of the three stages and is, perhaps, the most serious present barrier to more effective planning for water quality management through cost-benefit analysis.

In view of our earlier stipulation that water quality must enter the analysis of benefits explicitly, we must examine these three stages more carefully. (They are shown schematically in Figure 1.)

Stage 1. A single effluent discharge can contain many substances which affect water quality—for example, oxygen-demanding organic wastes (biochemical oxygen demand, or BOD), suspended solids, waste heat, or toxins. When these substances enter the waterway, they affect—in sometimes simple and in sometimes complex ways—such measurable components of water quality as dissolved oxygen, temperature, and chemical concentrations. For example, a nondegradable chemical substance will simply be diluted, and its concentration in the water body will be a calculable fraction of its concentration in the effluent stream. On the other hand, organic wastes affect water quality parameters in a more complicated way. As they are degraded by bacteria, they reduce dissolved oxygen levels to an extent and at a rate which depends on water temperature, wind, river flow rates, and other physical and biological characteristics of the receiving water.

Some of the physical measures of water quality, such as turbidity and smell, affect man's uses of the water directly. Also, these and other

Figure 1

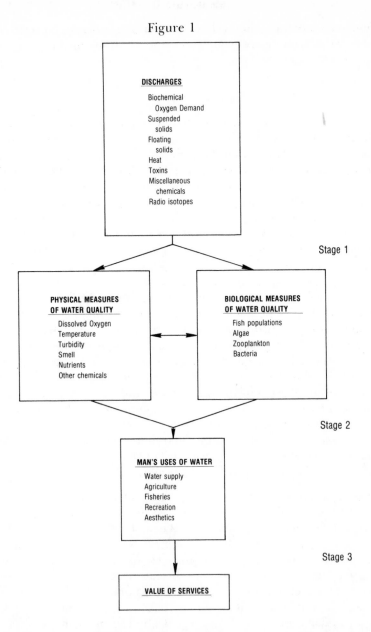

physical parameters affect stream ecology in complex and not always well understood ways. The populations and species distributions of fish, algae, zooplankton, and bacteria may all be affected, and not necessarily in the same direction, by changes in the physical-chemical parameters of water quality.

Even providing a descriptive characterization of the outcome of this first stage is a formidable task. Water quality cannot be represented by a single number on some scale but rather is an n-dimensional vector of the relevant parameters. Also, water quality varies across space and in time, further complicating the task of measurement and description. In subsequent discussion I will ignore these complications and, for expositional simplicity, will assume that there is a one parameter measure of water quality which fully describes the relevant characteristics of a water body. I will denote this measure as Q.

Stage 2. Man has many uses for his waterways. The beneficial uses of water which are affected by water pollution can involve withdrawal or diversion of the water or can be based on uses of the water body. Withdrawal uses include municipal and industrial water supply and irrigation for agriculture. In-stream uses include recreational activities taking place on or adjacent to the water body, and commerical fishing based on the biological and other productivity of the water body. Finally, aesthetic enjoyment can also be included in the category of in-stream uses to the extent that they affect individuals' utility.

MUNICIPAL WATER SUPPLY. Water pollution can affect municipal water supply either by raising the costs of treatment needed to provide safe and palatable drinking water, or by introducing harmful substances which are not removed in normal treatment processes. The effects of impurities in municipal water supplies can range from unpleasant taste and odor and damage to plumbing and appliances, to outbreaks of bacteria-related diseases and chronic long-term health affects due to the cumulative ingestion of impurities. On this latter possibility, there is very little information. The possible long-term adverse health effects of continued ingestion of heavy metals such as mercury, of inorganic salts, and of nondegradable or persistent hydrocarbons such as DDT and PCB have not, in my view, been adequately investigated.

INDUSTRIAL WATER SUPPLY. Where industrial activities withdraw water for in-plant use, pollution may either degrade the productivity of that water, or impose higher treatment costs. However, it is widely acknowledged that normally these effects are not economically significant. Either water quality requirements are so stringent as to require the full treatment of water drawn from any source, or quality requirements are so low as to permit the use of water from almost any source with no treatment beyond filtering out dead fish.[1]

AGRICULTURAL WATER USE. In the arid West, run-off and return flows from irrigated fields carry with them dissolved salts leached from the soil. The consequence of irrigation in upstream areas, therefore, is degraded water quality downstream. Where there is repeated rediversion of the river flow for irrigation purposes, the increasing salinity reduces the productivity of the irrigation water in agriculture. Re-

[1] In fact, in some cases improved water quality can apparently impose negative benefits on industrial water users. See Federal Water Pollution Control Administration (1966), pp. 71-74.

duced productivity of agricultural lands is a cost of this form of nonindustrial water pollution.

COMMERCIAL FISHERIES. Water pollution can affect commercial fisheries' production in a number of ways. Pollution of rivers can reduce or eliminate the spawning runs of commercially valuable species such as salmon, shad, herring, or smelt. Toxic substances can affect the biological productivity of estuarine areas which are the foundation of the biological food chain for commercially valuable species. Toxic substances can themselves reduce or eliminate populations of such things as crabs and lobsters. And finally, contamination from chemicals and bacteria can render surviving commercially valuable species unfit for human consumption. Thus, swordfish and tuna can be taken off the market because of contamination, and shellfish beds can be closed to commercial harvesting.[2]

RECREATION. One study which attempted to quantify the benefits of water quality improvement concluded that recreation benefits were by far the most quantitatively significant (Federal Water Pollution Control Administration, 1966). Water-based recreation activities which can be affected by water pollution include swimming, sports fishing, hunting water fowl, boating and sailing, and water skiing. Recreation benefits can be measured in units of user-days.

NAVIGATION. The presence of corrosive substances can shorten the lives of and otherwise damage vessels and such structures as wharves and pilings. Also navigation can be made more hazardous and be otherwise impeded by the presence of floating materials.

AESTHETICS. Aesthetics refers to other ways in which the quality of the water body affects the welfare or utility of those who live and work around it. Rather than speak in generalities, let me give a specific example. I feel myself adversely affected when I can smell the Androscoggin River a mile away as I walk across the campus at Bowdoin College. I am further adversely affected as I drive along the river's shore each day to and from work and see large globs of foam and scum. Since aesthetic effects are often not associated with direct use of the water, they pose severe measurement and valuation problems. But to the extent that they involve utility gains and willingness to pay, they are nevertheless every bit as real, in an economic sense, as recreation use.

HUMAN HEALTH. To the extent that pollution affects health by contaminating drinking water supplies, health effects have already been included in the listing set out here. There can also be adverse health effects associated with the impact of pollution on commercial fisheries, and recreation. Alternatively, health can be viewed as another dimension of human welfare which is adversely affected by water pollution.

---

[2]For economic rather than biological reasons, I am making a distinction between commercial fishing and sports fishing. The latter will be considered in the next paragraph under recreation.

However, the determination of the benefits of improving health, whether through improving the quality of drinking water, abating air pollution, or improving nutrition, poses some rather unique and difficult problems of method and empirical technique. They will be considered separately below.

NONUSER BENEFITS. There are a variety of benefits to nonusers of water, including existence value, preservation value, and option value (Krutilla 1967, Weisbrod 1964, Cicchetti and Freeman, 1971). The strategy for measuring benefits advocated in this paper is to observe the ways in which changes in water quality affect other economic decisions of water users so as to infer values by these indirect means. By definition, this strategy is essentially irrelevant for nonuser benefits. However, benefits to nonusers might be ascertained by questionnaires.

Stage 3. The rest of the paper is about stage 3, the determination of the economic values attached to the services affected by water quality. But the discussion cannot proceed very far without some prior reference to the relationship between water quality and value. The value of water for any one use may depend on a number of characteristics. For example, the value of a recreation user-day is affected by fish populations and species distribution, algae levels, the number and type of bacteria present, temperature, smell, turbidity, and concentrations of toxic substances. To further complicate matters, an increase in the magnitude of one characteristic may affect one use favorably while affecting an alternative use in a negative way. For example, higher temperatures may make for better swimming while adversely affecting trout and salmon populations. Industrial discharges of acids may adversely affect recreation and fisheries while improving the value of water for industrial uses because of retarded algae growth.

The difficulties in tracing out the effects of a discharge on the many parameters of water quality and, in turn, their effects on man's use of water, and our scant progress in this realm to date, substantially limit our ability to do careful cost-benefit analyses of water quality improvements. This does not seem to be fully appreciated by many advocates of greater use of cost-benefit analysis in this field. It was in 1968 that Allen Kneese wrote:

> I believe that our limited ability to evaluate the recreational losses associated with poor quality water, or conversely, the benefits of water improvement, is an extremely important barrier to rational water quality management . . . The first (complexity) is the relationship between the level of various water quality parameters and the recreational attractiveness of the water resource. This relationship can be viewed as being composed of two linkages: a natural one and a human one. I think these are both about equally ill-understood. It is my impression . . . that the biological sciences are almost never able to tell us specifically what difference a change in measured parameters in water quality will make in those biologi-

cal characteristics of the water that contribute to its recreational value. . . .
Perhaps the undeveloped state of forecasting is a result of the fact that
biologists have seldom been confronted with the types of questions we
would now like them to answer. . . . There is also a human linkage that is
ill-understood. What quality characteristics of water do human beings find
attractive for recreation? This is still largely an area of ignorance. [Kneese,
1968][3]

We need a better understanding of what attributes of water quality
influence individuals' decisions on whether to use a water body for recreation (or for other purposes) and determine the frequency with which
they use that water body. We also need to know how quality affects the
satisfaction that they gain from that use, and how it affects the nature of
the activities they undertake.[4]

To summarize the discussion up to this point, in the water pollution
field, we are committing scarce resources to pollution abatement in an
attempt to buy increases in welfare. Our applied welfare analysis (cost-
benefit) is based on the premise that dollar measures of willingness to
pay can be used as valid measures of welfare. The successful application
of cost-benefit analysis to the water pollution problem requires techniques for determining the dollar values of the outputs (i.e., the willingness to pay for the uses man makes of improved water bodies). But an
important requisite to this analysis is the ability to predict in what ways
and by how much man's use of waterways will change as a consequence
of pollution control efforts.

The Benefits of Q: A Classification

AS THE TERM *net economic efficiency benefits* implies, benefits stem from
resource reallocations that improve the efficiency with which resources
are converted to want-satisfying goods and services. Consider an
economy where all goods are supplied in competitive markets. If an
improvement in the allocation of resources increases the available quantity of one good, other things constant, there will be a decrease in the price
of that good and an increase in consumer surplus. The benefits accrue to
consumers. But in some circumstances, some portion, perhaps all, of the
benefits will accrue to factor inputs in the form of factor rents. In this case
producers' surplus must be measured and included in benefits.

However if producer and consumer surplus measures are to be
employed for empirical estimation of benefits, two basic assumptions
must be made. The first is that all prices other than for the good in

[3]The Resources for the Future interdisciplinary research team working on comprehensive, regional residuals-management modeling now includes an ecologist. For a
progress report on this work, see Robert A. Kelly (1973). [Complete bibliographical information on books and articles referred to in abbreviated form can be found in the alphabet-
ical list of references at the end of this paper.]

[4]For a first step in this direction, see David (1971). See also Kneese (1968), p. 181.

question are held constant. This assumption is necessary since benefit estimation involves areas under the demand curve or above the supply curve, and these curves are uniquely defined only under the *ceteris paribus* conditions that all other prices are constant. The second assumption is that the benefits do not change the real income of the beneficiary. Recall that benefits are defined as the area under the income-compensated demand curve. If real incomes do change, the measured demand curve diverges from the income-compensated demand curve and benefit estimates are distorted.

In this section and the next, Q represents a single parameter measure of water quality, or an appropriately constructed water quality index, where Q is assumed to be related to the beneficial uses of water quality in a known manner. Q enters economic processes in many different ways so as to become an economic good with economic value attached to it.

The benefits associated with a change in Q are equal to the area under the demand curve for Q over the relevant range. But since Q is not a marketed good, there may not be any direct way of estimating the shape of this demand curve from observed data. However, there are several situations where the way in which Q enters economic processes permits us to estimate willingness to pay by indirect means. The purpose of this section is to provide a partial classification of the ways in which changes in Q affect human welfare. The classification is based on the features of simple economic models constructed to describe these processes, and it permits the identification of several techniques for measuring benefits indirectly. This classification is not exhaustive; but it seems to cover the important cases, and it leads to distinctions which are useful in the discussion of the various estimation techniques.

Q as an Input to Production

This category consists of those cases in which water is used in some way in the production of a marketable good and where the productivity of water depends in part on its quality. This can be represented by a production function for good X of the following form:

(1) $X = X(K, L, \ldots, Q)$

where K and L are capital and labor respectively, and where the marginal product of Q is positive.

In this case, since Q affects the production and supply of a marketable good, the benefits of changes in Q can be defined and measured in terms of changes in the market conditions for good X. Assume perfect competition. There are three possible cases. The first is where X is a constant-cost industry and the change in Q affects the cost curves of a significant portion of the producers in the market. There will be a fall in price and an increase in total quantity. Figure 2 shows the case. The change in Q shifts the supply curve downward. The benefit accrues to consumers in the form of a decrease in the expenditure on the original

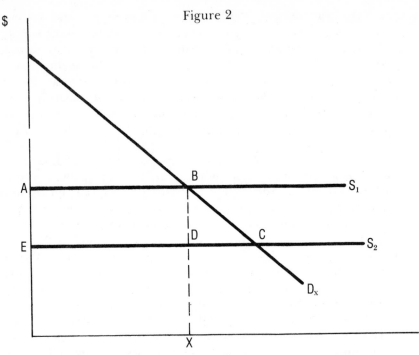

Figure 2

Benefits When Q is an Input in the Production of X: All Firms Affected

quantity demanded (equal to the area ABDE) plus the consumer surplus attached to the increase in the quantity (the area BCD).

The second possible case is where only one producer experiences a change in Q. In this case, the price of X does not change, and the benefit accrues to the owner of the fixed factor or the recipient of the residual income as a rent equal to the area ABCD in Figure 3. There is one special case in which this rent can be calculated in a straightforward manner from underlying production or cost data. When Q and some other purchased input are perfect substitutes in production,[5] and if the increase in Q lowers total cost without affecting marginal cost and output, the benefit is equal to the cost saving that the change in Q allows. This is an example of the so called "cost saving" or "cost of alternative" approach to measuring benefits to producers.

The third possible case is shown in Figure 4. Where the industry experiences increasing costs but the change in Q shifts cost and supply curves down to the right, there will be changes in both consumer and producer surplus. Benefits are equal to the net change in the sum of the surplusses, or ABCD in Figure 4.

[5]The analogous case for substitutes in consumption is discussed on page 80.

Figure 3

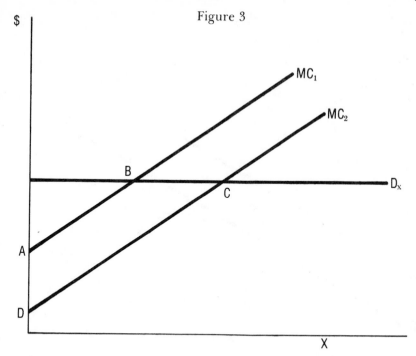

Benefits When Q is an Input in the Production of X: Only One Firm Affected

Q as a Complement to Other Goods

Karl-Göran Mäler has provided a rigorous proof that where a public good such as water quality is a complement to a private good in consumption, the benefits of increasing Q can be derived from information on the demand for the private good (Mäler, 1971). I will restate Mäler's results in a slightly different form so as to provide a clearer definition of the concept of complementarity, at least as it applies to the case under consideration.

Following Lancaster (1966), assume that individuals' utility functions are defined in terms of characteristics and/or consumption activities, rather than in terms of goods themselves. In this formulation of consumer theory, goods are inputs into a "consumption technology," the outputs of which are the characteristics of consumption activities valued by consumers. In symbols, this is described for a single individual by

$$(2) \quad U = U(\overline{V})$$
$$\overline{V} = V(\overline{X})$$

where $\overline{V}$ is a vector of characteristics and the second expression shows that the characteristics vector is a function of the vector of inputs of consumer goods. In Mäler's example the private good was a fishery

78

Figure 4

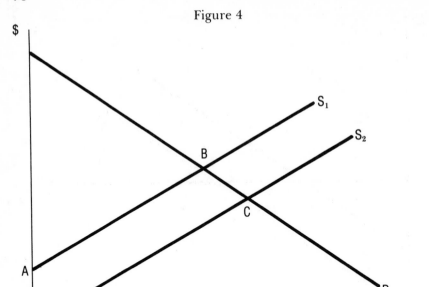

Benefits When Q is an Input in the Production of X: Producers' and Consumers' Surplus

where the unit of measurement was a day of fishing activity. The public good was the quality of the fishery, measured, perhaps, in terms of the number of fish caught per fishing day.[6]

The notion of complementarity employed by Mäler is embodied in the assumption that "those who do not use the fishery will generally be indifferent to quality changes" (p. 107). In other words, the marginal utility of the public good is zero if none of the complementary private good is being consumed. To put this into the context of the benefits of water quality improvement, let V_i represent a characteristic, "recreation experience," and let the "consumption technology" for recreation experience be

(3) $V_i = V_i (X_i, Q)$

where X_i is the number of days of attendence at a given recreation site (e.g., a lake), and Q is an index of the water quality of that particular site. Then Q is complementary to X_i if when $X_i = 0$, $V_i = 0$ and the marginal product of Q is 0 for all values of Q.

[6]Mäler specifically cites the early paper by Joe Stevens which is discussed below.

Figure 5

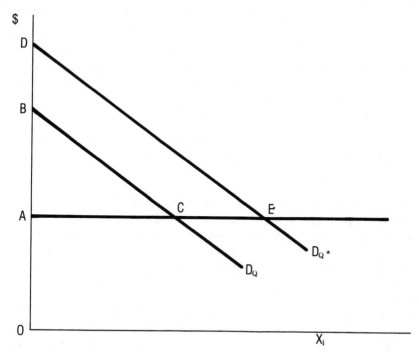

Benefits When Q is Complementary to X in Consumption

When the quality of the fishery is Q, the income compensated demand curve for fishing days is labeled D_Q in Figure 5. Assume that the price per fishing day is O-A.[7] The consumer surplus associated with the fishery is the area ABC under the demand curve. Now assume that quality improves to Q*. The increase in the quality of the fishery is assumed to increase the marginal utility of days spent at the site, thus shifting the income compensated demand curve outward to D_{Q^*}. The calculation of the benefits associated with this change is straightforward, and can be divided into three steps:

1. Given the old demand curve D_Q, postulate an increase in price from O-A to O-D. In order to leave the individual no worse off, he must be compensated by the area ABC.
2. Now postulate the improvement in quality and the shift in the demand curve to D_{Q^*}. Given the complementarity assumption described above, utility is unaffected; and therefore there is no need for compensation, either positive or negative.

[7]Of course, this price could be zero without affecting the analysis in any way.

3. Now postulate a return to the old price of O-A. The individual is made better off by the area ADE. In order to restore the individual to his original welfare position, he must be taxed by this amount.

The net effect of these three changes is a gain to the individual (in the absence of the taxes and payments described) of the area BCED. This is the benefit of the change in Q. But as Mäler points out, the identification of this area as the benefit of the improvement in Q would not be possible without the assumption of complementarity.

Q as a Perfect Substitute in Consumption

Define the utility function for an individual as follows:

(4) $U = U(X_j, X_i + aQ)$ $j \neq i$

where X_j and X_i are private goods which can be purchased at prices P_{x_j} and P_{x_i}, respectively, and Q is a public good, i.e., its consumption by the individual is equal to the total amount of the good which is supplied. The coefficient, a, represents the terms on which Q can be substituted for X_i. It is the marginal rate of substitution (MRS) between X_i and Q. If the MRS is known, the calculation of the benefits of a change in Q is straightforward.

The individual's indifference map for X_i, the private good, and Q, the public good, is shown in Figure 6.* The individual begins with Q^1 of the public good and X_i^1 of the private good. Now assume that the supply of the public good is increased to Q^2, moving him to indifference curve U_2. The individual could be taxed by an amount equal to $X_i^1 - X_i^2$ to leave him on the original indifference curve U_1. The dollar benefits of a change in Q, therefore, are:

(5)   Ben. $= P_{x_i} (X_i^1 - X_i^2)$
      $= P_{x_i} \cdot a \cdot (Q^2 - Q^1)$

Since Q and X_i are perfect substitutes, only knowledge of the objectively determined substitution coefficient is required to determine benefits. Where Q and X_i are less than perfect substitutes, i.e., where the indifference curves are convex, estimating benefits also requires more information on the utility function for Q and is tantamount to direct estimation of the demand function for Q.

Q as a Characteristic of Land

Economic theory has long recognized that the productivity of land differs across sites. The classical theory of rents holds that productivity

---

*An indifference curve shows consumption combinations of two goods which represent a given level of satisfaction. It is usually assumed that such curves exist over all levels of consumption for the goods in question; hence, in theory there is an infinite family of curves (an indifference map) of which only two are presented in Figure 6. [Editor's note.]

Figure 6

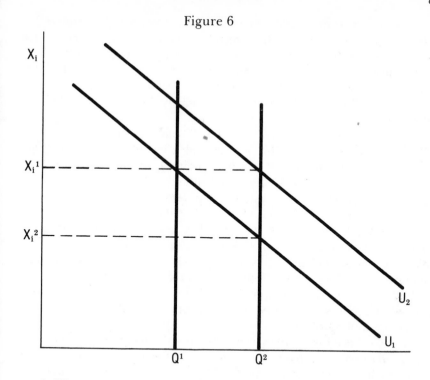

Indifference Curves U_1 and U_2 When X_i and Q are Perfect Substitutes

differentials will yield differential rents to land and therefore differential land values. Where land is a producer's good, competition and free entry are sufficient to assure that productivity differentials are fully reflected in the land rent structure. For any property where the land rent is less than the productivity, the activity occupying that land must be earning a profit. Some potential entrant will be willing to bid above the going land rent in order to occupy the site and reap the rewards of a superior productivity. It is this competition which bids up land rents to eliminate the surplus or profit. However, where the principal use of land is as a consumer good or an input into the household production function, the above result will not necessarily hold, primarily because productivity is not unambiguously defined in this case. A household may have a total willingness to pay in excess of the rent being charged for a particular site, and therefore be earning a consumer surplus. But so long as all individuals do not have identical utility functions, there may be no other household willing to bid up the land rent to eliminate the surplus.

Some environmental characteristics such as air or water quality may affect the productivity of land either as a producer's good or a consumer good. Where this is so, the structure of land rents will reflect these environmentally determined productivity differentials, perfectly where land is used as a producer's good, but only imperfectly where land is used as a consumer good. These results from classical rent theory have aroused considerable interest among economists in the possibility of using land rent or land value information as a measure of the benefits due to changes in environmental characteristics such as air or water quality. To illustrate the issues and problems involved, consider this example.

Suppose there are two lakes identical in all respects except that one lake has grossly polluted water (low Q) while the other lake is clean (high Q). Let p_1 be the rent of land around the dirty lake and p_2 be the rent of land around the clean lake. Clearly $p_2 > p_1$, since if $p_2 = p_1$, no one would wish to occupy land around the polluted lake.

Now suppose that the pollution is eliminated in the dirty lake so that both lakes have the same Q. There will be a single rental value for all land around both lakes. Let that be p_3 where $p_1 \leq p_3 \leq p_2$. The question is: Will the change in aggregate land rents equal the benefit of ending the pollution of the dirty lake? Lind (1973) has shown that under certain conditions the answer is yes. The problem is complicated by the likelihood that a change in the productivity of some sites will trigger a cycle of relocation among different sites by occupants. One of the major contributions of the Lind paper is to show that the benefits of eliminating pollution can be approximated by measuring the total net increase in the productivity of only those activities that locate on the affected land. In other words, as a first approximation, one need not be concerned with effects on activities locating on nonimproved land. Lind goes on to show that if it can be assumed that land rents fully capture productivity differentials, the benefits of pollution control can be approximated by the increase in the value of the affected land alone. Specifically, the change in the value of land directly affected by the project is an upper bound on the value of benefits, so long as rents are set so as to eliminate all surpluses.

Although this is an important result, I have argued elsewhere (Freeman, 1974a) that it is of limited practical significance for empirical benefit estimation. Briefly, if benefits are to be reflected in land rent changes, all other prices must remain unchanged. Lind's result requires the further assumption that rents eliminate all surpluses. But it is unlikely that both assumptions can be satisfied in practice. Suppose the new rent, p_3, is less than p_2. Then either those originally occupying the land around the clean lake are now earning surpluses, or there must have been resource reallocations, entry, and price changes to compete away those surpluses. One or the other of the two assumptions is violated; and rent changes do not measure benefits. The zero surplus assumption requires that $p_3 = p_2$. This in turn requires that those occupying land

around the two lakes have identical willingness-to pay-functions for Q.
But it will not, in general, be possible to make such a strong assumption.

Although land value changes cannot in general be equated with
benefits, Lind has shown that land value differentials existing before the
cleanup can be used to obtain approximate estimates of benefits. Specifically, an upper estimate of the benefit for improving any one site is
given by the difference between the present rent of that site and the
present rent of comparably improved sites. This is $(p_2 - p_1)$ in the above
example. This result is independent of the zero surplus assumption.

While Lind's result is an approximation which holds for nonmarginal changes in Q, I have shown that an individual's willingness to pay
for Q at the margin is equal to the marginal change in land rent in
moving from the site the individual occupies to one with marginally
higher Q (Freeman, 1974b). This follows from the conditions of household utility maximization which require that the household equate its
marginal willingness to pay for Q with the marginal purchase price for
Q.

Assume that Q is the only distinguishing characteristic of land and
that land rent (V) is an increasing function of Q. This rent function is
shown in Figure 7a. If the household is assumed to be a price taker in the
property market, it can be viewed as facing an array of alternative property value/quality combinations. The household maximizes its utility by
moving along the array in the direction of increasing Q until that point
where its marginal willingness to pay for an additional unit of Q just
equals the additional cost or marginal purchase price of Q.

Figure 7

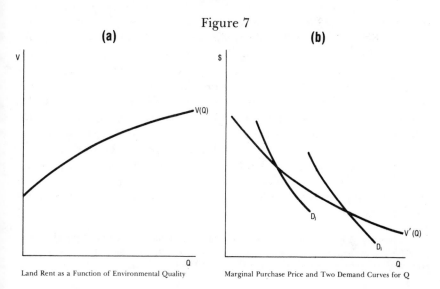

(a)

Land Rent as a Function of Environmental Quality

(b)

Marginal Purchase Price and Two Demand Curves for Q

Figure 7b shows the marginal purchase price function, which is the first derivative of the rent function V(Q). Also included are illustrative curves of the marginal willingness to pay (D_i and D_j) for two individuals. Although one point on each individual's marginal willingness-to-pay curve can be determined from land rent data by this method, these data will not reveal anything about the individual's willingness to pay for nonmarginal changes in Q.[8]

Techniques for Determining Values

In this section, several techniques for estimating values are described and related to the theoretical discussion of the preceding section. In most cases, these techniques have been developed to the point of empirical application. Where appropriate, references to examples are provided. The assumptions and conditions which are required to make the technique valid as an estimate of benefits in terms of the underlying theoretical welfare economics are stated and explained.

The Consumer Surplus Approach for Marketed Goods

Where water quality is an input in the production of a marketable good, but the change in water quality has such an extensive effect as to shift the supply curve for the marketable good, benefits must be inferred from changes in the price and marketed quantity of the good. Assume competition and a constant-cost industry. This case was briefly described and illustrated in Figure 2. The benefit of the water quality improvement has two components: the resource-saving on the original quantity demanded due to the price decrease, and the consumer surplus associated with the increased quantity demanded. If it can be reasonably assumed that all other prices are approximately constant and that the change is small relative to total income so that real income is approximately constant, then benefits can be measured by the appropriate areas under the demand curve. Application of this technique requires only that the analyst be able to calculate the cost change for the marketable good and that he have information on the demand function over the relevant range.

---

[8]However, I have shown that in the case of air pollution, it should be possible to use pooled data on land values from several cities to estimate the demand curve for clean air. A two-step procedure is required. See Freeman (1974b).

Net Factor Income

When Q is an input into the production of a marketable good, under certain conditions, the economic effect of a change in Q is to change the incomes of fixed factors of production. If the production unit in question is small relative to the market for the final product, and small relative to the markets for variable factors, it can be assumed that product and factor prices will remain fixed after the change in Q. The increased productivity accrues in the form of profit or in the form of a surplus income (quasi-rent) to the fixed factors of production. With sufficient knowledge of the operation of the production unit in question, it should be possible to use plant-engineering or economic data to estimate the changes in factor incomes in advance of the change in Q.

In the case of agricultural production, agronomic data may be used to determine the relationship between salinity of irrigation water and crop yields. Farm budget and farm practice studies can be used to determine the changes in factor inputs that would be associated with changes in the quality of irrigation water.[9] If the crop in question is one for which there is no federal intervention to support prices, the change in gross farm income is equal to the change in crop yield multiplied by the market price. Changes in expenditures for other inputs must be netted out to determine the change in net farm income. The only pitfall to be avoided is the possibility that the price used to value the output may overstate the true social value of the crop if price support policies are in effect.

The net income approach is also applicable to the case of commercial fisheries. Again, if the fishery in question is small relative to the total market for that fish product, it can be safely assumed that product price will be unchanged. Data on the relationship between biological productivity and water quality can be used to determine the increase in net physical yield associated with the pollution control program. If the fishery is being appropriately managed to maximize net economic yield, the benefit of the increase in water quality is equal to the market value of the increased yield, net of any changes in expenditures on other variable factors of production.

However, the assumption of maximizing net economic value is not likely to be valid. The economic benefit of the improvement in quality can be considered as an increase in the economic rent attributable to the fishery resource. But in an unregulated, free-access fishery, this rent accrues in the short run to the existing fishermen, and raises their income above their opportunity costs. The short run profits attract addi-

[9]The Bureau of Reclamation has long used this technique to estimate the benefits of providing irrigation water.

tional resources to the fishery, and ultimately this dissipates the rent to zero (Gordon, 1954; Crutchfield and Pontecorvo, 1969). In the absence of institutional reforms to improve the resource allocation in commercial fisheries, improvements in water quality will yield only transient benefits in the form of increased factor incomes. A policy of reallocating resources to improve water quality and thus improve commercial fishing in a physical sense will trigger an offsetting misallocation of additional resources to the fishery, ultimately resulting in zero benefits in this form.[10]

Cost Savings in Production

Where Q is a perfect substitute for other inputs in the production of a good, an increase in Q leads to a reduction in factor input costs. Where the change in total cost does not affect marginal cost and output, the cost saving is a true measure of the benefits of Q. Even where there is a reduction in marginal cost, if the reduction and corresponding increase in output are relatively small, this technique could still be used to provide a lower bound estimate of true benefits.[11]

In some circumstances, municipal water supply benefits might best be measured by this technique. If the water quality improvement simply results in a decrease in treatment costs (i.e., if there is no improvement in the quality of the delivered water and if there is no change in the price of the delivered water), benefits are measured by the reduction in treatment costs. But there are two other possibilities. Either quality or price (or both) might be affected by the reduction in water pollution. If the quality of the delivered water is improved and, for example, some individuals reduce their purchases of bottled drinking water and use more tap water for drinking and cooking, the situation is analytically equivalent to the complementary-good case described above (pp. 77–80). Water quality acts as a shifting parameter in the demand function for drinking water; and the benefit is the area between the demand curves. Where product price changes, the consumer surplus approach is appropriate.

The cost saving approach may also be appropriate for assessing industrial water supply benefits. Engineering data can be used to determine whether an improvement in water quality will result in lower treatment costs in a given industrial process. If, as seems likely, it can be

------

[10]However if the fishery is large relative to the market, so that the increased physical yield depresses the price, the above result does not hold. With the price decrease there is a consumer surplus benefit to consumers; but factor rents still sum to zero before and after the change.

[11]The cost saving approach is a variation of the cost-of-alternative-techniques approach employed to estimate the benefits of hydroelectric generation projects.

validly assumed that factor and product prices will not change, economic benefits can be readily calculated by this approach.[12]

Estimating the Demand for
a Recreation Site

Entry to a recreation site is a private good in the sense that it is divisible and that the exclusion principle can be applied. Therefore, since the water quality of that site can be considered as a complementary public good, the benefit of improving water quality there can be measured by estimating the demand curve before and after the quality change and determining the appropriate area between the demand curves. The Clawson-Knetsch (C-K) travel cost method can be used under certain conditions to estimate a demand curve for a given recreation site. At this point, I will first briefly describe the C-K method for estimating the demand curve;[13] then I will discuss several aspects of introducing quality differences into the analysis. The latter is essential for the estimation of the benefits of quality changes.

It must be emphasized that the C-K method is site-specific; it estimates the demand function for a specific recreation site rather than for a generalized recreation experience. While this may reduce the usefulness of the C-K method for some purposes—for example, predicting total recreation activity over some period of time—it is precisely this feature which makes the approach attractive for estimating the economic value of water quality improvement at specific recreation sites.

In implementing the C-K technique, two assumptions about the nature of visits to the recreation site must be made. The first is that the primary purpose of the trip is to visit the site. When there are multipurpose trips, at least some portion of the total travel cost is a joint cost which cannot be allocated meaningfully to the visit to the recreation site. The second assumption is that there are no close substitutes for this recreation site within the area being studied.[14] For a given recreation site, the surrounding area is divided into concentric circular zones for the purpose of estimating the travel cost from each zone to the site and return. The procedure is as follows:

1. Visitors at the site are sampled to determine their zones of origin.

[12]See Federal Water Pollution Control Administration (1966), pp. 71-74.

[13]For more detailed expositions of the method and discussions of problems, necessary assumptions, and applications to a variety of empirical problems, see Clawson and Knetsch (1966); Knetsch (1964); Cesaire and Knetsch (1970); and Cicchetti, Freeman, Haveman, and Knetsch (1971).

[14]For a more refined model that explicity incorporates possible substitution effects among sites, see Burt and Brewer (1971); and Cicchetti, Fisher, and Smith (1972).

2. Visitation rates defined as visitor days per capita are calculated for each zone of origin.
3. A travel cost measure is constructed to indicate the cost of travel from the origin zone to the recreation site and return.[15]
4. Socioeconomic data for each distance zone are gathered.
5. Visitation rates are regressed on travel cost and socioeconomic variables such as average income, median educational attainment, etc. The regression tests the hypothesis that visitation rates depend in part on travel cost.
6. The observed total visitation for the site from all travel cost zones represents one point on the demand curve for that site, i.e., an intersection of the present horizontal price line (either at zero price or the typical nominal entry fee) with the true economic demand curve.[16] To find other points on the demand curve, the C-K method assumes that visitors will respond to a $1 increase in admission price in the same way they would to a $1 increase in computed travel cost.[17]
7. To find the point on the demand curve for the site at an admission price of $1, use the estimated visitation rate equation to compute visitation rates and total visits for all time zones at the existing travel cost plus $1. These calculations are repeated for higher and higher hypothetical admission prices until the full demand curve is traced out.

This exercise has two basic purposes. One is to enable prediction of visits to the recreation site at alternative admission prices. The second is to determine the economic value or consumer surplus associated with that recreation site by calculating the area under the estimated demand curve (by integration).

To use the C-K method in estimating the benefits of a water quality improvement, we must have estimates of the demand curves for the

[15]Early studies used money travel cost, calculated in terms of cents per mile as a variable. Cesario and Knetsch (1970) have shown that the exclusion of time cost biases the estimates obtained by this procedure. For empirical estimates which do attempt to adjust for this time bias, see Cicchetti, Freeman, Haveman, and Knetsch (1971).

[16]If the site is characterized by congestion among users, the C-K approach is not valid. The observed point on the demand curve represents an intersection of the unknown demand curve with a supply/congestion cost curve of unknown shape. The absence of congestion is required for the establishment of the necessary a priori restrictions on the parameters of the model for proper identification of the demand curve. For discussion of the identification problem in connection with estimating the demand for recreation sites, see Cicchetti, Smith, Knetsch, and Patton (1972); and Cicchetti, Fisher, and Smith (1973).

[17]When time cost is included, it is converted to a dollar value by the choice of an appropriate value for the opportunity cost of time. Of course, estimates of the true demand curve are highly sensitive to the choice of this key parameter. Far more research needs to be done on how recreationists value their leisure time. In addition, some economists would argue that in principle time costs cannot be meaningfully or uniquely measured.

recreation site both before and after the improvement has taken place. One possible approach is to estimate the demand curve for a recreation site at a polluted waterway (assuming that recreation usage was greater than zero) before the pollution abatement project was begun, and then to re-estimate the demand curve after the project has been completed.[18] The new demand curve should be above and to the right of the old one; and following the method of Mäler, recreation benefits would be equal to the area between the two demand curves. This kind of ex post measurement would, itself, be valuable, particularly as a check on other methods for predicting ex ante benefits. But it does not meet the real problem posed by cost-benefit analysis, i.e., how to estimate benefits before the fact in order to plan for a more rational commitment of resources to water pollution control.

Analytically the problem can be stated as one of estimating a multivariate demand function of the form

(6) $X_i = X(P_i, Q)$

where there are no direct observations of changes in either independent variable. As described above, the derivative with respect to price can be inferred in the absence of direct observation of price changes by assuming that individuals respond to price changes and to travel cost changes in the same way. But the Q variable poses two problems. The first is to determine what empirical measure of Q is appropriate. The second is to determine a set of plausible assumptions on the basis of which the partial derivative can also be inferred indirectly.

Two studies of the recreation benefits of water quality improvement have dealt with these two problems in quite different ways. Stevens (1966) was analyzing a sports fishing resource (Yaquina Bay, Oregon), and defined Q as the number of fish caught per angler trip, or the success/effort ratio. Reiling, Gibbs, and Stoevener (1973) used subjectively determined variables for each of four activities (swimming, boating, water skiing, and fishing) to indicated the desirability of the site for that activity. They called these variables "use-intensity levels." Neither study had an adequate basis for relating changes in its measure of Q to changes in discharges or to specific abatement measures.

In both studies, demand curves for recreation sites were estimated using some variant of the C-K technique. Changes in their measure of Q were postulated, and new demand curves were calculated. The value of the change in Q could then be determined from the area between the two demand curves.[19] The question is: how much does the demand curve shift for a given change in Q?

[18]Of course, the estimations should control for other variables which change over time, such as income and population.

[19]Stevens erred in his definition of the change in welfare. See Stevens (1966, pp. 179-80; 1967; 1969) and Burt (1969).

Stevens estimated the relationship between total angling effort (X) and the success/effort ratio (Q) from time series data from other sport fishing areas along the Oregon coast. Regression analysis was used to estimate the coefficients and the elasticities of X with respect to Q for daily, weekly, and annual time periods. Since one might expect the daily and weekly data to be dominated by such factors as weather and fish movement, Stevens argued that the annual data might best indicate the possible longer term relationship between pollution-caused changes in the success/effort ratio and the resulting changes in total angler effort. Therefore, Stevens assumed that the elasticity of X with respect to pollution-caused changes in Q at Yaquina Bay would be the same as the elasticity estimated from annual data for another fishing area where pollution was not a factor affecting angler success. There are two critical questions with respect to Stevens' method. The first is whether the alternative sites chosen for estimating the effects of quality changes are in all other respects equivalent to the site being analyzed. In other words, can a relationship derived from one data base be applied in a different context? The second is whether the behavioral response to changes in angler success is invariant to the cause of changes in success.

To determine the relationship between use-intensity level and demand for the recreation site, Reiling et al. gathered data on recreation visits to four water-based recreation facilities in Oregon. The data for the four lakes were pooled in multiple regression equations in which a measure of participation was the dependent variable; and travel cost, on-site cost, and use-intensity levels were the independent variables.[20] These pooled regression equations yielded a single parameter for each of the independent variables. That parameter was assumed to be valid for each of the four lakes under consideration. To determine the demand curve for one lake, its site characteristics and travel cost characteristics were inserted into the general regression equation and values for visits from each distance zone were computed. To compute the benefits from a postulated change in water quality, higher values of the use-intensity variables were inserted into the equation and visitation rates were recalculated.

In summary, the C-K method is an operational technique which is theoretically valid for estimating the demand curve for a specific recreation site. The major considerations in applying the technique in general are the construction of the travel cost variable, and the difficulties when

---

[20]Apparently values for the use-intensity variables were assigned on the basis of actual levels of use or participation. This can be criticized on two grounds. First, use can be influenced by factors other than quality. Thus the use-intensity index may measure something in addition to or instead of quality per se. Second, if the use-intensity variable does reflect the number of users, the regression equations of Reiling et al. are, in effect, using a measure of visits to explain visits.

there are substitute or competing recreation sites. In applying the technique to the estimation of water quality benefits, the principle operational problems are the selection of the appropriate quality variables to include in the demand analysis, and how to determine the relationships between quality and use.* Recent investigators have experimented both with pooling data from several sites and with using observed relationships from other sites. These experiments must be considered as approximations whose validity and accuracy have yet to be determined.

Defensive Expenditures

Defensive expenditures are expenditures made to prevent or counteract the adverse effects of pollution. Assume for a moment that defensive expenditures are a perfect substitute for reductions in the level of pollution effects experienced. Then an individual can effectively purchase the optimal amount of Q through defensive outlays. And changes in these outlays, which accompany changes in Q, will reveal his marginal willingness to pay for Q.

In practice, defensive outlays which would be perfect substitutes for Q would be rare. There is no such thing as a perfect defense. There are some disutilities associated with pollution that cannot be prevented by further spending. Hence changes in defensive outlays are likely to give underestimates of the true benefits of the changes in Q. Nevertheless, recognizing this limitation, analysis of changes in defensive outlays related to water pollution could substantially narrow our range of ignorance about benefits. Candidates for study include expenditures on bottled drinking water, water softeners, and repairs to appliances and plumbing. In applying this technique to water pollution damages, the analyst must distinguish between defensive expenditures related to man-made pollution and expenditures for controlling naturally occurring substances.

Land Value Studies

It has been shown that an individual's *marginal* willingness to pay for an improvement in Q is equal to the *marginal* differential in land rents with respect to Q. Where land rents have been regressed on a number of explanatory variables including some index of water quality, this marginal differential is the partial derivative of the regression equation with respect to Q. Thus land value studies have a potential for providing

---

*Two additional difficulties should be mentioned: The first involves the fact that for recreation sites located in urban areas, travel time (and cost) are not likely to be significant considerations. The second factor involves the possibility that unaccounted benefits to nonusers may be present because the existence of an improved recreation site gives them the option of eventual use of the site. [Editor's note.]

information on the marginal benefits of water quality enhancement. However, it must be emphasized that land values reflect only those elements of Q which accrue exclusively to owners of specific sites. For example, where there is adequate public access to a water body and substantial recreation use by other than waterfront property owners, the recreation benefits to those users will not be reflected in waterfront property values.

Even where it can be shown that the benefits accrue primarily to property owners, there are several problems which limit the usefulness of the land value technique for estimating water quality benefits. As in all property value studies, the first question is how property value should be measured. Transactions are relatively infrequent and often inaccurately recorded; thus, using actual sale price is an unattractive alternative. Other choices include independent appraisals, owner self-reporting, and tax assessors' data. But each of these is subject to unknown degrees of error and possible bias. A second problem is the definition of the water quality variable. One study of lake front property in Wisconsin (David, 1968) classified lakes as having poor, moderate, or good water quality. This judgement was made by officials at state agencies. Whether the officials' criteria for classification coincided with those parameters which influence potential property owners' perception of water quality is an open question. An alternative approach would be to single out two or three quantitative measures of quality, and include them all in the regression equation.

A general problem facing all land value studies is the large number of variables which can legitimately be hypothesized to influence land values. The problem is to reduce this list to manageable size, to get accurate measures of those variables which are included in the study, and to obtain a sample large enough to support a thorough multivariate regression analysis.

I suspect that there are two reasons for the dearth of competent studies on the relationship between property value and water pollution.[21] The first is the amount of work involved in gathering data, etc., especially in view of the problems discussed here. The second is that property value studies measure only those benefits which accrue to land owners adjacent to the improved water body. But for major water bodies and major pollution control proposals, this is likely to represent a relatively small portion of total benefits. For small lakes and ponds where access to the water is largely limited to those who own adjacent land, land owner benefits will be a major fraction of total benefits. But by their nature, these latter cases probably do not warrant the large-scale com-

---

[21]David (1968) is the only published study attempting to relate water quality and property values systematically.

mitment of time and labor that a competent property value study requires.

The Interview Technique

In those instances where, for all practical purposes, improvement in water quality can be considered a public good, and where the use of alternative techniques such as those described above is inappropriate or not feasible, the possibility of simply asking people about their willingness to pay should be investigated. The problem with this approach is how to induce individuals to give unbiased answers to questions about their willingness to pay. Where individuals believe that their share in the cost of providing the public good will be in some way proportional to their revealed willingness to pay for it, they have incentives to understate their willingness to pay. On the other hand, where assurances are given that tax burdens will not be affected by the answers given, individuals have an incentive to overstate their willingness to pay in an effort to assure that the public good is provided, or that it is provided at a high level. Recently, two suggestions have been made for dealing with these incentives toward biased responses.

Peter Bohm suggests that the individual be left uncertain how, if at all, his response might affect his liability to pay and, further, suggests how the uncertainty should be conveyed:

> If the total willingness to pay exceeds the given construction costs, the plant will actually be constructed. Moreover, they may be called upon to pay exactly the amount they stated. They are, however, also informed that the actual payment may not be equal to this amount, that, for example, it may be proportional to the amount stated. Furthermore, it is possible that the costs of the plant will be partly or completely financed by the Federal government and thus only partially or not at all by the people in the region concerned. It is made clear that at the moment of the "referendum" the actual distribution of the payment is unknown. [Bohm, 1971, pp. 95-96]

Bohm then argues that the rational individual will be unable to perceive any clear advantage either to overstating or understating his willingness to pay, and therefore may decide that honesty is the best policy:

> The individual consumer knows that, if he states his true willingness to pay, he *may* end up in a situation in which he pays more than "necessary" for the given public good output. But neither can he neglect the fact that he may have *some* influence on the final decision of whether to produce the good or not *and* that he may have to pay very little, perhaps nothing, for this influence. It is hard to predict his behavior under these circumstances. . . . Once it is clear that there is no open and shut case for the individual when considering under- or over-statement of his preferences, the choice of strategy may well seem so complicated to him that he prefers to state his true maximum willingness to pay. [Bohm, 1971, p. 98]

Bohm's argument is more hopeful than rigorous on this point. However, it does suggest the possibility of designing questions in which the structure of incentives does not lead to strong biases in one direction or the other, and from which useful information can be elicited.

Tideman has provided a more carefully worked-out scheme in which the distribution of the tax burden for a proposed project is specified as part of the questionnaire, but individuals are asked to indicate their willingness to pay for marginal changes around the specified proposed level (Tideman, 1972). Tideman suggests that each individual be questioned as follows:

> The city is thinking about building a $50,000 park in your neighborhood. If the plan goes through, your share of the cost will be $28 per year. No final decision will be made until the value of the project is checked. In case a $45,000 park or a $55,000 park is built, how much should a $5000 change in the expenditure on the park be valued in terms of your benefits? Whatever figure you give will be used in adjusting your taxes. For instance, if you say that $5000 of park is worth $3 per year to you, your bill for the $55,000 park would be $31, and for the $45,000 park it would be $25.

> Would a respondent reveal his true marginal benefits? That would depend on his expectations. If he expected the level of expenditure to be revised upward, then it would be in his interest to understate his benefits, to avoid higher taxes. On the other hand, if he expected a downward revision then it would be in his interest to overstate his benefits, so that his taxes would be reduced by more than his loss of benefits. If the possibility of lower expenditure just balances the possibility of higher expenditure . . . then neither an overstatement nor an understatement of his benefits will involve expected gain. [Tideman, 1972, p. 113]

Tideman goes on to discuss a second question for use in estimating the elasticity of the marginal benefit schedules and the incentives it holds for understatement or overstatement of marginal gains and elasticities. He concludes that, at least under a fairly wide range of plausible conditions and assumptions, the incentives will not be biased and the rational individual will maximize his gain or minimize his potential loss by stating his true marginal preferences and elasticity. He concludes, "We could at least improve on the equilibrium from majority voting through the process described above, which is Pareto-efficient for almost all persons who reveal their true benefits" (Tideman, 1972, p. 117).

Tideman's suggestion is intriguing. But at the practical level it appears to require an individual of unusual sophistication to perceive accurately the incentive structure built into the questions. Nevertheless, the paper does show that there are real possibilities in more carefully designed questionnaire approaches to the problem of determining the willingness to pay for public goods.

There has been some experience with questionnaires in the recreation area. Davis utilized a questionnaire to determine the willingness to pay of recreationists using several public and private recreation areas in northern Maine (Knetsch and Davis, 1966). He questioned them while they were engaged in the activity, thus making the question less hypothetical, and, perhaps, making it possible for the person to envisage the possibility of being excluded from the recreation site because of nonpayment. It could be argued that this latter factor, the real possibility of exclusion, would minimize the incentives created by the free rider problem in the classical public goods case. Thus, the Davis approach might be appropriate for estimating the demand for existing recreation sites. It would be possible to use interview methods such as described by Davis as a cross-check on estimates of economic value or willingness to pay derived by other methods such as the C-K approach. However, the simple approach used by Davis appears to be less applicable to the case of evaluating potential improvements in water quality since it poses hypothetical, and perhaps difficult to visualize, alternatives to the individuals being questioned.

Health Effects

As discussed in an earlier section, it is not clear that water pollution has affected health and mortality in a quantitatively significant way. Nevertheless, for the sake of completeness in this survey, a brief discussion of techniques for dealing with the possible benefits of improving health and mortality is included here. There are two conceptually distinct but highly correlated effects to be considered: changes in the number of days a person is ill in his lifetime, and changes in his life expectancy. Changes in water and air pollution levels could affect either separately or, more likely, both in the same direction. Following the welfare theoretic framework chosen for this study, the conceptually correct notion of benefits is the willingness to pay: (a) to avoid a day of illness of some specified severity; and (b) to obtain an increase in life expectancy. (See Mishan, 1971.)

This simple statement hides a number of perplexing conceptual and empirical problems. For example, it seems likely that there are substantial externalities involved with living individuals in the sense that family members and close friends might be willing to pay some positive sum to increase that individual's life expectancy. Therefore, the marginal benefit of prolonging an individual's life expectancy by one day is equal to his own willingness to pay plus the (algebraic?) sum of the willingness to pay of all friends and relatives.

It is, perhaps, in the notion of willingness to pay for life expectancy where the problem of income distribution is posed in the most basic way.

Willingness to pay for additional life is constrained by ability to pay, i.e., wealth. Straightforward acceptance of the willingness to pay criterion in this case implies acceptance of the existing distribution of wealth.

What techniques are there for estimating willingness to pay? Can the questionnaire-interview technique be used? With respect to health benefits, properly posed questions may yield useful information, but the technique is probably more difficult to apply to the question of life expectancy. First, it is difficult for a person to interpret the meaning and value of changes in his own life expectancy. Furthermore, focusing on changes in the mean value of life expectancy ignores problems of variance and other higher order moments in the distribution and the possibility of risk aversion. An individual with a life expectancy of 30 more years could plausibly have a quite different willingness to pay for a one year increase in his life expectancy if it were achieved by marginally reducing the probability of his dying at any given age than if it were achieved by a substantial decrease in the probability of dying during a very narrow time span.

Are there any indirect methods for measuring willingness to pay? It has been suggested that the social value of willingness to pay for increases in life expectancy can be inferred from the preference revealed by society in the allocation of public resources to different life-saving activities—for example, highway safety, medical research, and nutritional aid for underprivileged children. But it appears that there is a wide variation in the implicit willingness to pay when comparisons are made across programs. Thus, they provide little guidance for resource allocation decisions (Lave, 1972).

Individuals also reveal implied trade-offs between life expectancy and other measurable economic variables. Individuals who continue to smoke despite the Surgeon General's warnings could be viewed as trading off the consumers' surplus associated with smoking against the now widely publicized reduction in life expectancy. An estimate of the consumers' surplus could provide an upper bound on the value of additional life expectancy for cigarette smokers as a group. However, this estimate would not be applicable to nonsmokers. For another example, individuals make implicit trade-offs between life expectancy and reduced travel time when they choose to drive to the store rather than walk. But with the latter the problem is not the same as with the pollution-health choice, since the variance of the probability distributions involved is quite different. Finally, this whole approach faces the difficulty that it does not capture any of the external benefits of an individual's life. Any estimates derived in this matter must be considered as biased downwards. In sum, it seems that there is no good reason to expect that the revealed marginal willingness to pay that could be inferred from examinations of choices such as this would yield consistent valuations of life expectancy.

Some empirical work concerning health benefits has been based on proxies for willingness to pay, such as medical costs and lost earnings (Ridker, 1967; Lave and Seskin, 1970). As proxies and in the absence of better information, there is some justification for using these measures, provided that the analyst is well aware of the limitations. If medical expenses were a perfect substitute for improvements in water quality in the production of human health, then reduced medical expenses could be a conceptually valid estimate of at least part of the benefits of pollution abatement.[22] However, it is clear that there are other disutilities associated with illness, so medical costs must be an understimate of the willingness to pay to avoid illness. The most obvious limitation of the lost earnings measure is that it places no value on the lives or health of those who are not working for reasons of age (retired), sex, or other factors. There may be other disutilities of illness, etc., not captured in an income measure, including the external effects discussed above. Finally, it is not clear that individuals would be willing to commit themselves to pay all their future earnings at any point in time to avoid death or illness at that time.

Finessing the Valuation Problem

Estimating recreation benefits or the value of a recreation resource involves estimating a fully identified demand curve for the recreation resource. An alternative approach involves less stringent data requirements, assumptions, and estimation techniques. These less stringent requirements can be considered advantages, but they entail the loss of our ability to infer values from the empirical analysis (Cicchetti, Smith, Knetsch, and Patton, 1972). This alternative approach is to estimate reduced form equations which could enable us to predict the increased participation in recreation one would expect if the supply of recreation sites were increased by, for example, the cleaning up of a river. If the value of a recreation day can be inferred from other sources, then the recreation benefits are approximated by the product of the increase in recreation days and the assumed value per day.

This type of model was used by Davidson, Adams, and Seneca (1966) as part of a study of the economic benefits of improving water quality in the Delaware estuary. These authors used data from the 1959 nationwide outdoor recreation survey to estimate equations for participation in boating, swimming, and fishing. One set of explanatory variables in these equations reflected the availability of water area for

[22]But see the discussion of defensive expenditures above.

water-based recreation. Actual socioeconomic data for 1960 and projected values for these variables for 1975 and 1990 were combined with
the estimated equations to predict recreation activity in the eleven
county area around the Delaware for those years. The variables for
availability of water area were those actually observed at the time of the
study, excluding the Delaware estuary itself. This gave an estimate of
recreational use assuming no improvement in water quality in the Delaware estuary.

Then it was assumed that the water quality of the Delaware would
be improved sufficiently to allow water-based recreation. This increase
in the availability of water area was used to predict new levels of recreational activity. The difference in the "with" and "without" predictions
was the output of the water quality improvement. However, it should be
noted that water quality parameters entered the analysis only through
judgment as to whether or not the river was suitable for recreation. The
model did not distinguish possible differences in water quality and the
quality of recreation among those water bodies included in the available
supply.

Finally, Davidson et al. made a series of "illustrative" calculations of
monetary benefits by applying arbitrary but "reasonable" dollar values
per recreation day to their estimated recreation outputs. The benefits
were then compared with the costs of achieving those outputs. They
concluded that a value of $2.55 per day in calculated recreational bene-
fits equalled the costs of the pollution control program.

In many instances the disparity between benefits and costs may be
so great that illustrative calculations like those above will point fairly
clearly and unambiguously toward one conclusion or the other (to undertake the project or not). In such cases, further efforts to obtain refined estimates of value may not be warranted.

The Water Resources Council (WRC) in its proposed guidelines
suggests the assignment of values per day as a technique for determining
recreation benefits (Water Resources Council, 1971). The Council suggests valuing "general" recreation days at between $.75 and $2.25 and
"specialized" recreation days ("those activities for which opportunities,
in general, are limited, intensity of use is low, and often they involve a
large personal expense by the user") at three to nine dollars. As indicated above, in some circumstances useful information can be gained by
trial calculations with tentative or arbitrary unit prices. However, the
danger in the WRC recommendation is that the suggested prices may be
misconstrued by some to be valid and accurate measures of value. The
application of unit values or average values does not give sufficient
weight to the concept of consumer surplus and total willingness to pay.
And the specification of a fairly narrow range of possible values does not
give adequate consideration to the wide range of possible different cir-

cumstances. For example, the total willingness to pay could be quite high where supply is restricted relative to demand, or where congestion exists, or where there are no close substitutes for the recreation resource in question. In sum, the use of arbitrary values can be useful in some circumstances, but they should be seen as a last resort and be used only in the most tentative way. Furthermore, every effort should be made to allow for specialized recreation resources, the availability of substitutes, and supply-demand considerations in selecting the arbitrary values to be used.

Ex Post and Ex Ante Analysis

THE WATER RESOURCE PLANNER who is trying to allocate scarce resources and is faced with a number of competing goals needs ex ante analyses of the effects of alternative resource allocations to guide his decisions. Ex ante analysis involves the prediction of the physical and economic consequences of resource allocation decisions on the basis of a model or a theory concerning the physical and economic processes involved. It involves visualizing two alternative states of the world, one with the project in question and one without, and then comparing these alternative futures in terms of some established criterion, such as net economic efficiency. Ex post analysis involves measuring the actual consequences of the project by comparing the observed state with a hypothetical alternative—the state of the world without the project. Ex post analysis, in effect, treats the project as a controlled laboratory experiment except that the control group is hypothetical rather than real.

Ex post and ex ante analyses are not competitive alternatives but should be viewed as complementary techniques for improving our knowledge. An ex post analysis of a project can be viewed as a check on the validity of the ex ante analysis. The ex ante analysis is a prediction of what will happen; the ex post analysis is a check of what actually did happen.[23]

It is particularly important that the economic analysis of water pollution control benefits include ex post analysis of existing projects because our knowledge of the physical and economic systems on which present ex ante analyses are based is extremely limited. It is necessary not only to develop more comprehensive models of the physical, biological, and economic aspects of the system but also to devote more effort to verifying these models through ex post comparisons of the predictions with observed results. For example, one of the most thoroughly studied river basins in the United States is the Delaware estuary. Strong policy

[23]For further discussion of this and related points, see Haveman (1972).

recommendations have been based upon analyses carried out with the basic dissolved oxygen model for the estuary. (Delaware Estuary Comprehensive Study, 1966; Kneese and Bower, 1968.) But serious questions have been raised about the validity of the model and therefore the solutions that have been reached; and it has been argued that inadequate effort was made to verify the model (Ackerman and Sawyer, 1972).

It must be emphasized that the ex post verification of the analytical models used in benefit estimation is not simply a comparison of actual results with predictions. Ex ante models are based upon some view of the future, some projections of economic magnitudes, such as population levels, real income, and price levels. Care must be taken in ex post analysis to sort out the effects of unforeseen developments, such as war or uncontrolled inflation, on the variables in question. For example, if the failure of income levels to rise on the projected path results in a shortfall of recreation benefits at a particular site, this is not a failure of the analytical model so much as a reflection on our inability to perceive the future. The real benefit of ex post analysis is in making the most of the opportunity to improve on the analytical models used.

Summary and Recommendations

THIS PAPER HAS REVIEWED THE NECESSARY ELEMENTS of a model of the production of the benefits of water pollution control. In particular, it has focused on the necessity for establishing the relationships between changes in abatement levels, changes in quality parameters, and changes in the uses man makes of water as a precondition to estimating the value of improved water quality. The paper has described benefits in terms of the different ways in which changes in quality and changes in use affect human welfare and willingness to pay; and it has used economic theory to determine the different ways of measuring or estimating benefits or values in each case.

With respect to techniques for estimating benefits, I offer two general observations on the present state of the art. First, the economic theory and analytical techniques for valuing benefits for most types of uses are relatively well developed. At least, this is the case for the most significant classes of user benefits.[24] The analytical framework and techniques for utilizing the net factor income and the complementary goods

---

[24]It also must be pointed out that we can say relatively little about the measurement and valuation of aesthetic effects, long-term ecological damages, and various forms of nonuser benefits such as option value.

and perfect substitutes approaches are now well established, and there are no major unsettled questions concerning the economic theory underlying them. This is also true, but to a lesser extent, for the land value approach and the C-K technique for recreation.[25] However, there are still major unresolved theoretical and analytical issues with respect to valuing health and life benefits and with respect to the proper design of questions for the interview-questionnaire technique for pure public goods. But these are areas which are relatively unimportant quantitatively in the water pollution area.

All this is not to say that estimation of benefits will be easy. The necessary economic data are not gathered on a regular and comprehensive basis. For the most part, each potential pollution control program must be evaluated de novo with primary data gathered specifically for that purpose.

The second observation is that this optimistic assessment refers to economic analysis and theory but does not carry over to the other aspects of the problem such as specifically relating abatement measures to quality changes and uses. In other words, the economic theory for the valuation problem is far better developed than the noneconomic aspects of the problem concerned with relating water quality parameters to human activities associated with water. We still do not know what indices of water quality are relevant to recreational uses of the water, or how people will respond to changes in different quality parameters. Nor do we fully understand the links between many quality parameters and discharge rates.

In conclusion, we are far from being able to apply the marginal techniques of optimization analysis in the pollution control area. And this is only partly due to the limitations of either economic analysis or economic data. More importantly, it is due to limitations of both theory and data in the realm of biology and ecology, and in the way they affect human uses of water.

This survey leads me to offer several recommendations for additional and renewed effort:

• A major effort should be undertaken to learn more about those biological and physical characteristics of water that contribute to its attractiveness as a recreational resource, and about how those characteristics are affected by different kinds of discharges.

• The Clawson-Knetsch approach to estimating the demand for recreation facilities should be applied to one or more water bodies with significant pollution problems. This work should focus on refining the model, developing more comprehensive and reliable measures for travel

---

[25]In the latter case it is the critical travel cost variable which poses the largest problems.

cost, and experimenting with different measures of quality as a demand curve shift variable.

• There should be a carefully constructed experiment with the interview-questionnaire technique for estimating willingness to pay. Preferably this should be coordinated with a study based on the Clawson-Knetsch technique in an effort to provide a cross-check or validation of benefit estimates obtained by different techniques.

• A careful and comprehensive program of ex post analysis should be developed and implemented. The period of the seventies will have been characterized by a major commitment of resources to water pollution control in every major river basin in the country. It would be penny-wise and pound-foolish not to plan now to allocate one or two percent of the total funds to be spent on pollution control toward measuring and evaluating what we will have bought with that massive expenditure.

• Finally, some effort should be made to define the relevant catagories of nonuser benefits and to begin the dual tasks of relating this class of benefits to water quality levels and of devising techniques to estimate monetary benefits.

REFERENCES

Ackerman, Bruce, and Sawyer, James. "The Uncertain Search for Environmental Policy: Scientific Fact Finding and Rational Decision Making along the Delaware River." *University of Pennsylvania Law Review* 20, no. 3 (January 1972), 419-503.

Bohm, Peter. "An Approach to the Problem of Estimating Demand for Public Goods." *Swedish Journal of Economics* 73, no. 1 (March 1971), 94-105.

Burt, Oscar R. "Comments on 'Recreation Benefits from Water Pollution Control' by Joe Stevens." *Water Resources Research* 5, no. 4 (August 1969), 905-7.

Burt, Oscar R., and Brewer, Durward. "Estimation of Net Social Benefits from Outdoor Recreation." *Econometrica* 39, no. 5 (September 1971), 813-27.

Cesairo, Frank, and Knetsch, Jack. "Time Bias in Recreation Benefit Estimates." *Water Resources Research* 6, no. 3 (June 1970), 700-4.

Cicchetti, Charles; Fisher, Anthony; and Smith, V. Kerry. "An Economic and Econometric Model for the Valuation of Environmental Resources with an Application to Outdoor Recreation at Mineral King." Xerox (1973).

Cicchetti, Charles, and Freeman, A. Myrick, III. "Option Demand and Consumer's Surplus; Further Comment." *Quarterly Journal of Economics* 85, no. 3 (August 1971), 528-39.

Cicchetti, Charles; Freeman, A. Myrick, III; Haveman, Robert H.; and Knetsch, Jack. "On the Economics of Mass Demonstrations: A Case Study of the November 1969 March on Washington." *American Economic Review* 61, no. 4 (September 1971), 719-24.

Cicchetti, Charles; Smith, V. Kerry; Knetsch, Jack; and Patton, R. A. "Recreation Benefit Estimation and Forecasting: Implications for the Identification Problem." *Water Resources Research* 8, no. 4 (August 1972), 840-50.

Clawson, Marion, and Knetsch, Jack. *Economics of Outdoor Recreation.* Baltimore: Johns Hopkins University Press, 1966.

Crutchfield, James A., and Pontecorvo, Giulio. *The Pacific Salmon Fisheries: A Study of Irrational Conservation.* Baltimore: Johns Hopkins University Press, 1969.

David, Elizabeth L. "Lake Shore Property Values: A Guide to Public Investment in Recreation." *Water Resources Research* 4, no. 4 (August 1968), 697-707.

David, Elizabeth L. "Public Perceptions of Water Quality." *Water Resources Research* 7, no. 3 (June 1971), 453-57.

Davidson, P.; Adams, F.G.; and Seneca, J. "The Social Value of Water Recreational Facilities Resulting from an Improvement in Water Quality: The Delaware Estuary." In Allen V. Kneese and Stephen Smith (eds.), *Water Research.* Baltimore: Johns Hopkins University Press, 1966.

Federal Water Pollution Control Administration. *Delaware Estuary Comprehensive Study: Preliminary Report and Findings.* Philadelphia, July 1966.

Freeman, A. Myrick, III. "Spatial Equilibrium, The Theory of Rents, and the Measurement of Benefits from Public Programs: A Comment." *Quarterly Journal of Economics* (1974a).

Freeman, A. Myrick, III. "On Estimating Air Pollution Control Benefits from Land Value Studies." *Journal of Environmental Economics and Management* 1, no. 1 (1974b), 74-83.

Gordon, H. Scott. "The Economic Theory of a Common Property Resource: The Fishery." *Journal of Political Economy* 62, no. 2 (April 1954).

Haveman, Robert H. *The Economic Performance of Public Investments.* Baltimore: Johns Hopkins University Press, 1972.

Kelly, Robert A. "A Conceptual Ecological Model of the Delaware Estuary." (Xerox), 1973.

Kneese, Allen V. "Economics and the Quality of the Environment: Some Empirical Experiences." In Morris Garnsey and James Gibbs (eds.), *Social Sciences and the Environment.* Boulder: University of Colorado Press, 1968.

Kneese, Allen V., and Bower, Blair T. *Managing Water Quality: Economics, Technology, Institutions.* Baltimore: Johns Hopkins University Press, 1968.

Knetsch, Jack. "Economics of Including Recreation as a Purpose of Eastern Water Projects." *Journal of Farm Economics* 46, no. 5 (December 1964), 1148-57.

Knetsch, Jack, and Davis, Robert. "Comparisons of Methods for Recreation Evaluation." In Allen V. Kneese and Stephen Smith (eds.), *Water Research.* Baltimore: Johns Hopkins University Press, 1966.

Krutilla, John V. "Conservation Reconsidered." *American Economic Review* 57, no. 4 (September 1967), 777-86.

Lave, Lester B. "Air Pollution Damage: Some Difficulties in Estimating the Value of Abatement." In Allen V. Kneese and Blair T. Bower (eds.), *Environmental Quality Analysis: Theory and Method in the Social Sciences.* Baltimore: Johns Hopkins University Press, 1972.

Lave, Lester B., and Seskin, Eugene P. "Air Pollution and Human Health." *Science* 169 (August 21, 1970), 723-33.

Lind, Robert C. "Spatial Equilibrium, The Theory of Rents, and the Measurement of Benefits from Public Programs." *Quarterly Journal of Economics* 87, no. 2 (May 1973), 188-207.

Mäler, Karl-Göran, "A Method of Estimating Social Benefits from Pollution Control." *Swedish Journal of Economics* 73, no. 1 (March 1971), 106-18.

Reiling, S.D.; Gibbs, K.C.; and Stoevener, H.H. *Economic Benefits from an Improvement in Water Quality.* Washington: Environmental Protection Agency, January 1973.

Ridker, Ronald G. *Economic Costs of Air Pollution.* New York: Praeger, 1967.

Stevens, Joe D. "Recreation Benefits from Water Pollution Control." *Water Resources Research* 2, no. 2 (1966), 167-82.

Stevens, Joe D. "Recreation Benefits from Water Pollution Control; A Further Note on Benefit Evaluation." *Water Resources Research* 3, no. 1 (1967), 63-64.

Stevens, Joe D. "Reply." *Water Resources Research* 5, no. 4 (August 1969), 908-9.

Tideman, T. Nicolaus. "The Efficient Provision of Public Goods." In Selma J. Mushkin (ed.) *Public Prices for Public Products.* Washington: The Urban Institute, 1972.

Water Resources Council. *Principles and Standards for Planning Water and Related Land Resources.* Federal Register 38, no. 174 (September 10, 1973).

Weisbrod, Burton A. "Collective-Consumption Services of Individual Consumption Goods." *Quarterly Journal of Economics* 78, no. 3 (August 1964), 471-77.

DISCUSSION

Discussant: *Charles J. Cicchetti*

One of the issues that Freeman avoids is the difference between using a price-compensated and an income-compensated demand curve. He thus avoids any possible problems of determining whether the ability to pay or the price-compensated demand curve should be used for cost-benefit analysis, or whether the individual himself has to be compensated, if he is going to suffer damages or receive benefits. That issue should have been included in his paper.

Another problem concerns the assumption that when there is a price or quantity change in the market, we consider it independent of all other prices and quantities in the economy. The assumption is often made, but it is not likely to be the case in many situations, and I think Freeman failed in not discussing that issue.

Some Institutional and Conceptual Thoughts on the Measurement of Indirect and Intangible Benefits and Costs

JOHN BISHOP and CHARLES CICCHETTI

OVER THE PAST FEW DECADES, economists and government officials have developed various procedures to measure the direct benefits and costs of water resource projects. Some of the procedures are straightforward, while others are much less simple and are often based on rather intricate sets of economic assumptions. It is very important to understand the basic concepts according to which direct benefits and costs are measured, but since they have been widely discussed elsewhere it seems repetitive to review them here. This paper will, instead, focus on the ways in which those concepts must be broadened to include *intangible* economic and environmental benefits and costs as well as social equity—as stipulated by the National Environmental Policy Act of 1969 and by its interpretation in the courts. One of the policy questions addressed in this paper is how these expanded objectives might be included in analyses of costs and benefits or of a project's environmental impact. In the discussion, we will concentrate wherever possible on water quality benefits and costs. We will, however, refer freely to other fields when doing so will better serve the point.

Intangible Benefits and Costs

EVEN WHEN THE OUTPUTS OF A PROJECT being evaluated are sold on the market, estimating the benefits or costs of that project is necessarily filled with extrapolations and assumptions. To determine benefits over time, one must estimate future relative prices, but they must be predicted on the basis of current prices and circumstances with possible adjustment for imperfections in the market.[1] Costs are often incorrectly estimated, especially when new technology is being implemented for the first time.

In principle, the benefit concept for outputs which are not traded in the market is the same as for outputs which are traded: aggregate willingness to pay, or, diagramatically, the area under the income-compensated demand curve. Empirical implementation, however, is more tortuous and may be impossible.

We prefer to think of intangibility as a characteristic of all unmarketed outputs, not a subset of them. The weaker the analogy between a nonmarketed or public good output and its best market counterpart, the more likely the analyst is to admit defeat and cover his retreat by proclaiming the output "an intangible." Once we start measuring the benefits of such nonmarketed outcomes as flood control benefits and recreation visits, intangibility defined in this manner is only a matter of degree. How far the valuation of nonmarketed outputs is carried depends on the state of the art and the skills and resources of the analyst. Generally, attempts at valuing nonmarketed outputs have been extremely ad hoc.

For private goods—provided the consumer knows what choices are possible and what each might mean to him—there is no reason why this must be so, because a theoretical basis for making extrapolations from marketed to unmarketed goods has existed for some time. The characteristics, or hedonic, approach to the theory of consumer choice introduced by Becker (1965), Lancaster (1966), and Muth (1966) provides a natural basis for using observable market prices to evaluate nonmarketed outputs.[2] In the new theory of demand, a consumer's utility function is defined over attributes—such as safety, style, comfort, popularity, solitude, or beauty—not over goods. Under this theory, time, learned skills, and marketed and nonmarketed goods and services are inputs for the production of these attributes. If the contributions of marketed goods and services to the production of attributes can be empirically specified, one can derive implicit prices for attributes. If it is also possible to specify empirically how the nonmarketed outputs of a project contribute to the production of these attributes, then one may estimate derived demand functions for these outputs as well.

In a very important paper, Sherwin Rosen (1973) has shown how the underlying structure of the attribute-production process can be identified when: (1) the bundle of characteristics inherent in a particular project or model "cannot be untied," (2) "a sufficiently large number of differentiated products are potentially available so that choice among

---

[1] For an application of this concept refer to A. C. Fisher, J. V. Krutilla, and C. J. Cicchetti (1972).

[2] For a review and extension of this early literature see Cicchetti and Smith (1973b).

various combinations of the characteristics is continuous for all practical purposes," (3) sellers are competitive, and (4) each consumer has perfect knowledge of each models' vector of characteristics and of the self-production function by which he creates his final outputs.[3] In our discussion of this issue we will not reproduce his elegant mathematical treatment but rather try to translate and summarize it in intuitive terms. Hedonic regressions of observed differentiated product prices on all of the products' characteristics provide summaries of currently observed implicit prices for attributes. There is no reason to expect that the implicit price of improvements in one attribute should be unaffected by the level of other attributes or constant over its own range of variation. Consumers, for example, choose a particular automobile so that any money-attribute trade-off equals the ratio of the marginal utility of the attribute to the marginal utility of money. For the group which includes all buyers of a particular model, the average valuation placed upon the attribute is an average of these ratios, weighted by the number of cars purchased.

This, however, does not tell us how to value nonmarginal changes in the level of an attribute. The average value of the attribute change will be overestimated if particular individuals receive nonmarginal increases in the attribute—and underestimated if nonmarginal decreases occur. The structure of demand is not identified by hedonic regressions unless all buyers have identical resources and preferences, a patently false assumption. Rosen proposes, however, the simultaneous estimation of the demand and supply of the characteristics of the product, using the marginal prices calculated from the hedonic regression as endogenous variables. To identify the demand requires that a number of variables measuring supply determinants—such as location, local factor prices, and measurable differences in the technology opportunities—be available.

Application of the new theory of consumer choice to the valuation of nonmarketed private goods like death postponement, health, and aesthetics is quite direct, although not necessarily simple. Difficulty in identifying the structure of demand creates no problem in many applications because most of the effects of a project occur in small doses to a multitude of people. As a result the observed implicit prices of an attribute captured by hedonic regressions are an unbiased estimate of the benefit received by the individual from the small change in that attribute. The impacts of water quality, air quality, or medical research on life-expectancy appear in individual utility functions as tiny movements in the "risk-of-early-death" attribute variable. However, in almost all cases the person whose death will be averted by the project is not and cannot be identified. In many circumstances (e.g., air pollution pro-

[3]See: Rosen (1973).

grams) it is not even possible after the fact to identify who suffered because a particular project was not undertaken. In these cases demand functions cannot be identified and additional theoretical efforts are necessary before unambiguous benefit measures can be derived.

If cost-benefit analysis is to become more than an academic exercise in hindsight, it must produce ex ante measures of benefits. If it is to become widely used its cost must be lowered. If large selections of potential projects are to be screened and compared, economic analysts must strive to develop techniques comparable to those used by accountants or structural engineers. Then different analysts would reach the same results. For valuing aesthetic experiences, the primary requirement is to de-emphasize the uniqueness of each experience. We must dissect the experience into component intellectual and emotional attributes, measure these attributes more or less objectively, and estimate the implicit prices of these attributes. Once this has been done well, the same techniques of measuring attributes and the same implicit prices can be applied to new project plans, and comparable analyses of competing projects can be obtained at minimal cost.

Adopting a characteristics approach to the valuation of outcomes which are not marketed accomplishes four objectives. It provides a system for extracting from observed prices the information necessary to place a value on outputs or outcomes that are never sold. It extracts the necessary information from already operating facilities or current market transactions making it possible to place values on future events that are currently no more than drawing board projections. It makes project evaluations comparable and lowers the cost of a systematic analysis of all options. Finally it ties the technique of measuring intangible benefits to the economic theory of the consumer. Although specific projects may not equally satisfy these objectives, we believe that professional cost-benefit analysis should begin with them.

In most cases, cost-benefit analysis must first measure those factors which can be measured in dollars. The remaining factors can then be measured either by a sensitivity analysis from which the value per unit can be inferred or by calculating "break-even" values. For those factors, which are not quantifiable in any monetary or physical form, such as aesthetic values, this information should be presented as a side display and contrasted to net measurable benefits (i.e., benefits minus costs) by the decision maker.

Nonuser Benefits

USER BENEFITS ARE CERTAINLY THE MAIN, but not the only, concern of those who practice the art of cost-benefit analysis. Below, we will discuss several types of real economic benefits which conceivably flow from the preservation of an environmental resource, but which do *not* accrue to users of the resource (i.e., to individuals who appear on-site to claim them). The point of this discussion is that persons or groups who do not appear on-site to "consume" the resource may nevertheless suffer economic injury from its conversion to other uses. Nonuser effects are by definition indirect, and usually they fall in the category defined above as intangible.

Several types of benefits can be attributed to nonusers. The reason these benefits should not be ignored in such national controversies as damming up Hells Canyon, constructing a ski resort in Mineral King Valley, or building the trans-Alaska pipeline is that benefits may be quite small per individual, but a large number of nonusers may be concerned about such projects and their concomitant damage to the environment. Accordingly, in estimating preservation benefits for such projects, nonuser benefits are likely to be important.

The first type of nonuser benefit is indicated by the amount that an individual would be willing to pay to preserve his option to enjoy some particular area some time in the future. If all potential users of Hells Canyon were certain they would visit it and each of these future users were counted, then one component of the benefits of preserving the canyon could be accurately measured. In most cases, however, more people will think they might want to visit an area than will actually visit it. Since most uncertain users would be willing to pay something to guarantee continued access to a specific preserved area, attributing value only to those who will actually use the area in the future is likely to underestimate total value. The factors which are important in estimating this value are: (1) whether the area is relatively unique (are there any substitutes) for the people it is likely to attract, (2) whether the development is reversible or the resource reproducible, and (3) the degree to which people facing an uncertain situation and risking the loss of a specific resource would be willing to pay to avoid this risk (are they risk averse). The name economists give to this nonuser benefit is "option value."

A simple example may help in understanding the importance of this concept. Suppose there is some amount that uncertain users would pay for preserving their present option to use a natural area. If they never actually visit the area, their willingness to pay will be unrecorded. Since most benefit estimates are based on forecasts of actual users and do not include those who think they might visit the area in the future, a major component of benefits would be incorrectly excluded. If the area in question is very popular, a large number of people might feel attracted to it, although each of them has a very small probability of visiting it. Each might be willing to pay to preserve the area in question although the amount would be small due to the low probability of use. But since there are large numbers of them, a significant component of the benefits of preservation would be lost if this information were ignored.

Suppose, for example, that an area would normally expect 100,000 visitors per season, each willing to pay ten dollars per visit to preserve the area in its present state. The annual preservation benefits attributable to users would be $1 million. Now suppose these 100,000 were drawn from one million people who had a 10 percent probability of use and that each person would be willing to pay $1.25 in order to avoid the risk of not having the option to visit the area. Under such circumstances the benefits of preservation are understated by $250,000 (or 1,000,000 people × $1.25 − 100,000 people × $10).[4]

There are two additional types of nonuser benefits. First, there are people who do not plan to visit an area but who derive value from the sheer knowledge that the area, which they may consider a "gift of nature," is preserved in its natural state. Many individuals contribute to action-oriented environmental organizations only to aid the effort to preserve such phenomena, knowing full well that they will not visit them. This "vicarious consumption" can also be designated "existence value." These contributions are analogous to those made to preserve the works of creative genius that were threatened by the floods of Florence a few years ago, by people who knew they would not themselves enjoy any more than photographic reproduction of those works.

A related type of nonuser benefit is the one derived by individuals who know that they themselves will not utilize such preserved areas, but who wish to preserve them for future generations. This "bequest motivation" is analogous to leaving personal fortunes in trust for one's children and grandchildren. Public trusts, whether determined by the executive and legislative branches, or won in judicial controversies, undoubtedly generate economic benefits for a considerable number of Americans.

An important characteristic of these nonuser benefits is that they may be exceedingly difficult to quantify directly in monetary terms. However, even if only the number of nonuser beneficiaries can be estimated this should be included in the analysis. If such an estimate of affected nonusers can be provided then it can be used to determine how large the value to these nonusers would have to be in order to justify nondevelopment.[5]

---

[4]Recently, Schmalensee (1972) has addressed the option value question and concluded that its sign may be plus or minus. By separating the problem in the way he does, the uncertainty and trade-offs of the early literature are lost. Cicchetti and Freeman (1971) have shown that if risk-averse demanders are faced with uncertain choice between (a) paying too much for the option that may not be used and (b) suffering the loss of not buying the option, then option value will be positive.

[5]For a complete discussion of nonuser benefits see Krutilla (1967).

Applying
the Hedonic Method

ALTHOUGH THIS PAPER DOES NOT INTEND TO REVIEW specific applications
in depth, it is important to note that hedonic principles, which view
consumers as producers, have been used by economists performing
cost-benefit analysis. Since the trail-blazing work by Marion Clawson,
economists have used location and distances traveled to measure the
economic benefits of a new recreation site. In some recreation studies,
water quality differences are used as explanatory variables in examining
recreational demand or in forecasting future recreational use. These
were early efforts; they were based on sound economic theory and were
applied long before theoretical developments about hedonic price indices and about consumers as producers.

The importance of the hedonic price approach is twofold: first, it
formalizes the inclusion of other important characteristics and factors,
when including such considerations is typically the exception rather than
the rule. Second, it permits the interpretation of empirical estimates in
terms of economic theory. This is necessary if such procedures are to be
properly applied to cost-benefit analysis.

To explain the above thoughts more fully we will review two recent
applications where hedonic pricing methods permitted an extension of
cost-benefit analysis into new areas.

In a recent paper Cicchetti, Fisher, and Smith (1972) simultaneously
estimate the demand for various skiing sites in California. The location
of other sites and therefore their relative prices are taken into account
explicitly by using a generalized least squares regression approach. The
benefits of new sites at various locations can be determined by simultaneously estimating the change in consumer surplus for the alternative
sites. The standard, but often unrealistic, cost-benefit assumption that
the price of goods in other markets is independent of the goods or
services provided by a given project has been explicitly dropped in this
study.

A second study by Cicchetti and Smith (1973) goes a step closer to
the hedonic price regression approach. In a recent study of the use of
wilderness for recreation, a specific measure of the willingness to pay for
wilderness use under different levels of congestion was made. Specifically, in a survey of wilderness users at a specific site, variations in price
were related to the length of the trip, the number of trails and camp
encounters, and the types of other parties encountered. In applying this
information the analysts examined: (1) the optimal level of use, (2) the
annual and present value of a wilderness, and (3) the benefits of alternative management values that could increase the level of use. The cost of

applying such in-depth analysis is not small, nor is the cost of basing decisions on incorrect assumptions.

The Valuation of Postponing Death

THE COST-BENEFIT LITERATURE has not yet settled upon a method of valuing the postponing of death. An ongoing debate has centered around the issue of whether the present value of the decedent's foregone consumption should be subtracted from the present value of his foregone future earnings. Fein and Klarman argue against subtracting consumption.[6] Weisbrod, Dawson, Reynolds, Thédié and Abraham, and Dublin and Lotka subtract the individual's consumption from his earnings to arrive at a "net benefit to society."[7] Their position is that "the individual would have produced but he would have consumed. What the nation has lost is the dead man's production (gross loss) less his consumption."[8] Under this approach extending the lives of the non-working poor, welfare recipients, and retirees is counted as a *cost,* not as a benefit of a health program. The proposition that society would be better off if certain people were dead seems counter to every ethical system we are aware of, including the one that is inherent in welfare economics. Certainly the man who dies loses. In one sense the very thing he loses is the consumption that the net-output approach subtracts. Despite the fact that the potential victim is alive at the time the analysis is undertaken, " 'society' is defined to exclude the individual whose life is being valued."[9]

Gross earnings seems at first glance a more acceptable concept. But that, too, leads the analyst into a number of counter-intuitive ethical positions. Extending the life of retirees and welfare recipients would produce zero benefit. This implies there is minimal benefit to doing research on diseases that attack only those above sixty-five. An illness that strikes during ones' vacation creates no losses. Preventing a death is

[6]Rashi Fein, *Economics of Mental Illness* (1958), p. 18; Herbert E. Klarman, "Syphilis Control Programs," in Robert Dorfman, ed., *Measuring Benefits of Government Investments* (1965), p. 379.

[7]R.F.F. Dawson, *Cost of Road Accidents in Great Britain* (1967); Louis I. Dublin and Alfred J. Lotka, *The Money Value of a Man* (1946); D. J. Reynolds, "The Cost of Road Accidents," *Journal of the Royal Statistical Society*, 109, (1956); Jacques Thédié and Claude Abraham, "Economic Aspect of Road Accidents," *Traffic Engineering and Control* 2 (1961); Burton Weisbrod, *Economics of Public Health: Measuring the Economic Impact of Diseases* (1961).

[8]Thédié and Abraham, p. 59.

[9]Weisbrod makes exactly this point though without our emphasis. *Economics of Public Health*, p. 35.

less important than preventing an illness that permanently destroys economic productivity (the illness requires costly medical treatment as well). Most of the users of the gross earnings measure admit there are other "intangible" costs of illness and death but feel that since they cannot be measured accurately they better not try. The decision to use gross earnings is not neutral among programs. However, it understates most of the benefits of programs which help the aged and those facing low wages. It overstates the benefits of reducing risks for people who enjoy risking their lives. While gross earnings may be a useful starting point for trying to measure the appropriate benefit concept, it is not itself that concept.

With Mishan we feel the appropriate benefit concept for impacts on mortality is the willingness to pay for reducing the risk of death that is calculated to occur.[10] In terms of the characteristics approach to utility theory, the private benefit is measured by the marginal price of reducing the risk of death for tiny changes in the individual's personal risk or by the area underneath the demand curve for risk reduction for large changes in personal risk. Indirect or externality benefits are analogously defined for other peoples' risk of death. In other words, a project is justified if it is "a potential Pareto improvement, one in which the net gains can so be distributed that at least one person is made better off, with none being made worse off."[11]

Applying the Pareto principle to death seems strange, for surely no sum of money could get most people to accept immediate *certain* death.[12] The outcomes we are evaluating, however, are not the certain death of named individuals.[13] Rather they are small changes in the risk of death experienced by many people. When valuing the risk, as opposed to the certainty of death, people consistently behave as if it has finite value. Steeplejacks, paratroopers, and mountain climbers all choose to risk their lives more than they would be obliged to in some other activity. Even the less adventurous demonstrate there is a trade-off when they smoke, pass a car on a two-lane highway or neglect to fasten lap and shoulder belts.

In most applications of cost-benefit analysis, private benefits are added to externality or indirect benefits to get social benefits. However,

---

[10]E.J. Mishan, "Evaluation of Life and Limb," *Journal of Political Economy* (July/August 1971), p. 692.

[11]E.J. Mishan, p. 692.

[12]This flows from the fact that we require time to enjoy the things money can buy. This is a central tenant of Becker's consumer-as-producer theory.

[13]Where certain named individuals are deliberately placed at high risk we are ethically required to ask for volunteers and all must be done to maximize the chances of survival. Shelling has a perceptive discussion of this in T. C. Shelling, "The Life You Save May be Your Own," in Samuel Chase, ed., *Problems in Public Expenditure Analysis* (Washington, D.C.: Brookings Institution, 1968).

when using a certain product involves the voluntary assumption of some risk and buyers are aware of this, the expressed willingness to pay for the product or service already accounts for the risk. Therefore estimated demand curves for a particular mode of transportation, for use of a wild river for whitewater canoeing, or for seaside swimming opportunities, do not require further adjustment to reflect the private risks the users are taking. The externalities one imposes on others are, however, legitimate adjustments to the estimated demand curves. Examples of such situations are taxes on cigarettes and laws mandating the use of safety belts or crash helmets. The justification for interference with free choice here must rest on: (1) the externalities one's own death imposes on others, (2) lack of knowledge of the risks on the part of the consumer, or (3) risks of death his foolhardiness imposes on others.

The private demand for risk reduction should be included in cost-benefit calculations, either when involuntary risks of death, injury, or sickness are involved or when the safety design of a facility is the issue. Air pollution, levels of background radiation, and unsafe drinking water all impose involuntary risks of death. Public programs designed to reduce these risks should include the private benefit of a reduction in the risk if people are willing to pay for a small reduction in the statistical probability of their own early death.

The second situation where it is appropriate to include the private benefit in calculations is where choices that have safety consequences are being made in the design of recreation facilities, automobiles, or highways. If measures of the total or social benefit of averting a death are available, then a determination can be made of how far risks should be minimized. In principle the private component of this benefit can be estimated by comparing the areas underneath structurally estimated demand curves for alternative designs. In practice, however, ex ante estimates are either not feasible or are equivalent to general estimates of the private benefit of risk reduction.

Placing a value on the benefit of a reduction in the risk of early death is not necessary for all policy decisions. Often impacts on mortality are minor, relative to the other benefits and costs of a project. Some policy makers may find being explicit about such matters "distasteful." In such circumstances, measures of (1) the expected number of deaths averted or caused and (2) expected reductions in the lifespan of those who die or live as a consequence of the project should be presented by the analyst to the decision maker as a collateral exhibit to the main cost-benefit analysis. By comparing estimates of effects on lifespan to the net benefits that have been expressed in money terms the decision maker can intuit whether one is large enough to outweigh the other.

The Indirect Benefits of Postponing Death

SOME OF THE BENEFITS OF REDUCING a person's risk of death or injury are not felt by that person. Some, though not all, of these external benefits are quantifiable. The indirect or external benefits are both financial and psychic[14] and they occur whether the individual voluntarily or involuntarily assumed the risk.[15] The financial costs imposed on others by an accident or death are: (1) the medical bills paid by insurance or charity (rational calculation of private risks should neglect these), (2) taxes that would have been paid out of earnings, minus the reduction in demand for government subsidy of private goods, (3) the earnings—including an appropriate valuation of a homemaker's time—which the decedent would have provided the family minus his consumption (this assumes a completely individualistic utility function in which the decedent values his own consumption only). On the other side of the ledger there is a financial benefit equal to the assets that would have been consumed during retirement and there is also the benefit arising from earlier inheritance by those who are party to the decedent's estate.

The psychic externalities of a death are the grief, guilt, and pain which relatives, friends, and possibly some strangers feel upon someone's death. Put into dollar terms it is the willingness to pay to reduce the risk of someone else's early death. This is definitely not an insignificant motive. Probably almost all parents will, if asked, say they would pay more to reduce their child's or spouse's risk of death than their own. Parents also act in ways consistent with a high subjective valuation on reducing such risks. The close supervision parents give their children is largely justified as a safety measure.

Even outside the extended family there are likely to be people who suffer a psychic loss from a death, and if the number of such individuals is large their aggregate willingness to pay could be quite large. It should be remembered in this regard that the circumstances surrounding a death can strongly affect the magnitude of the psychic loss others feel.

Publicity and blame can have powerful effects on the size of aggregate psychic externalities. Publicity not only increases people's knowledge of other deaths. When the deaths were potentially avoidable it carrys the moral message "How terrible! We should have done something to stop this." The politician, clergyman, reporter, or a Ralph

---

[14]Mishan, calls them "indirect or derivative risks," pp. 699-700.

[15]The size of psychic externalities may, however, be affected by whether the risk is voluntarily or involuntarily assumed.

Nader, who brings these issues to our attention is often trying to change our preference structure as advertiser's do for private goods. The fact that as a result we feel bad about something we would not otherwise have thought about does not delegitimize the feeling. Knowing that they will be told about certain types of deaths and that this will cause them sorrow, the public tries to reduce the number of deaths and to protect itself from feeling the deaths were avoidable by passing laws mandating certain safety practices.

If blame is fixed on or felt by someone other than the victim, the psychic externalities are larger still. Added to the victim's losses is the blamed person's feeling of guilt. Whether a person or institution is blameworthy is culturally, not absolutely, defined. Here again if the ecologists, or a Ralph Nader, can influence the conventions that ascribe blame, a real change in psychic externalities can be induced. We as economists might feel that a business should not be blamed for supplying people what they want. If, however, many believe that there is a moral responsibility not to pollute, etc., it is rational for companies and for cost-benefit analysts to take this psychic externality into account.

Measuring psychic externalities felt by those outside the extended family is likely to prove very difficult. As in the case of the bequest motive for maintaining wilderness areas and protecting endangered species, it is a pure public good. An individual's behavior cannot significantly effect the quantity of this public good so there is no private behavior that can be observed to measure how strongly a person cares about the matter. Questionnaires and observations of political behavior seem to be the only way to quantify it. It would seem that in the current state of the art the public's psychic externalities must be treated as a side exhibit.

With the elaborations outlined above, we generally concur with Mishan's use of willingness to pay for risk reductions as the ideal or conceptual measure for valuing the mortality impacts of government programs. We also have some sympathy for his evaluation of it as "the only economically justifiable concept."[16] However, its greater theoretical rigor and generality comes at the cost of making the concept considerably less specific and harder to measure. Even if we place additional restrictions on the situation—assume perfect information and an attribute utility function that values, along a common scale, all changes in personal mortality risk no matter what their source—we are still left with possibly insoluble measurement problems of valuing both private risk and psychic externalities. The problem of psychic externalities is not

[16]Mishan, p. 704.

unique to our perspective; however, the typical response has been either to deny their existence or to hide behind the label "intangible."

Applying the Hedonic Method to the Measurement of the Private Valuation of Risk

THE ATTRIBUTES APPROACH TO CONSUMER CHOICE not only provides a theoretical underpinning for measuring the price of an attribute; it shows how difficult an unbiased estimate is to obtain. (1) As discussed before, hedonic regressions which predict the prices of product models or brands with their characteristics as regressors measure only the value of tiny changes in risk. They overestimate the value of risk reduction if certain people experience large (inframarginal) reductions in risk. (2) The economy offers consumers a multitude of different trade-offs between money and risk, and between risk and other characteristics. People distribute themselves along an N-dimensional continuum according to their preferences. Obtaining an estimate of the trade-off that coal miners express when they choose between above and below ground work tells us about their preferences only. Other people are different and the people who face involuntary risks due to air pollution, and so on, are primarily "other people." Estimates of trade-offs must be obtained for everybody at risk including women and children. (3) The utility function is defined over a multitude of attributes. Certain attributes—risk of injury, *machismo*, and thrills—are often associated closely with the personal risk of death. The involuntary risks from air pollution, cancer, infectious diseases, or poisoned fish do not provide the risk bearer with *machismo* or thrills. Since a "pure" estimate of death risk aversion is required, these characteristics must be measured and controlled for when estimating the hedonic regression.[17] (4) We must believe that the individual has a reasonable idea what the risks of alternative choices are. One can hardly learn about these risks by personal experience. Learning from the experiences of others is unsatisfactory, for one is aware of the circumstances of only a very few deaths. Seeking out information on the risk of each of a multitude of choices is a costly process and often it will not be worth the effort. This does not imply that reductions in risk are not highly valued—only that the expected gain in allocational (between risk and money) efficiency is less than the costs of obtaining the information.

---

[17]Collinearity of characteristics is a very common problem in the estimation of hedonic regressions. It does not create serious problems for price index applications but it is critical for the application proposed herein.

Occupational and job choices seem to be the only situations where large numbers of people choose between situations with price tags on them (jobs) in a context where information on the risk of different jobs is potentially available.[18] There may be problems, however, in applying this approach to white-collar workers for people may not perceive any differences in risk among these jobs.

Even if this approach fails to provide us with good estimates of everybody's marginal evaluation of risk, it can provide a base by which a priori theoretical or questionnaire approaches can be validated. The questionnaire approach to measuring private death risk aversion involves asking questions like "Would you accept a short term job that would take no appreciable time, was not illegal or unethical, would receive no publicity, but carried a 1 percent risk of losing your life, if the pay were a 1 percent addition to the financial resources available for your consumption in every year of your life?" "Currently, your chance of dying this year is _____. Over your lifetime the additional pay would come to about _____. ($8000 for a college graduate). What is the minimum percentage addition to your consumption that would induce you to accept the risk?" Since an answer is not independent of the wording of the question, estimates derived from market analogs might be used to validate the design of the questionnaire.[19]

If we treat the family as a single consumption unit, it is possible, in principle, to estimate hedonic regressions for the combined private and familial externality willingness to pay for reductions in the risk of death. In practice, however, the consumer information on the safety characteristics of products and on the legality of potential alternatives is limited due to building codes, food and drug regulations, and so on, making it very difficult to get good estimates. Here again questionnaires may be useful sources of data.

The final approach that seems to have some promise is to generate results by placing a priori restrictions on the form of a person's utility function. Dan Usher has shown, for instance, that given (1) expected utility maximization and (2) utility functions that are additive over time at a given discount rate, r, and that have a constant elasticity of β, the

---

[18]Two examples of this approach are: Richard Thaler and Sherwin Rosen, "Estimating the Value of Saving a Life: Evidence from the Labor Market," presented at the Conference on Research in Income and Wealth, December 1, 1973; and, Richard Smith, "Compensating Wage Differentials and Hazardous Work," Technical Analysis Paper no. 5, Office of Policy Evaluation and Research, Department of Labor (August 1973).

[19]Because of the empirical problems with the hedonic regression method the analyst might have more confidence in an internally validated questionnaire method. In a number of experiments Bohm has found that questionnaire wording does not cause significant differences in average willingness to pay for public goods. P. Bohm, "Estimating Demand for Public Goods: An Experiment," *European Economic Review* 3 (1972).

trade-off between the consumption activity C(t) and immediate risk of death is

$$\frac{1}{\beta} \int_0^\infty C(t)S(t)e^{-rt}dt$$

where S(t) is one's probability of being alive at time t.[20] Note that if there is declining marginal utility ($\beta < 1$) the cost of a risk of death is greater than the expected loss of consumption. Usher's application was to the national income account, so he required that the consumption of goods and services be the only source of utility.

If one takes the broader perspective of Gary Becker's consumption activity approach, C(t) becomes redefined as a consumption activity whose price per unit time equals the use of purchased goods and services per unit time plus the opportunity cost of the person's time.[21] The expenditure constraint forces the sum of all expenditures to equal non-wage income plus wage income:

$$\sum_{i=1}^m p_i x_i = V + \overline{w}(T_w)$$

where x_i is a good or service purchased at price p_i, V is nonwage income, $\overline{w}$ is wage per unit time and T_w is time spent working. If work is valued for its money income only, the marginal wage equals the average wage, and, on average, people are able to adjust their hours of work (so that they are in equilibrium between time working and time consuming), then utility received over a time interval is $(V(t) + \overline{w}T)^\beta dt$ where T is the total time available for either consumption or work. The redefined trade-off becomes:

$$\frac{1}{\beta} \int_0^\infty (V_{(t)} + \overline{w}T) \, S(t)e^{-rt}dt$$

integrated over all usable time. Sleep time and eating time would seem to be the only subtractions necessary.

Over a given period of calendar time the imputed income will generally be about three times after-tax earnings, so this approach implicitly places a much higher value on avoiding a risk of death and indicates that so long as the marginal utility of consumption does not rise, the willingness to pay for a reduction in risk is three or more times that implied by the gross productivity approach conventionally used in cost-benefit studies. If utility rises as the square root of the consumption activity

[20]Dan Usher, "An Imputation to the Measure of Economic Growth for Changes in Life Expectancy," paper given at Conference on Research in Income and Wealth, November 4-6, 1971.

[21]Gary Becker, "A Theory of the Allocation of Time," *Economic Journal* (September 1965), pp. 493-517.

($\beta = .5$), an individual should be willing to pay about 6 percent of each future year's consumption of goods and services for the elimination of a 1 percent risk of immediate death. These numbers are, of course, a consequence of assumptions about utility functions that are a good deal more restrictive than those economists typically make. On the other hand, it could be argued that the alternative, statistical estimates of willingness to pay for small changes in risk, requires assumptions about the quality of information available or the relevance of an uninformed opinion—that no one but an economist would make.

The Valuation
of Good Health

THE TRADITIONAL METHOD BY WHICH HEALTH or morbidity effects of programs are valued sums the loss of output and the costs of medical treatment. Even when people are paid their marginal product, gross earnings will be a very inaccurate measure of the welfare loss individuals suffer from an illness whose treatment costs are covered by insurance. It asserts that none of the following impose a welfare loss on a family: a child's 103-degree fever, dysentery on a vacation in Mexico, breaking one's leg on the first day of a ski trip, a retiree breaking his hip. On the other hand, the man who has a slight touch of the flu, reports in sick, and uses the time to work on his stamp collection or catch up on his assignments in a night course, is asserted to have lost the welfare equivalent of one day's wages. Finally it is implicitly asserted that avoiding pain is not something people are willing to pay for. Many analysts admit the existence of additional "intangible" costs of illness and say they are calculating a lower bound estimate of benefit. However, cost-benefit analysis does not influence the size of an agency's budget nearly as much as it influences the priorities set within that budget. The "lost output" rule-of-thumb is not neutral with respect to illness or target population. It underestimates most greatly the costs or injuries and illness that produce continuous pain or high degrees of incapacity, that are associated with recreational activities, and that attack children or the aged. It overestimates (or underestimates least) the costs of mild infections that are enough to keep a person home without preventing him from working in bed or enjoying the leisure time he unexpectedly has.

The new theory of the consumer provides a way of conceptualizing the psychic costs of illness or injury. Valuing reductions in the risk of illness can be approached directly, in a manner completely analogous to our treatment of the risk of death. In principal, this approach is preferred, since it comes closest to recreating the outcomes of the government program being evaluated. However, the empirical problems (multicollinearity and imperfection of consumer information) are generally even more serious here than they are in valuing reductions in the risk of death.

An alternative approach involves estimating the willingness to pay to avoid an illness or to eliminate it once it is contracted. Then one can use the change in the probability of a period of illness times the estimated loss as the measure of the willingness to pay for a reduction in the risk of illness.

A serious illness or injury produces large (inframarginal) changes in the victim's well-being and use of time. Estimating willingness to pay for relief from the illness requires estimates of the slope of demand curves for the consumption activities that must be given up (skiing, eating solid food, etc.). Prices faced prior to the illness provide only one point on the curve. If this is taken into account in the calculation of the welfare loss suffered by the sick person, the use of the expected welfare loss as a benefit measure is no more than an application of standard cost-benefit discounting practice.

The impact of illness or injury on a person's welfare has three aspects: (1) pain; (2) cosmetic losses; (3) inability to allocate time in the most preferred manner. The experience of pain is notoriously difficult to relate. As a result, people who have experienced or are experiencing a particular type of pain would seem to be the only ones able to make valid consumer judgments about the value of relief from it. The price of the drugs and the services of medical personnel watching for side effects are only part of the costs of relieving serious pain. The other part is the lost or altered consciousness that the pain-killer produces. It is almost universally believed that there is no utility to be derived from living in a drugged state totally unaware of one's surroundings. This part of the cost of successful reductions in pain can be conceptualized as a disruption in the use of time (i.e., being unable to drive or read or converse). The sum of these costs provide an upper bound estimate of the willingness to pay when a drug is available but not used, and a lower bound estimate of costs when pain relief is medically impossible or interferes with treatment.

Cosmetic losses can result from the injury or illness itself, or as a side effects of radiation or chemotherapy treatments. Since private behavior is replete with expenditures wholly designed to increase one's own attractiveness, estimation of the implicit price of personal attractiveness should not be too difficult empirically.

Disruption of the use of time, is by far the most pervasive effect of illness and injury. Loss of earnings is only a part of this impact. We will illustrate this type of effect with a number of examples. A mild heart attack is likely to leave a white-collar worker's productivity unaffected. But strenuous recreation may very well be forbidden. So the man loses the consumer surplus he was receiving from skiing and playing tennis.

The less disoriented a patient is by the illness and the drugs he is given, the greater the number of consumption activities he may choose between. Since each activity produces diminishing marginal utility as more time is allocated to it, being able to do a variety of things means that no one activity must be driven to the point of zero return. While, for groups, the average wage will be strongly correlated with the marginal value of time, a number of factors can create systematic differences: taxation means that those receiving high wages will find that nonworking time will be worth 60 or 70 percent of the gross wage. For low wage workers the value of nonworking time will generally be closer to their gross wage. Enjoying the daily activities of one's job raises the marginal value of work time. Certain times of the day may be more valued than others. If opportunities to increase or decrease working time are restricted the marginal value of nonworking time can be greater or lower than the average wage. The opportunity cost of nonworking time is less than the average wage if the person moonlights at lower wages or *would* moonlight if a job that pays his average wage were available. We may assume it to be greater than the average wage if the worker is frequently late or absent.

While empirical estimation of the time disruption costs of different injuries and illnesses is a big job; it is not an impossible one. Abundant medical data on the physical effects (such as temperatures, respiration, sleep time, drugs administered) of each disease are obtainable. By correlating this objective information with time budget data (time spent reading, talking to others, playing games, working on hobbies) for both the period in the hospital and convalescing at home, it should be possible to construct profiles of the use of time for each type of illness or injury for specific categories of people. The final and biggest task is doing the necessary empirical work implied by Becker's theoretical insight. We must obtain structural estimates of the demand for each consumption activity that distinguish the behavior of different age, sex, and income groups. The extensive data-gathering and research activity needed for this will require many years. In the interim, a systematic application of rules-of-thumb to time budget data could greatly improve the quality of our estimates of morbidity losses.[22]

[22]An illustration follows. Assume time budget research shows that the Asian Flu knocks a person out for an average of 48 hours, and that the average use of that time is 16 regular sleep hours, 12 extra sleep hours, 2 regular prime time TV watching hours, 4 extra prime time TV hours, and 10 extra daytime hours plus 2 hours eating with no appetite and 2 hours conversing. Assume the following rules of thumb: (1) the base opportunity cost of work time is equal to the average wage, (2) the base opportunity cost of weekend, evening, and morning time is 110 percent the average wage, (3) while extra sleep helps the recovery it yields no current utility, (4) since TV was watched during all prime time hours, the marginal benefit of an extra hour declines from 110 percent of $\bar{w}$ to 55 percent, (5) since no daytime TV was watched before, its marginal benefit ranges from 66 percent to 0

Summary and
Conclusion

THE PRECEEDING DISCUSSION HAS POINTED OUT THAT, with respect to
environment protection, administrative agencies are required by national law to expand their concept and application of cost-benefit
analysis. Thus, the National Environmental Policy Act now mandates
that many impacts of government projects that were previously neglected as "intangible" be explicitly included in the planning process.

The underlying theoretical rationale of benefit measurement involves the concept of willingness to pay. Even when other techniques are
used in practice, they are validated by reference to willingness to pay.
Extending the scope of benefits to "intangibles" or indirect or externality
impacts does not change the underlying concept of benefits.

The problem that "intangibility" creates for the analyst is not a
result of the "vagueness," or "impalpability" or "indefinability" of the
benefit described. The benefit received from eating an apple is "intangible" in the same sense. The problem is a consequence of the lack of
markets in which the benefit is sold. Markets automatically provide the
analyst with an unambiguous measure of the trade-off people express
between money and the "intangible" benefit.

When analysts refer to something as intangible they mean hard or
impossible to measure. Thus, intangibility or difficulty of measurement
is in some degree a trait of all nonmarketed outputs. A number of ad hoc
procedures have been used in the past to place values on nonmarketed
outputs like death avoidance and good health. The lack of a strong
theoretical underpinning has contributed to the lack of consensus over
practices and has confused the discussion of the conceptual issues.

We propose that the characteristics approach to utility theory be
used to give a systematic theoretical basis to methods for using observable market prices to place values on nonmarketed outputs. By viewing

---

percent of $\overline{w}$, (6) the lack of appetite results in the 2 hours of eating time being valued at
only half its normal 10 percent of $\overline{w}$. Our estimate of losses is

$$(1-.33)\,\overline{w}10 + (1.1 - .825)\,\overline{w}4 + \overline{w}6 + 1.1\,\overline{w}6 + .55\,\overline{w}2$$

$$= (6.7+1.1+6.0+6.6+1.1)\,\overline{w}$$

$$= 21.5\,\overline{w}$$

In contrast, expected lost wages are approximately $(11.43)\,\overline{w}$ taking into account the fact
that only 5/7 of the days are working days. Admittedly the number arrived at is somewhat
arbitrary. However, as long as the time budget data is correct, it is a better estimate of the
true loss than lost wages. When used for comparing the morbidity costs of different
illnesses and injuries, errors introduced by choosing the wrong value for time spent in
specific activities will tend to cancel out since most illnesses produce similar adjustments in
the use of time.

the consumer's utility function as defined over a limited number of attributes or consumption activities rather than over an infinite list of unique products, we draw out the commonalities between things that are sold in markets and the free or public goods that are not. The trade-offs between money and a characteristic that people exhibit in their private behavior can be used to estimate the implicit price of attributes also provided by public projects. To consider time as an important input into the consumption activities that yield utility provides a basis for calculating the cost of loss or disruption of nonworking time. The attributes that often appear as intangibles in cost-benefit analysis—risk of early death, pain, personal beauty, option value, congestion, disruption in the use of nonwork time—were discussed and methods for placing money values on them were suggested and evaluated.

The characteristics approach is most helpful when private decisions can influence the level of one's own consumption of the characteristic. When the characteristic itself is a pure public good—as with the existence value or bequest motivation for preserving natural wonders, or with the psychic externalities to strangers of a death—there is no way of extrapolating from market behavior to the price of the pure public good. Thus if money values are to be placed on these effects, measurement must be based either on observation of political behavior or on questionnaires.

It was pointed out that intangible benefits are not distributed over project designs or program clientele proportionately to the more easily measured benefits. As a consequence, leaving them out inevitably distorts the planning process. Wherever possible, the agency or agencies should agree on standardized money values to be placed on nonmarketed outputs. Project selection, given one's budget constraint, is the primary application of cost-benefit analysis and this procedure produces comparable analyses at minimum cost. If standardized procedures are followed, a rough estimate of the correct concept is always better than an exact measure of the wrong concept.

In a few cases it may be necessary to handle a particularly unique psychic externality as a collateral exhibit. However, setting money values is generally preferred. A major benefit of cost-benefit analysis is the learning process induced by having to justify the value placed on particular hard-to-measure outputs. Making assumptions explicit facilitates the debate. The challenge for economist and practitioner alike is to expunge subjectivity to the greatest extent possible and at the same time increase the breadth and application of cost-benefit analysis. The courts will require this action and it is encumbent for economists to take an active role.

REFERENCES

Becker, Gary. "A Theory of the Allocation of Time." *Economic Journal* (September 1965): 493-517.

Cicchetti, C.; Fisher, A.C.; and Smith, V.K. "An Economic and Econometric Model for the Valuation of Environmental Resources." Presented at the meeting of the Econometric Society (Winter 1972).

Cicchetti, C., and Freeman, A.M., III. "Option Demand and Consumer's Surplus: Further Comment." *Quarterly Journal of Economics* (August 1971).

Cicchetti, C., and Smith, V.K. "Congestion, Quality Deterioration and Optimal Use: Wilderness Recreation in the Spanish Peaks Primitive Area." *Social Science Research* (March 1973).

Cicchetti, C., and Smith, V.K. "Interdependent Consumer Decisions: A Production Function Approach." *Australian Economic Papers* (1973b).

Fisher, A.C.; Krutilla, J.V.; and Cicchetti, C.J. "The Economics of Environmental Preservation: A Theoretical and Empirical Analysis." *American Economic Review* 62 (September 1972).

Krutilla, J.V. "Conservation Reconsidered." *American Economic Review* 57 (December 1967).

Lancaster, Kelvin. "A New Approach to Consumer Theory." *Journal of Political Economy* (April 1966).

Mishan, E.J. "Evaluation of Life and Limb: A Theoretical Approach." *Journal of Political Economy* (July/August 1971): 687-705.

Muth, Richard F. "Household Production and Consumer Demand Functions." *Econometrica* (1966).

National Environmental Policy Act, PL 91-190, January 1, 1970.

Rosen, Sherwin. "Hedonic Prices and Implicit Markets: Product Differentiation in Pure Competition." Discussion Paper Number 296, May 1973, Harvard Institute of Economic Research. Forthcoming in *Journal of Political Economy*.

Schmalensee, R. "Option Demand and Consumer's Surplus: Valuing Price Changes Under Uncertainty." *American Economic Review* 62 (December 1972).

Shelling, T.C. "The Life You Save May Be Your Own." in Samuel Chase, ed., *Problems in Public Expenditure Analysis*. Washington, D.C.: Brookings Institution, 1968.

Usher, Dan. "An Inputation to the Measure of Economic Growth for Changes in Life Expectancy." Paper given at Conference on Research on Income and Wealth, November 1971.

DISCUSSION

Discussant: *Allen V. Kneese*

I want to focus on the part of the paper that is devoted to methodology with respect to intangible benefits and the hedonic method. In reading this section, I have wondered what sort of genuine difference this new terminology makes in the situation. It seems to me that, as far as applied work is concerned, it may have very little relationship.

Looking for substitutes on the basis of which to evaluate public goods is not new. It has been done routinely for many years in the flood control area. Benefits are measured as the present expected value of the damages, and the damages can be substituted for by control efforts. If the design is appropriate, the marginal control cost will presumably be equal to damages and the substitution will be perfect. This is a routine method that has been around for a very long time.

DISCUSSANT: *Thomas D. Crocker*

I am perhaps somewhat less excited about the usefulness of the Lancaster formulation than the authors; that is, I question its usefulness for inferring the value of nonmarketed attributes from measures of willingness to pay for marketed attributes.

The first point is that if one is going to establish a hedonic index by regressing prices of the good or activity in question on its attributes, great care is required in defining the good itself. Otherwise, one is likely to be regressing expenditures; that is, price times quantity on attributes, which obviously gives relatively little information about willingness to pay. While this is not a problem with respect to many inputs or goods in household production functions, for something as intangible as recreation, health, avoidance of health risk and so forth, it seems to be a rather severe problem.

Second, many attirubtes have to be treated as what one might call nonseparable joint products. Thus, one cannot offhandedly assert that willingness to pay for the red color of a car is equal, at the margin, to willingness to pay for the red color of a chair.

DISCUSSANT: *Henry M. Peskin*

The thrust of this paper seems to be that if you *can* come up with a monetary measurement, then you have eliminated any philosophical question of whether you *should* come up with a monetary measure. I do not agree with this reasoning. That is, I can suggest some theoretical reasons why there might be no dollar value for certain intangibles. In that case, coming up with a monetary measure by whatever means—hedonic indices, or whatever—might be a disservice to the policy maker. You are falsely giving him a number to hang his hat on where in fact there is no justification for the number.

A Survey of Empirical
Benefit Studies

DENNIS TIHANSKY

INCREASED AWARENESS OF THE DEGRADATION OF WATER has promoted the evolution of decision making tools and of methods for assessing impacts. Traditionally, the economic evaluation of water pollution control has focused on costs. In recent years the emphasis has been transferred to empirical studies of the benefits realized by pollution controls.

The contributions of economists toward these ends are substantial. A large literature base is readily available on the principles and concepts necessary for the evaluation of benefits as they contribute to social welfare. Applying theoretical constructs to actual situations, however, is often difficult and time-consuming. In few instances is there a market price or an established monetary value which is a direct, or at least clear, representation of benefits. Some impacts of an aesthetic or intangible nature do not lend themselves to market parlance, but for making benefit comparisons they are monetized by assuming a shadow price or a weight proportional to some other benefits which have been monetarily expressed. Such comparisions are generally fraught with bias and, aside from being disputable in the moral realm, they are likely to underestimate real social values (Trice and Wood, 1958).

The analytical framework suggested by economic theory may be very complex or require too extensive a data base for implementation. Because externalities from water pollution are pervasive in society, it is argued (Kneese and d'Arge, 1969) that these effects should be examined in light of total residual flows. To do so normally requires a scope of analysis beyond the immediate interests of water resource planners and the data available to them. Instead, one can reduce information needs by estimating net benefits in a partial equilibrium context, but there may be less confidence in the solution. The fact that an estimate is only approximate in view of data restrictions does not brand it as useless. In some cases, knowing an order of magnitude is sufficient to justify an environmental policy.

This paper reviews an extensive compilation of major benefit studies. From an original collection of almost two hundred studies, approximately sixty are selected as contributing most to the conceptual

127

framework and empirical insights of benefit evaluation. Of the selected studies, less than 30 percent are theoretically valid, but even fewer seem cognizant of the applicability, let alone the existence, of welfare economics. (For a discussion of methodologies for benefit analysis, see the papers by Freeman and by Haveman and Weisbrod.) Studies pertaining to water quality as an input to production—e.g., with respect to industrial or municipal water supply—are generally acceptable in their conceptual approaches. But recreational benefit estimates are usually based on tourist expenditures rather than such welfare measures as willingness to pay or consumer surplus. If only the recreation studies are considered, fewer than 10 percent follow from theory. Yet at least one of them misapplies theoretic notions by measuring welfare gains incorrectly.

Although there is no theoretical basis for most benefit values, equally discouraging is the paucity of empirically derived damage functions. Only 20 percent of the surveyed literature derives dose-response relations from on-site data. With recreation studies, this proportion decreases to 10 percent. Practically all recreation loss curves are hypothetical, based neither on activity-day counts nor on public opinion surveys. Actual damage curves, in some studies, are based on a very small sample of observation points, seriously restricting their credibility. This is notably common in cost assessments of water supply treatment and land value changes, wherein two extreme water quality conditions determine a linear damage function. If the relation is actually nonlinear, then this simplistic derivation cannot identify the range of marginal damages necessary to determine optimal levels of water quality control. The above percentages on the adequacy of benefit estimates pertain to selected references. If the original studies are also included, these percentages would be smaller, roughly by a multiplicative factor of one-third.

There is a massive literature on water resource management, but this review considers only that portion emphasizing water quality. Certainly, this necessitates omitting many works that have contributed significantly to the development of methodology. However, most of these studies have been critiqued elsewhere (e.g., Hinote, 1969; James and Laurent, 1973; Loucks, 1972). The first section of this paper summarizes in historical context the impetus for benefit estimation. The second section outlines the state of the art on benefit studies and contrasts the empirical findings. The next section describes criteria for judging the adequacy or relevance of benefit assessments, after which the literature is evaluated according to these guidelines. Concluding remarks pertain to the utility of translating all benefits into precise monetary values. A multidisciplinary framework of analysis is recommended as the most viable alternative. In the Appendix, selected references are examined for methodological weaknesses. Empirical estimates are reviewed with an emphasis on their adaptability to local and regional planning.

It is important to note the distinction between critiques in the main text and those in the Appendix. While the text identifies only those studies deemed adequate in terms of the basic criteria, more extensive comments on all selected references are relegated to the Appendix. Here the studies are partitioned according to their focus on a specific pollutant, a particular water use, or a more comprehensive analysis. While the theoretical validity of each study is essential, the Appendix concentrates on the misapplication or misinterpretation of theoretical and empirical notions. That is, having chosen a methodology for benefit estimation, the empiricist next faces problems of gathering relevant data, formulating reliable damage relations from this data, and then interpreting them in a proper economic perspective. Even if he chooses a theoretically sound method, the resulting estimates may well be derived from faulty or unreliable information and thus have limited practical value.

An Historical Perspective

THE INITIATIVE FOR BENEFIT ESTIMATION in the United States originated in the mid-nineteenth century with a classic discussion on the utility of public water works (Dupuit, 1844). The Refuse Act of 1899 recognized the right of the public to intercede in cases where river dumping and waste discharges affect water quality. Several years later the River and Harbor Act of 1902 directed the Army Corps of Engineers to evaluate the "public necessity" of navigation improvements. Because of the local nature of many of the derived benefits, the federal government tried to limit its funding of these improvements. The updated River and Harbor Act of 1920 taxed local interests according to "a statement of special or local benefit which will accrue to localities affected by such improvement, and a statement of general or national benefits . . ." (Committee on Public Works, 1970). In spite of this decree, evaluation techniques by the Corps were very simplistic and pertained only to tangible effects. The Flood Control Act of 1936 expanded the scope of benefit analysis to include social security impacts on adversely affected persons, but the analysis was not obligatory and therefore did not achieve the desired objectives.

Most benefit estimates in this time period pertained to navigation or water resources planning, but not to water quality per se. Nevertheless, the techniques of estimation did prove of some use in the latter area.

The first nationally publicized benefit figures can be traced to the Special Advisory Committee on Water Pollution (1935). On the basis of questionnaire responses from each state, estimates were obtained on the economic effects of the destruction of fish and waterfowl by pollution. Although commercial fishery losses were of primary concern, several responses pertained to the decline of property values near polluted streams. Upon review of the estimates, the committee concluded that "the relevant data are so scattered, so incomplete, and, because of the intangible nature of some of the losses, so inconclusive, that it is impossible at present to state any figure as representing the national economic burden of pollution."

The next major effort was an economic study of acid mine drainage by the Ohio River Committee (1943). Only the treatment costs of water supply were compiled. More important benefits, e.g., recreation and aesthetics, were neglected because of problems of estimation. This general mood of pessimism on benefit quantification prevailed throughout most government agencies. By 1946, the Federal Inter-Agency River Basin Committee charged a subcommittee with the task of "formulating mutually acceptable principles and procedures for determining benefits and costs for water resources projects." The end-product of this appointment was the publication of the first economic guideline on project evaluation (1950). But this work focused on water resources rather than pollution control.

The first comprehensive assessment of water quality control pertained to the Ohio River Basin (Bramer, 1960). Several years later a large research team funded by the Federal Water Pollution Control Administration [FWPCA] (1966) quantified benefits along the Delaware River. By this time academicians had become involved in conceptual and theoretical issues. Various federal initiatives, such as the Water Quality Act and the Water Resources Planning Act of 1965, encouraged empirical approaches to water quality problems, but research programs did not really gain momentum until the early 1970s. The Federal Water Pollution Control Act, as amended in 1972, provided significant support and funding for applied research on "methods and procedures to identify and measure the effects of pollutants on water uses" (U.S. House of Representatives, 1972).

The number of empirical studies has grown rapidly to encompass a wide variety of water uses and receptors, pollutants, and regions. Whereas past estimates focused on river basins, recent efforts have included other types of watersheds, e.g., lakes, estuaries, and fjords. American in origin, benefit studies have also begun to appear in some of the more industrialized countries of Europe, notably Italy and England. Even developing nations, such as Singapore and Mexico, are now assessing damages because of their economic dependence on water-related activities.

State of the Art
on Empirical Estimates

FOR THE EMPIRICIST THE BASIC FRAME OF REFERENCE for benefit quantification is a geographic region, a specific recipient or water use, a water quality constituent, or any combination thereof. The most common objective is to predict total regional benefits over all water-related impacts. Such information, when compared to the costs of control, is useful in setting policies and in determining the overall impacts of environmental legislation.

Within this macro-setting, water quality criteria are designed to protect major water users. Streams are classified in many states according to their function in serving the interest of patrons. The sensitivity of these water uses to water quality specifications is reflected in the magnitude of benefits. The complementarity or perhaps incompatibility of activities, e.g., commercial fishing and navigation, underscores interdependencies among benefit recipients. Although water uses are generally responsive to a multivariate measure of water quality, legal standards on abatement pertain to individual water constituents. Consequently, there is a need for benefit studies relating to a single parameter.

Table I depicts the basic components or points of reference in any empirical study. Each element of the matrix links a water quality parameter (row headings) to categories of water use threatened with economic damage (column headings). The row and column headings are selected from recommendations on water quality criteria (e.g., FWPCA, 1968; McKee and Wolf, 1963; National Academy of Sciences, 1972).

The weighted pollution index in the bottom row is summed over standard parameters, such as dissolved oxygen, pH, total dissolved solids, iron and manganese, temperature, and total nitrogen. Among pollution variables the impacts of biochemical oxygen demand (BOD) levels on water use have been assessed most frequently. However, BOD is usually assumed to be a surrogate, or indicator, of the general level of water quality. On the other hand, other pollutants in the table can be identified more uniquely with water use responses.

Consistent definitions of each water use are necessary to compare benefit estimates by region or by pollutant within the same region. The "water supply" category pertains to treatment cost increases resulting from polluted water intake. "Navigation" is a catchall term for damages

Table I

STATE-OF-THE-ART MATRIX ON ECONOMIC DAMAGE ESTIMATES

Symbols

N — National estimate.   * — Generalized methodology.

R — Regional estimate.   • — Physical damages documented.

WATER QUALITY PARAMETER	Aesthetics	Agriculture	Commercial Fishing	Domestic Water Use	Ecology	Human Health (Ingestion)	Human Health (Water Contact)	Industrial Water Supply	Municipal Water Supply	Navigation	Power Generation	Property Value	Recreation (Water Contact)	Recreation (water non-contact)
Acidity	•	•	R •	•	•	•	•	R •	R* •	R •	•	•	NR •	
BOD	•	•	R •	•		•		•	R* •			R •	R* •	R •
Coliform Bacteria	•	•	R •	•		R •	N	•	•	•			R* •	•
Color	•		•	•				•	•				•	•
Floating Solids	•	•	•		•			•	R •			•	•	•
Hardness	•	•		NR* •	•			•	•					
Nutrients	•	•	•		•	•		•	•		•	•	R* •	•
Odor	•		•			•		•	•			•	•	•
Oil	•	•	R •	•	R •			•	•	•	•	R •	R •	•
Pesticides	•	•			•	R •		•	•				•	•
Radionuclides	•	•			•	•			•				•	
Sediment	R •	N •	R •		•			•	R •	R •		R •	•	
Taste			•		•				•					•
Temperature	•	•	•	•	•			•			•	R •		•
TDS & Salinity		R* •		NR* •	•	•		•	•		•		•	•
TSS & Turbidity	•	•	•	•	•	•		•	•	•	•	•	R •	•
Toxic Metals	•	N •		•	•	•	•	•	•			•	R •	
Weighted Index or General Pollution	R •		N* •		N •	N •		R* •		NR •	R •	NR •	NR* •	•

to bridges, piers, and ships, as well as boat accidents and water route closures. "Human health" can be affected by ingestion of drinking water or by bodily contact with beach water. Other categories are, for the most part, self-explanatory. It is obvious that some water use categories overlap, such as "recreation" and "aesthetics." Mutual dependencies of water uses preclude a complete partitioning of them. Nevertheless, for purposes of benefit estimation, they are usually considered distinct.

Based on a survey of relationships between pollutants and water use, most of the matrix elements can be adequately documented in terms of physical, chemical, or biological effects (Jordening and Algood, 1973). On the other hand, few of these effects have been translated into economic equivalents. The sparseness of economic estimates in the matrix suggests either the nonexistence of a literature base or, more frequently, the infeasibility of benefit quantification. Information in this table comes from both official and unofficial sources. To rely solely on the professional literature yields a partial, though well-tested, review. Many less publicized approaches, some being implemented at the policy level, also have much to offer.

The table contrasts the current state of knowledge on physical damage relationships with that of economic ones. The relative scarcity of the latter type follows for several reasons. Practically all economic estimates are based on a monetary weighting of physical damage measures, implying that the latter must be known beforehand. Some types of pollution may not adversely affect economic activity, or the pollution level may be of small enough consequence to be ignored. Many pollutants have significant impact only when there is a need for water of extraordinary quality, e.g., for special laboratory use. In normal ranges of pollution and for most water uses, corresponding economic effects may be negligible. Economic values have yet to be derived for some impacts in the aesthetic and ecological realm. But economic values need not be based on dollars or cash flows. The term *economic* includes social values and utility preferences as well. Nonmonetary values should therefore provide the multidisciplinary framework required for the enumeration of intangible benefits. There is also the extreme viewpoint of rejecting economic values in benefit analyses. This attitude is most frequent among ecologists unwilling to accept any "value system" on conservation.

Even more disconcerting is the widespread disagreement on what economic (monetary) values should be attached to certain unit damages. For example, most commercial fishing damages are measured in terms of total landing (dockside) revenue losses from closed areas, but at least one economist (Tybout, 1969) claims that net revenue losses (less production costs) are a better indicator of welfare losses. An even better measure, he contends, is a regional adjustment for resource and income shifts after fishermen find new employment.

Recreational values have perhaps received most attention in the literature. In virtually every instance, the procedure is to estimate the number of participants at a site before and after pollution control and to determine, by survey or other means, a typical user-day of expenditures. Economic factors may include travel costs alone or other expenses, e.g., lodging and admission fees, in addition to travel. Traditional supply-demand analysis suggests that consumer surplus is more appropriate, although most engineer-economists and water resource experts seem unaware of this concept and instead use expenditures. In a few instances, consumer surplus is rejected as seemingly impractical or superficial (e.g., Merewitz, 1966), although most studies show otherwise (e.g., Stevens, 1966).

The relative size of the region can significantly affect the scope of benefit estimation. In small lake and estuary studies, recreation benefits from pollution abatement are likely to compete with those in neighboring areas. But these substitution effects are seldom assessed. Large regional studies (of a major river basin, for example) are less inclined to ignore these effects. Yet even these estimates may inadequately reflect national impacts.

Input-output tables are often constructed in regional studies for predicting the consequences of local expenditures on the rest of the economy. Income multipliers are calculated to show the magnitude of intersectoral transactions. Almost all empirical studies, however, define regional boundaries in terms of watersheds rather than economic juris-dictions. Problems thus arise in computing input-output relations for an open economy, especially for small regions highly dependent on trade.

For small, local regions the most frequent benefit estimates pertain to commercial fishing. Information on fish catches and dockside revenue is readily available for all major harbors. Recreation demand and other important benefit statistics, on the other hand, are generally lacking. Coupled with the problem of determining substitute activities in nearby regions, benefits of these water uses are difficult to estimate at the local level. For larger watersheds, especially river basins, subbasins, and large lakes, recreation benefits are assessed most often. In some studies these impacts alone are considered sufficient to justify water quality enhance-ment.

Five benefit categories have been assessed in a fairly consistent man-ner: municipal and industrial water supply, commercial fishing, domestic water use, and navigation. They generally entail a straightforward calcu-lation of treatment costs, sales revenue, or expenditures on marketable items. Unfortunately, these water uses seldom in themselves yield the majority of benefits. Leisure activities along water bodies generally constitute the largest economic gains. But recreation, while experiencing phenomenal growth in popularity, eludes precise evaluation. This appears to be even truer of aesthetics. To date, any attempt to quantify this attribute has resulted in a relative rather than an absolute measure, e.g., as a fixed proportion of total monetary benefits.

Most empirical studies are region-specific and cannot be easily generalized to other areas. The number of water use categories is usually incomplete for want of data. Some water uses are deliberately omitted, particularly domestic water supply and navigation channel dredging. To include these uses as pollution-related begs the question of segregating man-made from natural water quality problems. For example, hardness and total dissolved solids in domestic water sources may be traced to natural processes of erosion and ground water conditions. But to negate man's influence on the presence of such constituents is clearly in error. Important land uses, such as agriculture and deep-well waste disposal, can sharply increase erosion and ground water infiltration rates. For most regions there are no reliable methods of apportioning ambient waste concentration by emission sources. To determine economic losses attributable to human activities is thus a matter of conjecture.

The distribution of water quality benefits among water uses varies by region. There are differences in causal factors—pollution levels and mixtures of pollutants, income and occupation of residents or tourists, availability of water areas and capacity for uses, climate, and, above all, values assigned to each benefit unit, e.g., dollars per recreation user-day. As a result, empirical studies cannot be expected to yield similar results across regions. Table II compares relative benefit estimates from selected studies, in which complete abatement is assumed in each region. Partial abatement is likely to modify these relative impacts, as water use intensities respond uniquely to environmental alterations.

Among these estimates, recreation is most often the leading contributor to total monetary benefits. Property values are also significant. Commercial fishing benefits are relatively small in most instances, but they are easy to measure and thus often appear in empirical studies. Ecology has yet to be measured extensively in economic terms. There are a few attempts (e.g., Spencer, 1970) to evaluate noncommercial fish kills for the purpose of discouraging pollution in sport fishing areas. The usual technique is to place a fixed value on dead fish. In all cases, the value is more dependent on potential revenue than on ecological risk. With the exception of very few studies, water uses are not covered comprehensively. As a result, the total values of calculated benefits are biased downward.

Relative magnitudes alone, as illustrated in Table II, cannot justify the enforcement of water quality standards. The impetus for water quality legislation hinges on the absolute dollar value, and primarily on rec-

Table II

Distribution of Total Economic Damages from Water Pollution

(percent)

REGIONAL STUDY	Agricultural Water Supply	Commercial Fishing	Domestic Water Use	Ecology	Human Health	Industrial Water Supply	Municipal Water Supply	Navigation	Property Value	Recreation
Annapolis R. (Nova Scotia) (Bradfield, 1970)							1.5		32.1	66.4
East Norway (Nedenes, 1972)						17.9	10.7			71.4
Italy (Scaiola, 1971)				33.2		26.0	4.9			35.9
Maumee River (Ohio) (Matson & Bennett, 1969)			25.4			21.5	30.0		1.1	23.0
Ohio River Basin (Bramer, 1960)	.001	0.1	15.6			30.5	8.5	0.2	1.6	43.5
Onandago Lake (N.Y.) (Faro & Nemerow, 1969)		1.4					18.3[a]		26.8	53.5
San Diego Bay (FWPCA, 1969)						26.1	12.9[b]	13.1[c]	14.7	33.2
San Francisco Bay (FWPCA, 1967)								7.2	71.4	21.4
Seneca Creek (Md.) (Brandt et al, 1972)		19.4				10.2	4.8		39.2	26.4[d]
United States (Fisher, 1972)		1.7			5.8				30.8	61.7

[a] Includes industrial water supply.
[b] Includes hospitals, schools, and other institutions.
[c] Primarily naval and maritime activities.
[d] Primarily aesthetics enjoyment by tourists.

reation benefits. Table III lists several published values of such benefits, given complete removal of pollutants. The most important activities include sport fishing, boating, and swimming, although the latter has been found insensitive to changes in water quality in at least one study (Davidson, Adams, and Seneca, 1966). Land-based activities, such as beach picnicking, are dependent on water quality to some extent, but few studies provide estimates for this category.

Table III

Economic Damages to Recreation from Water Pollution

($ million)

REGIONAL STUDY	Boating	Camping	Fishing	Swimming	Water Skiing	Total
Eastern Norway (Nedenes, 1972)	$ .12	$ .03	$ .03	$ .12		$ .30
Italy (Scaiola, 1971)			35.0	122.6*		157.6
Maumee River (Ohio) (Matson & Bennett, 1969)						7.7
Ohio River Basin (Bramer, 1960)						120.0
Onandago Lake (N.Y.) (Faro & Nemerow, 1969)			2.7	0.9*		3.6
San Diego Bay (FWPCA, 1969)	2.3		1.0	2.3	0.4	6.0
Seneca Creek (Md.) (Brandt et al, 1972)	0.6		0.7			1.3
Upper Klamath Lake (Ore.) (Stoevener, 1972)						3.9

*Includes swimming, boating, camping, and water skiing.

Aside from the availability of water use data, benefit estimates crucially depend on the method of evaluation. The lack of unanimity among empiricists on the most appropriate methodology precludes a straightforward comparison of these estimates. On the one hand, traditional welfare economics relies on supply-demand analysis, e.g., consumer surplus, to evaluate benefits. Because many social costs of pollution are not recorded in market transactions, the economic analysis may be based on willingness to pay or more subjective ratings of consumer preferences. On the other hand, there are opponents who question the practical meaning of a demand or supply function and may even deny its existence, except at the equilibrium point. For them, a better index of social value is usually direct expenditures or revenue generated by a water user. This assessment technique avoids the derivation of demand schedules, but its validity in measuring net welfare is challenged convincingly by noted economists. To complicate the analysis, there are "induced" and "secondary" benefits in addition to direct impacts of water quality enhancement.

In the next section, four basic criteria are defined as indicative of the adequacy or acceptability of benefit estimates. These guidelines pertain to the conceptual validity of the methodological framework, the derivation of damage relations, the empirical soundness of data, and the general applicability of the estimation procedure to other studies. Other criteria are also discussed, such as the assessment of uncertain data by sensitivity analysis, the comprehensiveness of the benefit categories, and recognition of the distributional aspects of benefit gains. The most useful benefit studies are then identified in light of these criteria.

Evaluation of the Literature

IT FREQUENTLY HAPPENS THAT POLICY MAKERS ACCEPT benefit estimates without questioning their worth. For cost-benefit comparisons, these estimates are useful only if they attempt to measure changes in human or social welfare. In evaluating the adequacy of a study, one should assess the following criteria: (1) Is the estimate derived from the appropriate conceptual framework of welfare economics? (2) Is the relationship between water quality and personal utility changes well established? (3) Does the study make explicit estimates of monetary benefits? (4) Is the method of benefit calculation applicable or easily generalizable to other empirical studies? The last property of technique transferability is desirable, but lack of transferability does not invalidate the analysis, provided the other criteria are met.

A review of the literature indicates that most studies are woefully inadequate in satisfying any of these guidelines. Only two studies (Appalachian Regional Commission, 1969; Tihansky, 1974) in the reference section fulfill all objectives, but they fail to distinguish the benefits of controlling man-made versus natural pollution sources, which may be an important delineation for the design of water quality standards.

The first criterion of theoretical adequacy is met by fewer than 30 percent of the references. But most of these pertain to values of water withdrawal use. Cost functions are derived from detailed surveys of industrial or municipal intake water treatment (Appalachian Regional Commission 1969; Eliassen, 1962; Greeley and Hansen, 1969; Ohio River Committee, 1943). Unfortunately, these end-of-the-line treatment costs are not sufficient in measuring total economic impacts. Other factors not considered include materials recovery, alternative production technologies, and intermedia transfer problems (see the paper by Hanke and Gutmanis). Economic damages to household items from usage of mineralized water supplies are assessed (Black and Veatch, 1967; Metcalf and Eddy, 1972; Tihansky, 1974). These calculations require standard engineering data on chemical costs, depreciation rates, maintenance expenses, and other input requirements. To date, there are only two systematic analyses of property value changes near polluted water (David, 1969; Dornbusch, 1973). However, the first reference uses subjective and uninformative water quality observations in damage assessments, while the second compiles better data but then derives a hypothetical model not adequately fitting these observations. Health benefits of purer drinking water are assessed as the potential savings in medical costs and lost wages (Lackner and Sokoloski, 1973; Liu, 1972). But neither of these studies recognizes the limitation of these values as proxies for willingness to pay.

From Table II, the largest benefits are associated with recreation, but fewer than 10 percent of these studies are theoretically valid (Davidson, Adams, and Seneca, 1966; Reiling, Gibbs, and Stoevener, 1973; Sampedro, 1972; Tybout, 1969). Only one of these references (Reiling, Gibbs, and Stoevener, 1973) derives actual willingness-to-pay curves; others assume a range of hypothetical values for each recreation day. A main reason for conceptual deficiencies is that the wrong talents are employed: the ratio of economics to engineering analysis is far too low. But even when economists were the authors, several studies failed to mention welfare concepts and, in at least one instance, applied the concepts incorrectly (see Appendix, Figure 1). In a few cases (Stevens, 1966; Stoevener, et al., 1972; Tihansky, 1973), consumer surplus was recognized as the most appropriate measure, but the major effort and time required to collect data and formulate demand schedules precluded this approach. The deficiency of these studies, then, is one of omission rather than negligence.

The second criterion, the formulation of realistic damage functions, is useful to decision makers evaluating the marginal cost-benefit trade-offs over feasible ranges of control. Fewer than 20 percent of all references contain empirically tested functions. Water treatment cost functions were formulated for industrial and municipal plants (e.g., Appalachian Regional Commission, 1969; Water Resources Engineers, 1969), but these analyses are based on a small sample set. Consumer cost curves for domestic water use are based on a variety of sample sizes (Black and Veatch, 1967; Tihansky, 1974), depending upon the household item affected. Generally there are many observations of losses in the service life of an item from contaminated water supplies but few (usually two) estimates for operation and maintenance costs. Most of these observations pertain to highly mineralized water, so damage curves for improved water quality must be extrapolated with little confidence.

Only 10 percent of recreation benefit studies derive actual demand schedules. Regression equations were formulated from extensive data surveys (Davidson, Adams, and Seneca, 1966; Megli, Long, and Gamble, 1971; Robert R. Nathan Associates, 1969; Stevens, 1966). In most of the literature, however, such curves are hypothetical and thus lack practical justification (e.g., Barker and Cramer, 1973; Faro and Nemerow, 1969; Stone and Friedland, 1970; Water Resources Engineers, 1969). The primary obstacle to formulating damage curves in any benefit study involves high expenses and personnel needs for interviewing, collecting, and analyzing the data. In many instances, the lack of information on water quality trends makes such derivations impossible.

In satisfying the third criterion, practically all studies provide dollar estimates of benefits. The 5 percent not doing so are very comprehensive, and include intangible and aesthetic effects which the authors are reluctant to monetize (e.g., Battelle Memorial Institute, 1972; Nedenes, 1972; Nighswonger, 1970). Although these qualitative approaches may provide guidance to some policy makers, they are less relevant to most decisions that rely primarily on dollar comparisons. A deficiency with the majority (almost 60 percent) of studies providing dollar estimates is that they are not location-specific. Notably in the recreation literature (e.g., Bureau of Outdoor Recreation, 1968), it is common to use standard economic values of a user-day, as recommended in federal or state government documents. The accuracy of these estimates in reflecting social welfare impacts is disputable.

A fourth attribute is the availability of an explicit set of equations (Barker and Cramer, 1973; Faro and Nemerow, 1969; Meredith and Ewing, 1969; Minnehan, 1968; Orlob, et al., 1970; Water Resources Engineers, 1969), a computer program (Tihansky, 1974; Water Resources Engineers, 1972), or other more comprehensive models (Battelle Memorial Institute, 1972; Nedenes, 1972; Stone and Friedland, 1970) to assist in the general calculation of benefits. Such explicit rules are valuable to planners making cost-benefit estimates. Only 15 percent of the references present generalized procedures for benefit estimation, but some of these (e.g., Orlob, et al., 1970; Water Resources Engineers, 1969) are specific to regional water uses or pollution problems. For example, while a damage curve relating municipal water treatment costs to acidity is meaningful in the Appalachian states, it is far less important in areas without severe mine drainage problems.

Aside from these general criteria, there are features relevant to individual studies. Approximately 25 percent of the studies are defined to be comprehensive (e.g., Battelle Memorial Institute, 1972; Nedenes, 1972; Water Resources Engineers, 1972), but in fact they seldom include all receptors listed in Table I. Moreover, all regional estimates published to date ignore secondary or indirect benefits. To obtain direct benefit estimates is difficult enough, in view of the various conceptual approaches and data observations required to evaluate each benefit category.

Another valuable feature of benefit studies is the use of sensitivity analysis when uncertainty exists in the model specification or the data base. Only two studies (Davidson, Adams, and Seneca, 1966; Sampedro, 1972) purport to measure the range of potential benefits, given different values of a recreation day. Causal factors underlying many benefit responses are not well understood. For example, seasonal catches of commercial fish fluctuate with climatologic as well as pollution changes, but the effects of these phenomena are difficult, if not impossible, to segregate.

The distributional aspects of benefit estimates are rarely addressed. One study (Appalachian Regional Commission, 1969) assesses the impacts of industrial water quality on major sectors of the economy, while a more recent report (Roberts, Hanemann, and Oster, 1974) surveys willingness to pay for recreational opportunities in the Boston area.

Concluding Remarks

IN THE LITERATURE THERE ARE VARIOUS ECONOMIC BENEFIT ESTIMATES of water quality enhancement, although most of them focus on limited objectives, such as the value of a single water use in a small region, e.g., commercial fishing losses in New Haven Harbor, Connecticut (Federal Water Quality Administration, 1970). This paper surveys only those studies pertaining to a major watershed, presenting a comprehensive analysis, or offering a unique methodology.

In practically all of these references, the determination of benefits is less than satisfactory from a conceptual or theoretical point of view. With recreation, for example, the correct measure is the amount people are willing to pay for cleaner water. But the marketplace does not accurately reflect this value. Typically there are no charges, or perhaps minimal fees, for the use of recreational waters, particularly those maintained by public authorities. The expenditures of a recreationist pertain to travel, lodging, etc., but not to direct enjoyment of water. In such cases, the empiricist derives an index which, although inadequate in reflecting welfare, nonetheless provides a quantitative assessment useful to the planner. For recreation demand, total user days is a common index, which is converted into a monetary equivalent by assuming a typical expenditure rate per user day. While this measure is easier to quantify and to interpret than willingness to pay or consumer surplus, it is clearly unacceptable as a surrogate measure of personal utility or welfare. Yet, because the estimation of net welfare often requires extensive data gathering, the theoretically valid concept is unlikely to receive widespread support among empiricists, many of whom seem to lack understanding of the principles of economics.

Common among earlier benefit studies was their complete reliance on dollar values. Intangible benefits were ignored, although recent approaches incorporate them as a weighting of monetary benefits. There is still an overwhelming tendency among empiricists to impute pecuniary values to "nonmonetary effects," primarily to preserve a common denominator in cost-benefit analyses. The dollar unit, however, is only one economic measure of worth. Criss (1971), for example, states, "The growing social consciousness and concern with human well-being has resulted in numerous water resource use and control programs, the results of which must be measured not in the customary monetary terms, but rather in terms of social and human welfare." To be comprehensive, benefit estimates should not be restricted to monetary values. It is indisputable that money is spent in pursuit of intangibles, but their direct value measure is, to a large extent, based on subjective preferences. The importance of ecological and social values should be weighted (by expert staff opinion, or whatever) into an integrated, system-analytic framework. In short, the scope of benefit studies should be multidisciplinary to satisfy the needs of more water resources planners.

Even if empiricists agreed unanimously on a correct method of evaluation, there would still remain sources of uncertainty in the estimates. Economic values of externalities are not recorded accurately in market transactions, and hence such estimates are often obtained through personal opinion surveys and interviews. But responses to questionnaires may be biased, and furthermore they represent a subset of the population base. Even more fundamental is the prediction of physical or behavioral reactions to water quality control. These responses for most water uses are projected as likely occurrences, rather than as observed from past experiences. Few case studies have traced the impacts of long-term water quality changes, probably because of the lack of interest in pollution control until recent years. These shortcomings seem to indicate that a point estimate of benefits is misleading. A more realistic estimate is a confidence interval or range of potential values. This implies that the justification of environmental programs cannot be based on a single cost-benefit ratio. In cases where the interval includes likely ratios above as well as below unity, decisions are not clear-cut and become more dependent on the likelihood of either occurrence.

The general tenor of this literature survey has been to describe theoretical and empirical weaknesses in the techniques of benefit quantification. The Appendix extends this critique by describing errors in applying methodology. These negative comments should not be construed as pessimism about the utility of benefit studies. In each study, the critique was purposely focused on questionable assumptions which the unwary empiricist might take for granted in his own calculations. In the past, benefit analysis has been rejected because of its frequent misapplication in trying to justify the worth of water resources projects. If a favorable cost-benefit ratio does not exist by straightforward (honest) calculations, then the benefit figures should not be juggled to yield a desired outcome. Certainly, this misapplication does not negate the worth of benefit evaluation.

A thorough analysis of water quality impacts is practically impossible in any regional empirical study because of the great range of water-related impacts, let alone the pervasive effects on the rest of the economy and society. Contributing to this limitation is the usual reliance on a common unit of measurement, preferably the dollar. If a benefit category is not directly valued in this unit, then some "convertibility factor" is often imputed to achieve a consistent result. To force such an evaluation runs counter to the opinions of many water resources planners. Water quality enhancement is only one facet, albeit an important one, of social welfare maximization.

Hopefully, the multidisciplinary approach will become more popular in benefit studies. It has already been implemented successfully in regional planning. For example, legislation in Norway requires that water quality control be integrated into an overall perspective of socioeconomic problems. To this end, Nedenes (1972) and other government officials devised and currently use a resource development plan to evaluate environmental quality in a more comprehensive setting. For optimal planning of water resources in Kansas, Nighswonger (1970) developed an assessment model to classify streams for their most promising uses. The Regional Water Quality Control Board of California's Central Coastline sponsored an evaluation (Orlob, et al., 1970) of alternative plans to comply with proposed water quality standards. Such empirical benefit studies have gained momentum in both the United States and other nations, as a means of sensible planning for the preservation of environmental values.

Appendix:
Water Quality Benefit Studies

IN THIS APPENDIX, representative studies are surveyed as they contribute to the development of methodologies. The order of presentation is based on the frame of reference outlined in Table I. Thus, empirical results are partitioned according to (1) specific water receptors and uses, (2) single pollutants, or (3) regional estimates of a more comprehensive nature. In cases where (1) and (2) occur simultaneously, the study is reviewed with respect to the more important category.

Water Receptor Studies

Recreation

Many water pollution control projects are justified on the basis of recreational values. In an historic study of the Delaware estuary, Davidson, Adams, and Seneca (1966) develop a method of estimating social benefits. Multiple regressions predict the impact of a host of decision variables on recreation demand. It is assumed that abatement makes available water surface areas previously closed to activities, and benefits are thus increased in proportion to the added capacity. A range of arbitrary dollar values per user day is chosen to estimate the present discounted stream of total benefits. This dollar range is not justified by the authors, although it appears to cover the range of typical expenditures by outdoor recreationists. Whether it is a reasonable approximation to average consumer surplus is unexplored. The crude assumption that recreation demand will increase in proportion to water availability is convenient but remains to be proven. In actual case studies along the Great Lakes (Bureau of Outdoor Recreation, 1967), polluted beach closures reduced participation rates but did not deter at least one-fourth of the original attendees from swimming (even two or three years later), despite posted warnings of contamination.

Some variables in the multiple regressions are interpreted subjectively, and hence the extrapolation of these results to other regions would be difficult. For example, one index ranging in value from 1 to 5 rates the adequacy of recreational facilities. Just how this index relates the site's physical capacity to water quality is unanswered. In spite of these weaknesses, the study provides insights into causal factors of recreational popularity. Option demand, for example, is found to be more sensitive to the extent and quality of facilities than to the size of the water area. Swimming is shown to be nonresponsive to water quality changes, although this conclusion runs counter to practically all other published results.

Another report on the Delaware River Basin was prepared by Dutta and Asch (1966). Their objective was "the development of the most applicable techniques for measuring in dollar terms the value of various levels of water quality." The general problem of benefit measurement is addressed, although recreation benefits are emphasized because they are thought to be decisive in justifying the net worth of pollution control. It is concluded that withdrawal uses of water, such as water supply treatment, are easier to evaluate in monetary terms than instream activities, since the former are based on current market or engineering conditions, while the latter involve less objective changes in future demand. The measurement of recreation demand is viewed as formidable, necessitating a fairly elaborate and detailed model with many causal factors. Numerical examples of benefit numbers for the basin are hypothesized, but no actual estimates are attempted, even for water withdrawal uses.

The consumer surplus method of estimating recreation benefits is applied by Reiling, Gibbs, and Stoevener (1973) for Upper Klamath Lake, Oregon. Importantly, certain pollutants are segregated with respect to their economc effects. Direct benefits are first estimated for algae control and then cumulatively for temperature reduction and beach debris removal. Oxygen deficiency, a primary determinant in numerous benefit studies, is not considered here because recreationists are only peripherally aware of it. A regional input-out analysis measures indirect (or secondary) benefits, but since these values are based on sales transactions, they cannot be compared consistently with the direct benefits of added consumer surplus. Moreover, the assumption of fixed technical coefficients in the input-output analysis is inconsistent with respect to changes in water quality and consequent shifts in water use input needs. Throughout the study it is recognized that changes in the use of this lake will probably affect activities on nearby lakes, but the empirical results ignore substitutability effects.

Stoevener, et al. (1972) conducted a similar, though more detailed, study of Yaquina Bay, Oregon. The analysis is limited to fishing and its impact on the bay economy. Net benefits are related to the shift in demand as water quality enhancement increases angling success. From this study, Stevens (1966) extracted his contribution to the demand analysis. Figure 1 describes his method of estimating benefits. The original demand curve, CF, for a polluted fishery shifts to curve, DG, after abatement. The rectangular area, ABEF, represents his estimate of benefits, defined in economic jargon as the maximal revenue earned by a nondiscriminating monopolist. In a note of criticism, Burt (1969) argues that this methodology is logically invalid. A better measure of net welfare changes, he claims, is the area, CDGF, bounded by the two demand

146

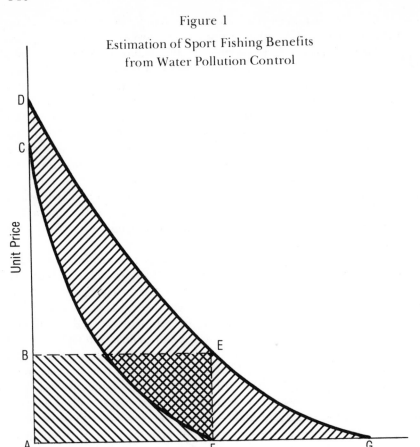

Figure 1

Estimation of Sport Fishing Benefits
from Water Pollution Control

Unit Price

Total Angler Effort

contours. This area corresponds to the amount that people are willing to pay for cleaner water.

More popular than the consumer surplus method of quantifying recreation benefits is the expenditures method. Megli, Long, and Gamble (1971) regress regional income from recreation on numerous socio-economic and environmental variables in the state of Pennsylvania. Three water quality variables are most explanatory: pH (a surrogate of acid mine drainage), dissolved oxygen, and temperature. Linear damage functions are found to provide a better model fit than quadratic functions or linear spline curves. In spite of sophisticated sampling and data analysis, the regressions have a statistically poor fit to observations and are also beset with multicollinearity problems. The dependent variable, regional expenditures, is neither defended as the appropriate measure of benefits nor observed with sufficient accuracy. Personal expenditures are surveyed at each sample area, but they are converted to regional income by an interindustry analysis of merely one county. Because this county is largely rural, its economic structure is likely to be different from that of a more urbanized area, such as Philadelphia County.

FWPCA (1969) estimates direct recreational benefits realized after several years of waste abatement in San Diego Bay. Benefits are also predicted through the year 2000 on the basis of population and recreation demand projections. It is difficult to accept these estimates with confidence, since they are extrapolated well into the future. In addition, to assume that sport fishing benefits are worth a flat $1 million is unwarranted without relating this value to actual participation rates.

In attaching a pecuniary value to recreation, practically all studies assume a fixed value, or fixed range of values, per unit of activity, regardless of the level of water quality. The Bureau of Outdoor Recreation (1968), on the other hand, recognizes that expenditures may vary according to one's reactions to water conditions. As a general guide in evaluating recreation days, the bureau recommends low, medium, and high estimates. Low values, $0.50–$1.00, are likely if pollution seriously limits the direct use of water. Medium values, $0.75–$1.25, occur if water is generally adequate for direct use but restricts some activities during brief or infrequent periods. The highest range, $1.00–$1.50, is expected when there are no damaging effects of water quality, and the water body is capable of supporting all forms of recreation. These values pertain to rural portions of Appalachia. For more affluent areas of the country they may underestimate actual spending patterns.

Property Value

The sensitivity of residential property values to water quality is investigated by David M. Dornbusch and Co. (1973). Observations are selected from five regions in different parts of the United States. For each region, separate multiple regressions on land value changes quantify the influence of a number of factors, including distance of property from the shoreline, zoning restrictions, and proximity to special land developments and urban areas. Excluded from all regressions is the water quality variable, since it is assumed to change uniformly near all property in each region. But given local pollution levels, this study extrapolates national property value losses from a hypothetical function relating percentage of land value changes to water quality. Unfortunately, the case study results do not substantiate this hypothesis and, in fact, reveal by themselves an unrealistic downward trend in damages as water becomes more degraded. The authors contend that the number of case studies is too small to confidently represent average conditions, and hence they resort to hypothetical estimates. There are other problems with the national extrapolation. Although water quality data for case studies refer to specific constituents, national estimates are based on a different pollution duration-intensity factor, since a more disaggregate index is not readily available. Obviously, the nonisomorphic nature of these data bases casts suspicion on the extrapolated results.

In another land value study, Beyer (1969) substantiates the above relative damage estimates. The values pertain to residential areas along the Rockaway River in New Jersey, but are based on opinions from real estate agents. David and Lord (1969) conduct a more quantitative analysis along artificial-lake shores in Wisconsin. The regression results, however, are not easily implementable by environmental policy makers. The water quality variable in the model is subjectively equated to 0 if pollution is negligible, or to 1 if it is moderate. How to translate this integral value into continuous, multidimensional measures of pollutant levels is unresolved.

Domestic Water Supply

There is only one national estimate of water supply benefits to domestic users. Tihansky (1974) derives individual functions relating physical damages for each type of household appliance and water distribution facility to mineral and other constituents of water. Data for these functions are extracted from other studies of household damages (e.g., Black and Veatch, 1967; Metcalf and Eddy, 1972). These impacts are translated into ranges of economic losses from operating problems and equipment depreciation in a typical household. Then they are aggregated at the state and ultimately the national levels. Average household damages are found to be significantly dependent on the source of water supply, partitioned into publicly treated surface water, publicly treated ground water, and private well water. The most economically damaging constituents of water are hardness and total dissolved solids. Because these constituents are partly natural in origin, their impacts are difficult to interpret in evaluating the abatement of man-made pollution. In few watersheds can man-made and natural mineral loads be segregated because of the lack of time-series data. The notion of a typical water quality level as used in this study is disputable, because of geographic variations of natural water quality and alternative options of water supply treatment. Although dependent on a larger data base, a histogram rather than a point estimate of water quality would provide a more respresentative sample of household conditions.

Commercial Fishing

Although commercial marine fishery losses from pollution are estimated frequently for small coastal areas, estuaries, and river stretches throughout the United States, there are only several national estimates. Bale (1971) estimates total national losses of (dockside) revenue from

DDT, mercury, and pathogenic organisms. Fish kills are evaluated by assigning an arbitrary price per fish, assuming that roughly two-thirds of reported kills are commercial species. This assigned unit value may be quite modest, and, moreover, because fish kills are not carefully monitored, damage estimates are highly conjectural. Economic losses to the shellfish industry are calculated by assuming that of all species, only clams and oysters are reduced in catch since they are practically immobile. The reduction in potential supply is assumed proportional to shellfishing areas closed by pollution. Potential revenue gains from cleanup are calculated from the original price of shellfish, which is incorrect since an increase in the national supply should lower prices.

The Council on Environmental Quality (1970) also estimates national revenue losses of commercial marine shellfishing, but this value is four times as large as Bale's estimate because it includes all species caught. This estimate is probably too high because some species, such as lobsters and crabs, are more tolerant of existing water quality than clams and oysters. Weddig (1972) calculates the impact of mercury scares on inland and estuarine fish catches in the United States. He assumes, without any apparent validation, that roughly 1.5 percent of potential domestic supply is thus lost. A more interesting part of this estimate pertains to a potential coho salmon fishery in Lake Michigan. Since it "closed before it began," the real dollar losses to private fishermen are assumed to be nil. This viewpoint overlooks the social cost of unfulfilled opportunity and lost option demand.

Human Health

Economic assessments of human illness from drinking water pollutants are highly subjective for several reasons. Epidemiologic and other health data are not compiled in detail for acute clinical illnesses, and, in fact, do not exist for subclinical and low-level illnesses (such as common diarrhea) as related to water quality. Waterborne pollutants are transmitted not only through drinking and culinary use, but also through water-based recreation, agriculture, and food, e.g., shellfish. To trace the extent of these complex effects is currently infeasible. Some enteric viruses harmful to man cannot be accurately monitored in water, so the origin of illnesses is difficult to identify. The economic value of human life is also a moot topic. There are many cost components to evaluate, such as medical needs, productivity losses, premature burial and lost wages in the case of early death, and psychic costs. Whether, for instance, a man's contribution to the work force should be measured by total wages or the residual wage net of consumption expenses is not resolved among empiricists.

Lackner and Sokoloski (1973) estimate national benefits that would accrue from the reduction of waterborne communicable diseases, such as infectious hepatitis and typhoid fever. Reported outbreaks of diseases attributed to water are evaluated monetarily on the basis of ten days of lost income and five days of hospitalization. Nonreported outbreaks, valued in analogous fashion, are assumed to be fifteen times as large as reported ones. Benefits for a third category of illness, gastroentiritis, are measured by assuming that one million cases occur per year at a unit social cost of $100. As a fourth category, dental expenses are expected to decline with improved fluoridation of public water supplies. These benefits are calculated for all individuals drinking water with minimal fluoride levels.

The direct benefits of communicable disease control are then multiplied by a "contagion multiplier" of five, indicating the highly communicable nature of each illness. These estimates, in turn, are doubled to account for the national economic impacts of personal expenses. Obviously, both the economic data and illness rates in these calculations are very subjective, implying that extreme caution must be exercised in basing policy on the results. At least one of the estimates is questionable. It is not clear how the lack of beneficial fluorides in water is considered a pollutant.

Liu (1972) estimates national health costs from viral pollutants in water. Assumptions on annual health impacts include 2 million work days lost from acute gastroentiritis and diarrhea at an average lost wage of $30 a day; 1,000 deaths from infectious hepatitis at $100,000 for the loss of a productive life; and 10,000 cases of spontaneous abortions at $300 per individual. Similar calculations are made for seven other disease categories. Although the assumptions are based on an extensive search of the literature and are more detailed than others in the literature, they remain "guesstimates." Yet they do recognize that economic impacts differ according to the type of associated illness.

Aesthetics

Aesthetic benefits of water quality improvement continue to elude direct quantification. Practically all studies of aesthetics consider water quality only peripherally. Concepts and individual attitudes associated with aesthetics relate not only to conditions of the physical environment, but also to political, social, and cultural conditions. The basic acceptance of what constitutes aesthetic values differs among empiricists as well as government agencies.

Visual aspects of streams are rarely assessed in economic terms, but Orlob, et al. (1970) derive a benefit function for wild river conservation. For the Shagit River in Washington, intangible values are estimated as a multiple of monetary impacts, such as tourist expenditures. The function for river basin development impacts is expressed as,

$$B_m = \frac{a}{2} m^2 - acm + B_o,$$

where B_m is the apparent dollar value of "nonmonetary" benefits from the development; m, the level of development ranging in value from 0 to 10; B_o, the total value of the river in its wild or natural state; a, the coefficient measuring the degree of curvature of this function; and c, another coefficient defining the economically optimal level of development. The estimates of these coefficients are arbitrary and differ according to personal biases. The authors admit that "no claim is made that the values assigned to the various subfactors are correct in the absolute." Perhaps more questionable is the assumption that pecuniary values do indeed make sense. To translate nonmonetary benefits into absolute dollar equivalents belies the definition of "nonmonetary."

Nighswonger (1970) is more comprehensive in evaluating aesthetics along streams in Kansas. His method is to inventory specific environmental attributes of an area, including such factors as road accessibility, campsite facilities, land area, and water pollution discharges. All attributes are weighted independently within a preassigned interval, and their sum provides an index of visual quality. Pollutants are subdivided according to their source into three groups: industrial, municipal and domestic, and agricultural. Only one (and the same) scalar value is assigned to the presence of each pollutant source, regardless of incremental changes in damages. Obviously, that restriction fails to recognize differential impacts of water quality constituents.

Krutilla (1971) proposes still another method, relying on direct monetary values rather than relative weightings. Benefits are estimated over time for two water uses in Hells Canyon: the development of a hydroelectric project at a dam site and the "preservation alternative" (maintaining the area for recreationists). Although the author believes that consumer surplus is an appropriate measure of recreation benefits, he also claims that the lack of data and "insufficient time" preclude such an analysis. Instead, economic damages from nonpreservation are approximated by the expected loss of tourist revenues. The current value of recreation is adjusted by future growth, making it favorable to the hydroelectric project. The recognition of these dynamic impacts makes the study particularly valuable.

This technique of using a surrogate measure of benefits is quite common in studies on recreation and aesthetics. But it evades the more complex task of quantifying impacts directly. Most measures also neglect the welfare impacts of substitutes for water uses. In this study, the comparison of alternative benefits is not entirely consistent. On the one hand, the hydroelectric project is weighted against options for power production. On the other hand, recreation losses (nonpreservation) are projected without consideration of welfare gains from substitute choices, such as staying at home. Finally, it is not obvious that the construction of a dam for power generation destroys the "preservation alternative." Indeed, this type of water impoundment may attract more tourists into the region.

Single Pollutant Studies

Oxygen Deficiency

Biochemical oxygen demand (BOD), a measure of dissolved oxygen deficiency, is the most popular water quality parameter in empirical studies. Its use is justified on both theoretical and practical grounds. More transport and diffusion models have been designed to account for, or incorporate, this pollutant than for any other. BOD readings are common in water quality stations throughout major watersheds and have been found to correlate highly with survivability of aquatic life.

The best known river basin studies relate monetary benefits to the reduction of BOD. For the Delaware estuary, FWPCA (1966) sponsored a number of interrelated studies. The primary outcome was the derivation of a large scale model relating alternative BOD treatment levels to the waste transport and the assimilative capacity of the estuary. Within this framework, treatment costs are minimized, subject to water quality constraints describing various intensities of beneficial water uses. Problems of estimating monetary benefits are thus purposely avoided by excluding benefit measures from the objective function. The authors claim that only a few benefits can be quantified. Benefits of industrial water supply, they argue, are inversely related to dissolved oxygen levels, because dissolved oxygen corrodes intake facilities.

In an extension of this model, Tomazinis and Gabbour (1967) estimate the economic impacts of pollution control on specific pastimes —boating, fishing, swimming, and beach picnicking. Like Davidson, Adams, and Seneca (1966), they assume that demand increases linearly with the supply of clean surface water. But they assume a narrower range of economic values per user day, comparable to anticipated per capita expenses for travel, board, and lodging. Moreover, there is a different economic value associated with each activity. The assumption of a fixed range of economic values does not allow for the case where water quality enhancement leads to congestion of recreational zones, thereby lowering the activities' social desirability and worth.

Minnehan (1968) derives a single equation model to predict total benefits from recreation allowing for congestion costs. However, his attempt to verify the model with data taken from studies of the Delaware estuary and San Francisco Bay (FWPCA, 1966; Stanford Research Institute, 1966) is not convincing.

Acidity

The first study of acid mine drainage control dates back to 1943 (Ohio River Committee). Detailed questionnaires to selected industries provided estimates of excess treatment costs and boiler replacement damages from acidic water intake. Economic damages to such structures as bridges are estimated subjectively. Bacterial pollution of public water supplies and recreational areas is identified as the most serious problem, but no attempt is made to quantify these impacts. Despite its inadequate coverage, this study is unique in recognizing the geographic variations of impacts. It concludes, "While the intensity of stream pollution is severe in numerous localities, serious damage is confined to a few places where water use is extensive."

Because acids seriously degrade more than half of Appalachia's streams, the Appalachian Regional Commission (1969) sponsored a major study of economic damages. For industrial water users, linear damage curves relate pH levels of intake water to costs of lime neutralization. Unfortunately, field observations are not used to test the validity of the linearity assumption. Navigation costs include accelerated corrosion and maintenance of bridges, piers, and ships. By means of interviews, human reactions to acidic water are examined. Recreationists are found to be tolerant of pH content as low as 5.5 or even 5.0, but reduce their activity rates at lower levels. Because highly acidic streams in most of this region are intermeshed with cleaner water bodies, the likelihood of substitution between areas discourages estimates of this benefit category.

A companion report, however, by Robert R. Nathan Associates (1969) includes the substitutability question in separate models predicting total user days and probability of participation. The model determinant, recreational water area, changes with the extent of pollution. Another variable, recreation land area near water bodies, is assumed to reflect potential demand for substitutes to water-based activities. Whether water enthusiasts do turn to nearby land activities is questionable and should be investigated more thoroughly. Furthermore, the choice of boundaries for the land area variable appears to be somewhat arbitrary. In converting demand predictions into monetary values, this study assumes a typical range of total expenditures per user day and then extrapolates local values to the regional economy by an appropriate income multiplier. The use of aggregate demand and expenditure rates for all recreation conceals differential economic impacts of specific activities as well as their unique sensitivities to water quality.

Salinity

Serious economic and social impacts on the agricultural sector can occur if irrigation water is too saline. In Australia, Callinan and Webster (1971) estimate farm production losses and the social costs of uprooting rural community life by forcing farmers to move. They assume that the economic cost of re-establishment is a fixed amount per capita, and that the corresponding "social cost" is one-fourth of the economic value. These cost estimates are not convincing because they are never defined explicitly. In addition to salinity, plant damages may be caused by other factors, such as poor farm management, whose effects are not considered in the study. Finally, farm crop losses are only an immediate effect, and may not reflect long-term damages. Crops more tolerant of salinity can be substituted in irrigated fields, thus reducing initial income losses.

Vincent and Russell (1971) present a more comprehensive analysis of saline water uses. Economic losses pertain to decreased agricultural crop yields, municipal and industrial water treatment costs, corrosion of water supply intake pipes, and reduced palatibility of drinking water (which bears on the consumers' willingness to pay for taste). Because sufficient information on salinity effects is lacking, the authors derive estimates by applying Bayesian probability theory. Expected values and probabilities of damage levels are both assigned judgmentally, thereby reducing the precision of the answers. Moreover, damage estimates are expressed as single points rather than with confidence intervals, the latter being more realistic with stochastic data.

Although constrained by information deficiencies, this study provides insight on the optimal strategy for environmental controls. Over the feasible range of salinity concentrations, total regional costs and damage curves are drawn. The economic optimum is shown to be that control level at which the slopes, e.g., marginal changes, rather than the intercepts of the curves, are equal.

In a theoretical decision model, the U.S. Environmental Protection Agency (1972) extends this concept of optimality to identify the least-cost solution for salinity control. In response to saline irrigation water from the Colorado River basin, farmers have five options, varying from no remedial action (with reduced crop yields) to maintenance of past crop yields (with increased water requirements). Nonlinear economic damage curves are formulated for each action, but they are based on minimal data and incomplete surveys of farmers' preferences for action. As a hedge, all damage estimates are made on the conservative side.

Another study of this region, by the Bureau of Reclamation (1969), derives what appear to be high damage values. On the basis of "no remedial actions," crop losses are evaluated, given the assumptions that the highest valued crops are destroyed and soil leaching conditions are extreme. Estimates of regional crop damages are based on fixed prices and thus overlook the possibility that large-scale changes in supply may significantly affect the equilibrium price.

Coliform Bacteria

Only one comprehensive analysis relies on coliform bacteria as the basic pollutant. Water Resources Engineers (1969) combines both tangible and intangible benefits in an economic framework, applied near the Monterey Bay area of California. Primary benefits for nine water uses, including wildlife preservation and scientific beach research, are evaluated over a fifty-year period. These values are converted to regional as well as national impacts by subjectively determined multipliers. They are further adjusted by a "constituent-importance factor" dependent on the level of water quality. This factor ranges from a low value for mild synergistic effects of pollutants to a high value for hazardous conditions. As water quality improves, this value decreases and thus yields economic benefits. Because this adjustment is intuitively derived, the benefit estimates are difficult to interpret. Moreover, the choice of coliform bacteria as the only water quality variable is questionable, since it usually relates directly to only a few water uses, e.g., human health and commercial fishing. Yet high coliform counts can sometimes indicate the presence of other damaging pollutants.

Sediment

Sediment from erosion is identified as a primary determinant of damages, particularly in rural areas. Brandt, et al. (1972) assess economic detriments along the Potomac River stretch north of the District of Columbia. Impacts are distinguished as on-site (soil losses at the discharge source, for example) and off-site (as with effects on downstream water users). It is recognized that not all effects are detrimental; for example, turbidity in municipal water supply absorbs certain foul-tasting and odor-producing constituents. Hypothetical linear damage functions relate sediment loads to chemical treatment costs of the water supply. Perhaps the most unique element of this analysis is its evaluation of aesthetics. A certain percentage of annual visitors to the Washington area are assumed to be concerned about the visual aspects of sedimentation. To each visitor, it is worth a "conservative amount (twenty-five to fifty cents) to alleviate this problem. These monetary values provide a range of total aesthetic benefits. Unfortunately, the estimates are purely hypothetical. Furthermore, the population base in this study is limited to tourists, whereas aesthetics is also of concern to residents.

In a more comprehensive geographic analysis, Stoll (1966) has estimated annual sediment damages for the United States. Damage categories include reservoir capacity losses, inland navigation route blockages, obstruction of irrigation canals, excess turbidity in public water supplies, and commercial fishery losses. The validity of such estimates is suspicious because accurate data on sediment discharges, transport, and subsequent effects are practically nonexistent in most regions of the country. For some water uses, the cost of dredging is estimated as a surrogate for damages, whose direct quantification is far more complex.

Multi-Pollutants

There is only one regional analysis which partitions total damages into specific water quality components. For the Ohio River Basin, Bramer (1971) ranks the most economically important parameters as suspended material, hardness, toxicity, BOD, and oil, with smaller contributions from other parameters. Although the author claims that these estimates follow from a previous study of water use benefits (Bramer, 1960), this is not obvious. Furthermore, to disaggregate the effects of parameters implies that synergistic and complementary aspects are ignored.

Comprehensive Studies

The previous studies pertain to individual pollutants or beneficial water uses. For the geographic areas of concern, these single indices appear self-sufficient in measuring most of the monetary benefits of pollution control. In most regions, however, the analysis is far more complicated. A large number of water uses and benefit recipients may compete with, or complement, one another. Each recipient is uniquely sensitive to components of water quality. For instance, a pollutant (such as hardness) detrimental to one user group (households) may have little or no affect on another group (tourists). Besides the multidimensional nature of water quality and receptors, there are also differential welfare impacts within the region. Urbanites, for example, exhibit a set of water use preferences distinct from those normally found among their rural counterparts.

In an historically important comprehensive study, Bramer (1960) has evaluated water uses in the Ohio River Basin. Monetary benefits are calculated in the most simplistic manner, by assuming that abatement will increase current expenditures or revenue by a fixed percentage. For example, participation rates in all recreational activities are assumed to rise uniformly by 50 percent if tertiary controls are enforced. Most of the physical damage estimates are crude and subjective, whereas economic values of physical damages are documented from other sources. The author recognizes an often neglected benefit—the distribution and use of domestic water supply. He attempts to measure man-made damages by including only residents using surface water sources while ignoring those using ground water. But such a dichotomy is unrealistic, since nature clearly affects surface water, e.g., by erosion, and man contaminates subsurface aquifers, e.g., by deep-well waste disposal. Matson and Bennett (1969) use the analytical framework developed by Bramer to study the Maumee River basin (in Ohio, Indiana, and Michigan). While the methodology is the same, relative benefit values differ in response to the uniqueness of each basin.

The U.S. Army Corps of Engineers (1963), in their early study of the Potomac River basin, identified the most damaging pollutants as temperature, pH, BOD, and coliforms. Thermal pollution control yields water supply benefits, including savings in pumping costs and reduced maintenance of cooling water systems. Acid reduction results in savings by reducing chemical requirements for abatement. BOD is assumed detrimental to leisure activities, but the benefits of control are estimated indirectly as reductions in treatment costs. Only in the case of fishing are benefits measured directly as additional state income from fishing license sales and expenditures, but these values are incorrect in theory.

For Onandago Lake, New York, Faro and Nemerow (1969) calculate separate benefits for recreation, water withdrawal and waste disposal, bordering land costs, and instream water uses, such as commercial fishing. There are several unique contributions from this study. Recreational benefits (personal expenditures) are netted or adjusted by the cost of maintaining parks and shorelines. Some water uses, such as sport fishing, are related to a pollution index rather than a single, less informative water quality parameter. Property values along Lake Onandago are compared with those along a cleaner nearby lake to estimate differential impacts of the two water quality levels. Yet despite an extensive compilation of data related to water use, the benefit estimates contain elements of uncertainty. Recreation demand, as a case in point, is hypothetically related to the pollution index. Commercial fishery losses are assumed directly proportional to the decline in catch of certain species over the past half-century. But this assumption ignores the possibility that consumer preferences for fish products and substitutes have changed. Flood damages incorporate the effects of siltation on reservoir capacity, but this may be primarily of natural rather than man-made origin.

For the purpose of setting standards, this study has limited application. Benefits are calculated at only two water quality levels, and a straight line is drawn through them to represent the total damage curve. Benefit estimates at other water quality levels must therefore be interpolated, or, less confidently, extrapolated, from a relation whose constant slope may indeed be unrealistic.

Another report (Sampedro, 1972) assesses the minimal benefits required to exceed annualized costs of pollution abatement in Biscayne Bay and nearby canals of Miami, Florida. Potential benefits are identified as increased marine animal production, decreased health hazards, and greater use of the bay for recreation and visual appreciation. The empirical values of these impacts are based largely on hypothetical data and questionable assumptions about maximal demand for water-related activities. Health benefits are arbitrarily set at $10 million annually, despite disagreement from public health officials contending that no dangers currently exist. The incremental value of commercial fish yields from pollution control is assumed to equal $3 million, on the grounds that total current revenue from bay catches is also this amount. Obviously, there is no evidence for these estimates.

Recreational benefits are calculated to be one-third of the difference between actual participation rates and maximal capacity estimates. This adjustment factor is not justified, and, to make matters worse, the capacity constraint is assumed identical for all beach resorts, in spite of their apparent distinctions in land and water development. Although these empirical flaws cast suspicion on the benefit estimates, this study recognizes the best theoretical approach. The gross expenditure and travel cost methods of measuring welfare losses are rejected in favor of willingness-to-pay measures. Moreover, the report recognizes the dynamic impact of water uses by calculating the present value of recreation benefits.

The only attempt to eliminate double counting of benefits is by Bradfield (1970), in an economic model of the Annapolis-Cornwallis River systems of eastern Canada. Recreational values are assumed to be partially reflected by property value changes. To determine the amount of overlap, this study divides recreational expenditures into local and tourist trade, and then excludes the former component as being capitalized into land value. This partitioning is not intuitively obvious, however, because both components increase the demand for waterfront acreage.

The calculation of recreation benefits in this study is based on two demand groups. The first group, current participants, lengthens the duration of visits after water quality enhancement and hence increases revenues to the local economy. The other group, new participants, is a source of additional revenue. The latter benefits to the local economy are measured as a fixed economic value per acre-foot of cleaner water. This unit value is quoted from another study of competing water uses by Wollman (1962). Whether recreationists do in fact increase regional income depends on what they spent on previous land-related activities. To apply Wollman's values (derived for U.S. streams) to a Canadian region ignores possible variations of consumer demand and value judgments.

In an investigation of the San Joaquin River basin in central California, FWPCA (1967) ranks pollutants according to relative impacts on each water use. For example, ambient concentrations of dissolved oxygen and chlorophyll are graphed against relative declines in sport fish populations. But damage functions are all derived from "expert opinion" instead of field surveys. However, they are expressed as confidence intervals to account for uncertainty of the mean estimates. Total sales revenue or expenditures are used as the economic measure of welfare. But there is no adjustment for costs of recreational facilities to arrive at a net income figure.

Tybout (1969), on the other hand, argues that gross benefits bias the analysis in favor of environmental controls. Revenue losses to commercial fisheries, for example, should reflect reallocation of resources, e.g., manpower shifts from fishing to other pursuits. In surveying Lake Erie pollution problems, the author finds reliable estimates of past attendance losses along polluted beaches. Thus, unlike most recreation studies that hypothesize changes in future demand, this analysis presents actual results. Ranges of economic values for each activity day foregone are documented from other well-known surveys (e.g., Clawson, 1966) to yield probable bounds of benefit estimates.

Social costs of the famous Santa Barbara oil spill are evaluated in a comprehensive manner (Mead and Sorensen, 1970). Damages are separated into control costs (beach cleanup, oil well control efforts, and oil collection), tourism and local recreation losses, declines in property value, declines in commercial fish harvests, and intertidal plant and animal damages. In one of few published analyses of substitution effects from pollution, this study surveys tourist demand and finds that Santa Barbara resort losses are offset by gains in neighboring resort areas. Net recreation losses are thus found to be negligible. Shortcomings of this conclusion are that it is based on expenditures rather than more appropriate economic welfare concepts, and that it ignores the social or psychic cost of transferring to a substitute resort area.

Particularly ambiguous is the assessment of ecological damages along intertidal areas. Economic loss bounds of $1,000 and $25,000 are chosen without any apparent justification. Bird losses are evaluated by attaching an arbitrary social value of $1.00 per life lost. Other marine ecological changes, such as decreases in the diversity of species and changes in the habitat that favored more tolerant species, were either too subtle to estimate or not significant enough to claim as damages. Despite the uncertainties inherent in some of the estimates, this study is an important forerunner in attributing damages to oil pollution.

An alternative to direct treatment of pollutants is low flow augmentation. Pyatt, Grantham, and Carter (1969) and Merritt and Mar (1969) purport to measure monetary benefits in Connecticut and Oregon streams, respectively, but both studies calculate avoided costs of municipal treatment. Such a proxy measure, so common in cost-benefit analyses, neglects the real issue of quantifying benefits (or damages foregone) and hence is equivalent simply to a cost analysis. Furthermore, by using avoided costs as a welfare measure, any project is justifiable since the resultant "cost-benefit" ratio is always unity.

National Estimates

Comprehensive benefit studies of water quality enhancement on a national scope are few in number. Practically all rely on extrapolations of a subregional analysis because of the enormous task of collecting statistics on a region-by-region basis.

A cursory review by Fisher (1972) provides such estimates for several benefit categories: recreation, bordering land values, commercial fishing, and human health. Municipal and industrial water withdrawal benefits are considered minimal. Wildlife values are also excluded, since they are either included under recreation and commercial fishery categories or ignored because exotic wildlife species cannot be priced. Per capita recreation benefits to the country are assumed identical to those in the Delaware estuary study (FWPCA, 1966). To assume that benefits in the highly polluted and industrialized Delaware estuary are identical with those in the nation as a whole seems indefensible. Property value estimates are calculated as 50 percent of recreation benefits, in accordance with the same proportional estimates in the Onandago Lake analysis (Faro and Nemerow, 1969). Both the extrapolation from this small-lake study and the assumption of fixed proportionality are intuitively disturbing. Moreover, part of the health cost estimate is just as subjective. EPA estimates (Lackner and Sokoloski, 1973) of waterborne communicable diseases are corrected for miscalculations, and then this total is increased to a rounded figure to account for viral and low-grade illnesses. The credibility of this procedure is questionable, but the final estimate cannot be challenged without more research results on health effects.

Kimball (1972) of the National Wildlife Federation also estimates annual water pollution damages. But his figure is not disaggregated, although it supposedly includes aesthetics impacts. There is no discussion of the method of evaluation nor of the input data needed for the analysis. The author claims that the value is derived by a "federal investigation team," whose identity must remain anonymous. In an attempt to unveil this lid of secrecy, a press agency (Benjamin, 1972) stated that the estimate is "built upon" information from the annual report of the Council on Environmental Quality (1971). Although this report reviews damages from air pollution, it has no such coverage for water. Hence, the source and validity of the Federation estimate remain uncertain.

For Italy, total losses from water pollution are projected over 1970 to 1985 by Scaiola (1971). Direct damages to recreation, and to municipal and industrial water supply are calculated at a social discount rate of 4 percent, but excluded are intangible and secondary damages (the latter defined to include property value changes and transactions among economic sectors). Recreational losses pertain to seaside tourism, lake activities, and both marine and inland sport fishing. Most river-based activities are not considered because of the paucity of relevant data. Although several types of recreational impacts are recognized—varying from the psychic costs of reduced enjoyment of water and the destruction of natural resources to losses from nondevelopment of polluted shores—quantification is limited to a few categories: the cost of cleaning beaches, the value of activity days lost, and reduced value per current activity day. The first category is less indicative of benefits than costs. The latter categories are more meaningful, and, unlike most hypothetical estimates in the literature, they are based on survey responses from independent tourist agencies and seaside resorts. The translation of lost activity days into monetary terms involves a "suitable correction of market prices." It is incorrect, however, to assume that private recreation fees and charges can be applied to public activities. These are two distinct recreational products; the former has the comparative advantage of privacy and usually better facilities.

In extending Scaiola's study of Italy, Cannavo, et al. (1970) published perhaps the only monetary estimate of ecological damages. Their results are prefaced by the following realization: "The evaluation of the economic damages caused to the ecologic patrimony presents great methodologic difficulties on the theoretic level. They concern chiefly both the inadequacy of the present knowledge of the consequences caused by pollution to the various ecologic systems and the impossibility of attributing a monetary value to the natural patrimony." Their estimate is limited to several natural products—fish, game, and wood (forests)—which have an obvious market price. It thus overlooks the more basic ecological concerns of the dynamic balance between natural habitat and living communities, both plant and animal. Instead, products are evaluated in terms of their replacement cost, as determined from questionnaires mailed to public administrators in each Italian province. Losses of salt-water fish, on the other hand, are assumed proportional to their wholesale price.

Of all marine water areas, estuaries are perhaps most vulnerable to pollution along the United States coastline. Wasserman (1970) documents case studies in which economic losses from man-caused activities are traced to the estuarine environment. Examples include shellfishing closures, waterfowl kills from oil spills, shoaling of a major harbor from debris, and swimming opportunities foregone from contaminated beaches. In quantifying losses, this study neglects to distinguish that portion of total losses due to man-made pollution. A second problem is the assignment of a dollar value to unit damages. For example, duck kills from oil pollution are valued as the replacement costs for game breeding. This value is probably inadequate to assess the recreationist's dissatisfaction with the reduced probability of a successful hunt.

The majority of benefit estimates focus on inland pollution problems. There is one somewhat extensive study of marine waters by Tihansky (1973). Water uses most frequently impaired are listed, but only three—commercial fishing, recreation, and navigation (shipping accidents)—are evaluated in a monetary framework. State-by-state information on water use impacts is compiled, and economic values are aggregated into a national estimate. Data for some coastal areas are lacking or highly suspect because of incompleteness, thus making final estimates vulnerable to error. Assumptions on water quality detriments to water uses are also questionable. For recreation, the percentage of beaches closed because of pollution is considered directly proportional to the relative loss of activity days. Similarly, losses in commercial fish catches are assumed proportional to shellfishing area closures. The economic magnitude of fishery benefits is based on the gain of consumer surplus resulting from a shift in the supply curve. Contrary to economic theory, the supply shift is determined by increases in potential catch (implying a horizontal shift on the demand-supply graph) rather than changes in marginal costs of fishing (a vertical shift). The lack of data on marginal costs prohibits the latter method of empirical analysis.

Generalized Regional Estimates

Many regional studies derive water quality benefit functions, which serve several general purposes. First, the functions depict water use values over a range, and not just a single observation, of water quality potentials. Second, for beneficial uses they reveal the degree of linear or nonlinear demand sensitivity to specific water constituents. Third, they describe relations, in brief mathematical notation, that may be too difficult to comprehend otherwise. Fourth, and perhaps most essential to empiricists, they provide a well-defined framework for the analysis of damages in other regions. The studies outlined below incorporate benefit relations in a system-analytic context. Some of them are more comprehensive than others in recognizing noneconomic (e.g., ecological) value systems. In all cases, the analysis is applied to a specific region.

The most detailed mathematical model of benefit evaluations is formulated by Meredith and Ewing (1969). Economic benefits to major water uses are linearly related to water quality changes. The complexity of model equations is illustrated in the following sample calculation,

$$WS = \sum_i \sum_j \sum_k \sum_m (C_{ijkm} N_{ijkm}(L1_{ijkm} - L2_{ijkm}))$$

where WS represents the total regional cost increment of water supply treatment; C_{ijkm}, the unit cost of treatment, at the ith water supply distribution station, of the jth pollutant in water whose source is the kth sector of the mth watershed; N_{ijkm}, the quantity of water intake; and $L1_{ijkm} - L2_{ijkm}$, the change of intake water quality after standards are enforced.

Typically, the large number of variables in this model preclude its use since input data are difficult to obtain. Although linear dependencies between water quality and uses are hypothesized, actual functions are not derived since they probably differ by region. Because pollution indices are added linearly, synergistic effects are neglected. Only small variations in water quality can be assumed, since long-term effects, although considered important, are excluded from the model. It is assumed, for instance, that municipal water suppliers reduce expenditures for chemicals but do not alter control technologies in response to cleaner water.

Greeley and Hansen, Engineers (1969) apply the above model for estimating benefits of dredging along Lake Michigan shores near Chicago. Annual cost savings of municipal water treatment are extremely small, but no other water uses are evaluated. The study was terminated because of insufficient data on recreation and other categories likely to yield the largest impacts.

Stone and Friedland (1970) develop a unique formula to assess both economic and social values of clean water. The formula is expressed as,

$$B_i = \sum_j a_{ij}W_j ,$$

where W_j is a weighted index of benefits incurred by water use j; a_{ij}, a weighted estimate of the impact of alternative treatment method i on water use j; and B_i, a normalized scale (in percent) of benefits resulting from each treatment method i. Every factor under the summation is assessed independently by attitude-scaling questionnaires. First, all water uses are weighted in economic importance (W_j^{econ}), such that the sum of weights is 100. Next, they are weighted analogously in social importance (W_j^{soc}). The economic and social values are then combined by relative scale, y, as follows,

$$W_j = W_j^{econ} + y \cdot W_j^{soc}$$

This model is applied to San Diego Bay, but there are several biases in the empirical estimates. Although value judgments of water use implicitly depend on water quality conditions at the time of the attitude survey, they are weighted with respect to future control strategies. The survey is submitted to public officials, who, while supposedly knowledgeable on water resources planning, represent only one segment of the bay community. Also, the socioeconomic integrator, y, is arbitrarily assigned values between 0.6 and 1.0. But for compatibility with other estimates, this integrator value should be obtained through a survey. The relative importance of water uses is not defined in terms of geographic scope. For example, expenditures on tourism can be measured either solely for bay activities or for their multiplier effects on generating revenue throughout the regional economy. But such distinctions are not recognized in the study. Finally, even if the formula is correctly applied, the solution vector would be difficult to implement by policy makers. The normalized socioeconomic ratings (in percent) of water uses are not directly comparable to control costs (in dollars). And the transformation of social values into a monetary equivalent is nonobjective.

In an assessment for the state of Illinois, Barker and Cramer (1973) present a simplified model of four major water uses: municipal and industrial water supply, recreation, and commercial fisheries. Benefits are distinguished as potential (based on future controls) and fulfilled (as a result of past efforts), contrary to the usual practice of evaluating only potential impacts. Each water use category is assessed uniquely. For example, municipal water supply benefits are calculated as the cost reduction of add-on technologies to control specific waterborne constituents. For recreation, a hypothetical nonlinear damage curve relates water-area closures to a pollution index, as defined by Sumitomo and Nemerow (1969). The damage function approach is rejected for commercial fishery losses since correlations between fish species and the pollution index are found inconclusive (probably because data are not compiled accurately). Instead, past volumes of catch are compared to current levels to estimate lost revenue.

This study recognizes that more than one pollutant is responsible for damages. However, the application of merely one weighted index does not reflect differences between water use intensities. Indeed, this may partially account for inconclusive tests relating the loss of fish species to the fixed pollution index. Hypothetical assumptions for recreation demand also lower credibility of the estimates. Economic values per unit-change in water use are selected without apparent validation. To assume, for example, that unit costs are identical for all conventional water supply treatment add-on facilities simplifies the analysis but is hardly plausible without conducting a survey of actual costs. Nevertheless, the model is valuable in accounting for synergistic effects of pollutants and in identifying alternative strategies of water supply treatment.

Water Resources Engineers (1972) designed a computer program to calculate benefits in any region, and applied it to the Santa Ana watershed and Monterey Bay areas of California. Water use categories include municipal and industrial water supply, domestic water use, agricultural water supply (for irrigation), fish and wildlife preservation, and recreation. Eight water quality measures must be supplied as input data, varying from hardness and total dissolved solids to dissolved oxygen levels. Damage functions relate specific water constituent concentrations to the demand or supply of water-dependent activities. These formulations are nonlinear and, in most cases, are derived from field and published surveys.

There are several interesting facets of the computerized algorithms. With some water use categories, benefits are calculated in a theoretical decision framework. For example, municipal water can be supplied from several sources, such as surface water or ground water, or it may be reclaimed. Among these options, the program selects the least damaging source. Pollutants are considered jointly in terms of their damaging effects. In the case of agricultural crop damages, either excessive chlorides or total dissolved solids is the major pollutant, whichever has the greater relative concentration. Their sum, on the other hand, would overestimate damages. The computer program also has the flexibility of evaluating single-period or multi-period benefit streams.

This model cannot be applied to other regions without a few modifications. Damage estimates for commercial fisheries pertain to conditions along the California seacoast. Some pollutants neglected in the model are detrimental in certain areas. Acid mine drainage, for example, is a serious problem in Appalachia and parts of the Ohio River Basin. Finally, the optimal selection of water supply sources may be economically but not politically feasible. Important surface water areas may be privately owned and hence off limits to municipalities.

Battelle Memorial Institute (1972) designed a complex environmental assessment model of impacts likely to result from water quality management. The model is used to evaluate water resources planning in the Bear River basin (of Idaho, Utah, and Wyoming). Based on a multidisciplinary state-of-the-art survey, this methodology weights environmental damages with and without water quality controls. The difference between these scores is a measure of realizable benefits. Major impact component of the assessment include ecology, environmental pollution, aesthetics, and human interest (including health and cultural patterns). In addition to scaling impacts according to relative importance, the analysis flags those disturbances that irreversibly affect the environment or cause a major distribution of resources.

Water quality damages are included within a larger environmental pollution category. For each major pollutant, the ambient concentration is transformed nonlinearly into an environmental quality index, which represents a surrogate for the accumulated effects of the pollutant. However, the transformation is too general to depict impacts on specific water uses such as recreation. Application of the method to a certain region, therefore, does not take into account effects on specific activities or land values. In addition, the impact measure cannot be easily translated into monetary terms. Economic impacts, as dictated by the original objectives of the assessment, are integrated with ecological, social, and physical and chemical factors. However, they play a secondary role and are replaced by a numerical rating scale.

Another model with water quality impacts assessed in a comprehensive setting is formulated by Nedenes (1972). It is designed to compare water resources planning strategies with other media-related aspects, e.g., urban development, nature conservation, agriculture, and forestry. Competing uses of water in each subregion are ranked according to "goal preference weights," determined by political experts. The resulting weight is distributed among subregions where uses predominate. Finally, the weight in each zone is adjusted by a "goal achievement coefficient," ranging from 0 percent for avoidance of a water use to 100 percent for complete fulfillment. Total benefits are then calculated by summing adjusted weights over all uses.

Not all important environmental impacts are evaluated in this model. Aesthetics and conservation of nature are excluded because water resource managers (in Norway) felt unable to weight them. To choose among optimal strategies, benefit weights must be translated into monetary constraints. Each "unit of benefit" (e.g., user day of recreation) must have a certain dollar value to exceed corresponding costs of control.

Unlike most benefit studies, this model emphasizes the trade-offs between water-related activities and other environmental values. For instance, if recreation is assumed to be the only source of benefits, the need for high quality waters might restrict urban development along water bodies. However, with the inclusion of the latter as a factor of regional concern, pollution control strategies may indeed change. There are several difficulties in deriving empirical estimates. Defining the boundaries of each subregion is challenging because impacts extend beyond the local area. Some water management goals, such as improvement of visual quality, elude simple quantification and thus vary in relative importance among planners. Even tangible benefits, such as sport fishing, are not adequately priced in the marketplace and therefore remain subject to personal evaluation.

DISCUSSION

Discussant: *Robert H. Haveman*

In presenting this paper, the author said he had observed about sixty studies, many of them done by contractors for EPA and other government agencies and some of them going back one, two, or even three decades. To me, the situation that he is reporting on is extremely discouraging, because, in my view, what has passed for benefit estimates in these studies forms a catalog of what not to do in cost-benefit analysis.

Someone once decried the fact that there were thousands of studies done for government agencies that were buried in the executive branch files, and that they should all be revealed. I once believed that too, but now I think they probably should remain buried. The private contract research industry has been referred to as the "flim-flam industry"—I now understand that term.

Changes in expenditures generated by improvements in water quality are taken as benefit estimates and treated as appropriate measures of willingness to pay. Changes in the revenues obtained from fishing licenses are taken as benefits from improvements in water quality in the same light as willingness-to-pay-estimates. Changes in *gross* outputs—such as a reduction in total gross commercial fish sales—are taken as the cost of deterioration in water quality, when it is the *net* output that is relevant. Benefits on a before-after basis are accepted as unquestionable benefits in a with-without framework. Estimates of only a *few* components of benefits are passed of as *full* benefit estimates. Benefit estimates of overlapping components are aggregated. User benefits and property values are aggregated, although they are measuring the same thing in many cases.

Only regional benefits are considered when nonregional beneficiaries are also affected by the project. There is acceptance of secondary benefits without questioning whether or not there is a basis for their even being secondary benefits in particular cases. Arbitrary estimates of the dollar value of user days, such as those used in the Bureau of Reclamation, are presented as unequivocal benefit estimates.

All of these are examples of what *not* to do in cost-benefit analysis.

REFERENCES

Appalachian Regional Commission. *Acid Mine Drainage in Appalachia,* vol. 1. Washington, D.C., 1969.

Bale, H. E., Jr. *Report on the Economic Costs of Fishery Contaminants.* Washington, D.C.: National Marine Fisheries Service, Department of Commerce, 1971.

Barker, B., and Cramer, P. "Water Quality Conditions in Illinois." Appendix A, *State-Wide Water Resource Development Plan 1972.* Springfield, Ill.: Illinois Department of Transportation, Division of Water Resource Management, 1973.

Battelle Memorial Institute. *Environmental Evaluation System for Water Resource Planning.* For Bureau of Reclamation, U. S. Department of the Interior. Columbus, Ohio: 1972.

168   *I. Benefit Measures*

Benjamin, S. *Anti-Pollution Measures May Change the Economy.* United Press release. New York, N.Y.: August 26, 1972.

Beyer, J. *Water Quality and Value of Homesites on the Rockaway River.* New Brunswick, N.J.: Water Resources Research Institute, Rutgers, The State University, 1969.

Black and Veatch, Consulting Engineers. *Economic Effects of Mineral Content in Municipal Water Supplies.* Kansas City, Mo.: May, 1967.

Bradfield, M. *Benefit-Cost Study of the Annapolis-Cornwallis River Systems.* Halifax, Nova Scotia: Dalhousie University, 1970.

Bramer, H. C. *The Economic Aspects of the Water Pollution Abatement Program in the Ohio River Valley.* Ph.D. thesis. Pittsburgh, Pa.: University of Pittsburgh, 1960.

Bramer, H. C. *Economically Significant Physicochemical Parameters of Water Quality for Various Uses.* Pittsburgh, Pa.: Mellon Institute, 1971.

Brandt, G. H., et al. *An Economic Analysis of Erosion and Sediment Control Methods for Watersheds Undergoing Urbanization.* Midland, Mich.: Dow Chemical Company, 1972.

Bureau of Outdoor Recreation. "Recreation and Aesthetics." Appendix F, *Development of Water Resources in Appalachia.* Washington, D.C.: U.S. Department of the Interior, 1968.

Bureau of Outdoor Recreation. *Water-Oriented Outdoor Recreation in the Lake Ontario Basin.* Ann Arbor, Mich.: Federal Water Pollution Control Administration, 1967.

Bureau of Reclamation. *Value of Desalted Water for Irrigation.* Denver, Colo.: U.S. Department of the Interior, 1969.

Burt, O. R. "Comments on 'Recreation Benefits from Water Pollution Control' by Joe B. Stevens." *Water Resources Research* 5, no. 4 (1969), pp. 905-7.

Callinan, B. J., and Webster, R. G. "Economic and Social Aspects of Saline Water Use and Management," In *Salinity and Water Use,* edited by T. Talsma and J. R. Philip. New York: Wiley-Interscience, 1971.

Cannavo, P., et al. *Pollution and Ecologic Patrimony.* ISVET Document No. 33. Rome, Italy: 1970.

Clawson, M., and Knetsch, J. L. *Economics of Outdoor Recreation.* Baltimore, Md.: Johns Hopkins University Press, 1966.

Committee on Public Works. *Laws of the United States Relating to Water Pollution Control and Environmental Quality.* Washington, D.C.: U.S. House of Representatives, 1970.

Coomber, N. H., and Biswas, A. K. *Evaluation of Environmental Intangibles.* Bronxville, N.Y.: Genera Press, 1972.

Council on Environmental Quality. *Environmental Quality.* Second Annual Report. Washington, D.C.: 1971.

Council on Environmental Quality. *Ocean Dumping: A National Policy.* Washington, D.C.: Government Printing Office, 1970.

Criss, R. R. "Socio-Economic Accounting Applied to Water Resource Planning." *Water Resources Bulletin,* August 1971.

David, E. L., and Lord, W. B. *Determinants of Property Value on Artificial Lakes.* Madison, Wisc.: University of Wisconsin, Department of Agricultural Economics, 1969.

David M. Dornbusch and Co. *Benefit of Water Pollution Control on Property Values.* San Francisco, Calif.: 1973.

Davidson, P.; Adams, F. G.; and Seneca, J. "The Social Value of Water Recreational Facilities Resulting from an Improvement in Water Quality: The Delaware Estuary." In *Water Research,* edited by A. V. Kneese and S. C. Smith. Baltimore, Md.: Johns Hopkins University Press, 1966.

Dupuit, J. "On the Measurement of Utility of Public Works." *International Economic Papers* 2 (1844).

Dutta, M., and Asch, P. "The Measurement of Water Quality Benefits." Report prepared for Delaware River Basin Commission, Bureau of Economic Research, Rutgers, The State University. New Brunswick, N.J.: May, 1966.

Eliassen, R., and Rowland, W. F. "Industrial Benefits Derived from Improved Raw Water Quality in the Contra Costa Canal." Stanford University, Institute of Engineering-Economic Systems, September 1962.

Faro, R., and Nemerow, N. L. *Measurement of the Total Dollar Benefit of Water Pollution Control.* Syracuse University, Dept. of Civil Engineering. Research Report no. 10, January, 1969.

Federal Inter-Agency River Basin Committee, Subcommittee on Benefits and Costs. *Proposed Practices for Economic Analysis of River Basin Projects.* Washington, D.C.: Government Printing Office, 1950.

Federal Water Pollution Control Administration. *Delaware Estuary Comprehensive Study: Preliminary Report and Findings.* Philadelphia, Pa.: U.S. Department of the Interior, July 1966.

Federal Water Pollution Control Administration. *Effects of the San Joaquin Master Drain on Water Quality of the San Francisco Bay and Delta.* San Francisco, Calif.: U.S. Department of the Interior, January 1967.

Federal Water Pollution Control Administration. *The National Estuarine Pollution Study,* vol. 2. U.S. Department of the Interior, November 1969.

Federal Water Pollution Control Administration. *Water Quality Criteria.* Washington, D.C.: National Technical Advisory Committee, April 1968.

Federal Water Quality Administration. *New Haven Harbor: Shellfish Resource and Water Quality.* Needham Heights, Mass.: U.S. Department of the Interior, August 1970.

Fisher, A. C. "The Evaluation of Benefits from Pollution Abatement." Washington, D.C.: Environmental Protection Agency, Office of Planning and Evaluation, 1972.

Greeley and Hansen, Engineers. "Study for Determination of Benefits from Improved Great Lakes Water Quality." *Dredging and Water Quality Problems in the Great Lakes.* Buffalo, N.Y.: U.S. Army Corps of Engineers, June 1969.

Hinote, H. *Benefit-Cost Analysis for Water Resource Projects.* Nashville, Tenn.: University of Tennessee, June 1969.

James, L. D., and Laurent, E. A. "Economics." Annual Literature Review, *Journal of Water Pollution Control Federation* 45, no. 6 (June 1973), pp. 1414-21.

Jordening, D. L., and Algood, J. *State of the Arts: Estimating Benefits of Water Quality Enhancement.* Manhattan, Kansas: Development Planning and Research Associates, 1973.

Kimball, T. L. "Air, Water Pollution Now Cost U.S. $28.9 Billion a Year." *National Wildlife Federation,* February-March 1972.

Kneese, A. V., and d'Arge, R. C. *Pervasive External Costs and the Response of Society.* Washington, D.C.: Resources for the Future, July 1969.

Krutilla, J. V. *Evaluation of an Aspect of Environmental Quality: Hells Canyon Revisited.* Washington, D.C.: Resources for the Future, 1971.

Lackner, J., and Sokoloski, A. A. *Safe Drinking Water Act of 1973: Estimates of Benefits and Costs.* Washington, D.C.: Environmental Protection Agency, 1973.

Liu, O.C., *Proposal on Research of Enteric Virus-Related Diseases.* Narrangansett, R.I.: Environmental Protection Agency, 1972.

Loucks, D. P. *A Selected Annotated Bibliography on the Analysis of Water Resource Systems.* Ithaca, N.Y.: Cornell University, December 1972.

Matson, J. V., and Bennett, G. F. "Cost of Industrial and Municipal Waste Treatment in the Maumee River Basin." Presented at the American Society of Mechanical Engineers—American Institute of Chemical Engineers Joint Conference on Stream Pollution and Abatement, New Brunswick, N.J., 1969.

McKee, J. E., and Wolf, H. W., eds. *Water Quality Criteria.* Sacramento, Calif.; The Resources Agency of California, State Water Resources Control Board, 1963.

Mead, W. J., and Sorensen, P. E. "The Economic Cost of the Santa Barbara Oil Spill." Santa Barbara Oil Symposium, University of California, Santa Barbara, Calif., December 16-18, 1970.

Megli, L. D.; Long, W. H.; and Gamble, H. B. *An Analysis of the Relationship Between Stream Water Quality and Regional Income Generated by Water-Oriented Recreationists.* University Park, Pa.; The Pennsylvania State University, Institute for Research on Land and Water Resources, 1971.

Meredith, D. D., and Ewing, B. B. *Systems Approach to the Evaluation of Benefits from Improved Great Lakes Water Quality.* Proceedings of 12th Conference, Great Lakes Research, Buffalo, N.Y., 1969.

Merewitz, L. "Recreation Benefits of Water Resource Development." *Water Resources Research* 2, no. 4 (1966), pp. 625-40.

Merritt, L. B., and Mar, B. "Marginal Values of Dilution Water," *Water Resources Research,* December 1969.

Metcalf and Eddy, Engineers, *The Economic Value of Water Quality.* Palo Alto, Calif.; January 1972.

Minnehan, R. F. "A Test of the Hypothesis That Water Pollution Control Is Worth What It Costs." Proceedings of the Water Resources Seminars, 1967-1968, Newark, Del.; University of Delaware, 1968.

National Academy of Sciences. *Water Quality Criteria.* Washington, D.C.: Committee on Water Quality Criteria, December 1972.

National Commission on Water Quality. *Assessment of Modelling and Indirect Benefit Indicators.* Washington, D.C.: 1973.

Nedenes, O. *Water Management—Cost Benefit Analysis.* Presented at the 6th International Conference on Water Pollution Research. San Francisco, Calif.: 1972.

Nighswonger, J. J. "Methodology for Inventorying and Evaluating the Scenic Quality and Related Recreational Value of Kansas Streams." Topeka, Kansas: Report no. 32, March 1970.

Ohio River Committee. "Report upon Survey of the Ohio River and Its Tributaries for Pollution Control." *House Documents,* vol. 19, pt. 1, no. 166: 78th Congress, 1st Session. Washington, D.C.: 1943.

Orlob, G. T.; Sonnen, M. B.; Davis, L. C.; and Norton, W. R., *Wild Rivers: Methods for Evaluation.* Walnut Creek, Calif.: Water Resources Engineers, October 1970.

Pyatt, E. E.; Grantham, G. R.; and Carter, B. J. *A Model for Quantifying Flow Augmentation Benefits.* Gainesville, Fla.: University of Florida, September 1969.

Reiling, S. D.; Gibbs, K. C.; and Stoevener, H. H. *Economic Benefits from an Improvement in Water Quality.* Washington, D.C.: Environmental Protection Agency, 1973.

Renshaw, E. F. "Value of an Acre-Foot of Water." *Journal of the American Waterworks Association* 50 (1958), p. 304.

Robert R. Nathan Associates. "Impact of Mine Drainage on Recreation and Stream Ecology." Washington, D.C.: Appalachian Regional Commission, June 1969.

Roberts, M. J.; Haneman, M.; and Oster, S. *Study of the Measurement and Distribution of the Costs and Benefits of Water Pollution Control.* Cambridge, Mass.: Harvard University Press, 1974.

Sampedro, P.M. *Costs and Benefits of the Abatement of Pollution of Biscayne Bay, Miami, Florida.* Coral Gables, Fla: University of Miami, Sea Grant Technical Bulletin no. 24, December 1972.

Scaiola, G. "Public Intervention Against Pollution: Estimates of the Economic Costs and Benefits Related to a Project for Eliminating the Principal Forms of Atmospheric and Water Pollution in Italy." *Rapporto di Sintesi* (June 1971), pp. 137-73.

Special Advisory Committee on Water Pollution. "Report on Water Pollution." Washington, D.C.: National Resources Committee, Water Resources Section, July 1935.

Spencer, S. L. "Monetary Values of Fish." Montgomery, Ala.: The Pollution Committee, American Fisheries Society, 1970.

Stanford Research Institute. *Recreation and Fishery Values in the San Francisco Bay and Delta*. SRI Report No. 5838. Palo Alto, Calif.: October 1966.

Stevens, J. B. "Recreation Benefits from Water Pollution Control." *Water Resources Research* 2, no. 2 (1966), pp. 167-82.

Stoevener, H. H., et al. *Multi-Disciplinary Study of Water Quality Relationships: A Case Study of Yaquina Bay, Oregon*. Corvallis, Ore.: Oregon Agricultural Experiment Station, February 1972.

Stoll, J. B. "Man's Role in Affecting Sedimentation of Streams and Reservoirs." Proceedings of the Second Annual Water Resources Conference, Chicago, Ill., 1966.

Stone, R., and Friedland, H. "Estuarine Clean Water Cost-Benefit Studies." Presented at 5th International Water Pollution Research Conference, San Francisco, Calif., 1970.

Sumitomo, H., and Nemerow, N. L. "Pollution Index for Benefit Analysis." Syracuse, N.Y.: Syracuse University, Department of Civil Engineering, 1969.

Tihansky, D. P. "An Economic Assessment of Marine Water Pollution Damages." Third Annual Conference International Association For Pollution Control, *Pollution Control in the Marine Industries,* Montreal, Canada, 7 June 1973.

Tihansky, D. P. "Economic Damages from Residential Use of Mineralized Water Supply. *Water Resources Research* 10 (1974), pp. 145-54.

Tomazinis, A. R., and Gabbour, I. "Water-Oriented Recreation Benefits: A Study of the Recreation Benefits Derivable from Various Levels of Water Quality of the Delaware River." Philadelphia, Pa.: Institute for Environmental Studies, University of Pennsylvania, February 1967.

Trice, A. H., and Wood, S. E. "Measurement of Recreation Benefits." *Land Economics* 34 (1958), pp. 196-207.

Tybout, R. A. "Economic Impact of Changes in the Water Resources of the Great Lakes." Proceedings of Conference on the Economic and Social Impact of Environmental Changes in the Great Lakes Region, State University College, Fredonia, N.Y., November 7-8, 1969.

U.S. Army Corps of Engineers, North Atlantic Division. *Potomac River Basin Report*. Baltimore, Md.: February 1963.

U.S. Environmental Protection Agency. "Physical and Economic Impacts." *The Mineral Quality Problem in the Colorado River Basin*, Appendix B, Regions VIII and IX. Denver, Colo.: 1972.

U.S. House of Representatives, 92nd Congress, 2nd Session. *Federal Water Pollution Control Act Amendments of 1972*. Conference Report no. 92-1465. Washington, D.C.: 1972.

Vincent, J. R., and Russell, J. D. "Alternatives for Salinity Management in the Colorado River Basin." *Water Resources Bulletin* 7, no. 4 (August 1971), pp. 856-66.

Wasserman, L. P. "Economic Loss of Our Estuarine Resource Due to Pollutional Damage." Livingston, N.J.: Infinity, Ltd., 1970.

Water Resources Engineers. *Evaluation of Alternative Water Quality Control Plans for Elkhorn Slough and Moss Landing Harbor*. Walnut Creek, Calif.: September 1969.

Water Resources Engineers. *A Quality-Use Benefit Computer Program for Evaluation of AMBAG Management Plans*. Walnut Creek, Calif.: 1972.

Weddig, L. J. "Effects of Pollution on the Commercial Fish Industry." Washington, D.C.: National Fish Institute, 1972.

Willeke, G. *Effects of Water Pollution in San Francisco Bay.* Ph.D. dissertation, Stanford University, 1969.

Wollman, N., et al. *Value of Water in Alternative Uses.* Albuquerque, N.M.: University of New Mexico Press, 1962, pp. 220-82.

PART TWO

COST METHODOLOGY AND MEASUREMENT

Costs of Water Quality Improvement, Transfer Functions, and Public Policy

ALLEN V. KNEESE

Introduction

A SUBSTANTIAL AMOUNT OF WORK HAS BEEN DONE on estimating the costs of water quality improvement. The work ranges from engineering and econometric studies of how costs vary with degree of waste removal in treatment plants of a given hydraulic capacity, through detailed studies of costs for particular industries, to estimates of the cost of water quality improvement for particular water resource regions and for the nation as a whole. Studies have differed greatly in sophistication and in the care with which they state their assumptions and interpret the results.

Results of different studies can vary for a number of reasons. These include the projected level to which waterborne residual discharge is to be reduced, either for individual outfalls or for larger areas; the way water quality objectives are taken to be achieved, whether by uniform reduction of discharge at all sources or by some other criterion; and according to the range of technological options for water quality improvement that are considered admissible.

Implicit in some of the more popular discussions of pollution control is the impression that there is one single figure which represents the costs of "cleaning up." A more sophisticated view has it that costs will vary depending on the degree of water quality improvement we want, and that they may get very high as we approach zero discharge.

There are potentially an infinite number of cost functions (for individual outfalls, for regions, and for the nation), depending on how we choose to constrain the technological and policy options available for water quality management. How one or another investigator sets the constraints will often depend on what kind of water quality improvement strategy he thinks can be successfully implemented.

175

That it makes a tremendous difference how the problem is constrained in the costs obtained has been amply demonstrated, as I illustrate below. An exception would appear to be "zero discharge," which was stated as a national goal in the 1972 amendments to the Federal Water Pollution Control Act. But even here the results are strongly dependent on what is considered permissible discharge to other environmental media—the atmosphere and the land. If an effort were made to approach zero discharge to all environmental media simultaneously, costs would tend to approach infinity. The following sections document these assertions and also explain the relationship of cost estimation to transfer functions, which translate amounts of discharge into environmental conditions.

Concepts
of Costs

THE CONCEPT OF COST itself merits a brief discussion. In ordinary discourse, cost is considered the monetary outlay required to purchase something. To economists, who usually have a more fundamental notion in mind when they use the term, cost means "opportunities foregone." If a consumer spends some of his budget for liquor, he cannot spend it for milk. Unless his income is so high that he can satiate all his wants, he will always forego utility from the consumption of some things when he consumes something else. If he calculates rationally, and the goods and services he consumes are divisible, he will spend on each item or service until a situation prevails in which he obtains the same utility from an additional dollar spent on every item he consumes. In this way opportunity cost plays a role in optimizing behavior.

The reasoning is very similar in production situations where a producer who wishes to maximize his profits owns a resource which is fixed in supply and can be used in the production of multiple outputs. At the optimum, the profit derived by devoting an additional unit of the resource to an output must be equal for each output to which that resource can be devoted. If, on the other hand, the producer can buy units of the resource on the market, he will buy them until the profit from devoting an additional unit to each output is reduced to zero.

Cost-benefit analysis usually assumes that competition exists in all markets supplying inputs to the public good which is to be provided. Thus the prices of inputs can be taken to represent the social opportunity costs of undertaking the activity. Cost-benefit analysis has usually been applied to the provision of public works—especially major reservoirs for power, flood control, irrigation, and navigation. The context is similar when a public investment in a publicly supplied treatment plant or reservoir for water quality improvement is contemplated. The question here too is whether the cost of inputs from the private sector is outweighed by the benefits of using them for public works. In such instances, if the markets in which the inputs would otherwise be used are not competitive, then an estimate of costs derived from the price of such inputs will not reflect social opportunity costs.[1] According to standard economic analysis, in a monopolistic situation the price of a good is higher than its marginal cost. This means that the marginal users of that good will be willing to pay more to retain marginal resources (those used for the last units actually produced) in that use than is registered by the market price of such resources. In such a case the market price of these resources *understates* their value in alternative uses. This does not mean that it might not be desirable to estimate market costs, say for budgetary reasons, but they will no longer bear a strict relation to social costs and will indeed underestimate them.

In other situations, market prices may *overstate* the value of marginal units of resources in alternative uses (i.e., social opportunity costs). This could be so if these resources would otherwise be unemployed or underemployed—another type of market imperfection. Wages may be rigid downward (say because of union restrictions or minimum wage laws) even in otherwise competitive markets, and in such cases, since capital investment cannot adjust instantaneously to declining markets, capital resources may be diverted to a public project, or for that matter to a (subsidized) private one at a social opportunity cost less than that implied by the market price of capital.

The literature has taken account of these types of market imperfections in cost-benefit analysis, and some limited efforts have been made to quantify them. But for the most part, cost-benefit analysts have chosen to stay with the assumption that markets are reasonably competitive and have thus avoided the considerable conceptual problems and complications in quantification attendant to abandoning this assumption.[2]

Since by far the largest source of wastewater discharges in the United States is private industry, including agriculture, major conceptual aspects of cost-benefit analysis in this area will occur in a context different from that in which cost-benefit analysis has been conventionally applied. What is involved is calculating the social costs of laying restrictions on the use of certain (environmental) resources by private interests. Costs associated with restricting waste discharge from an industrial source could be calcu-

[1] In this discussion, I leave aside evaluation difficulties associated with the noncorrespondence of private and social rates of discount.

[2] For studies that have analyzed and to some extent quantified the types of imperfections just mentioned, see John V. Krutilla and Otto Eckstein, *Multiple Purpose River Development: Studies in Applied Economic Analysis* (Baltimore: Johns Hopkins University Press, 1958); Robert H. Haveman and John V. Krutilla, *Unemployment, Idle Capacity, and the Evaluation of Public Expenditures* (Johns Hopkins University Press, 1968); and Charles W. Howe and K. William Easter, *Interbasin Transfers of Water: Economic Issues and Impacts* (Johns Hopkins University Press, 1971).

Figure 1

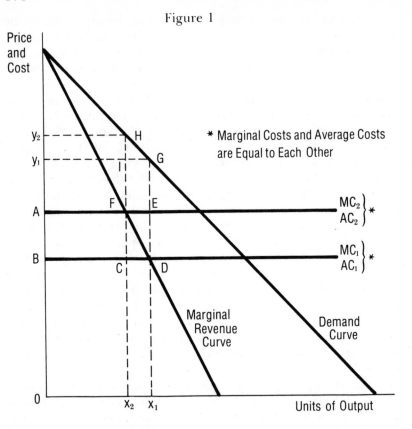

lated in numerous ways. Perhaps the standard way, however, is to calculate the additional cost per unit of output associated with the controls and then multiply this unit cost by the number of units of output to get a total cost. Even if such a unit cost is calculated correctly, in the sense that it reflects the way in which the private industry would actually adjust (more is said of this later), such a calculation may not reflect opportunity costs accurately if the effect on unit costs, and therefore on the number of units it is profitable to produce, is large or if the affected industry is monopolistic. Let us look at the second case first. This case can be illustrated by a standard chart, which can be found in any economics principles book, depicting the price output solution in a monopolistic industry. To avoid additional complications, it is assumed that the good in question is produced at constant unit cost.

In the initial cost situation represented by MC_1 and AC_1, depicted in Figure 1, output is x_1 and price is y_1. Now assume required discharge controls shift unit costs to MC_2 and AC_2. The corresponding price and output are given by y_2 and x_2. The standard procedure would calculate the projected total cost of the controls as the area ABDE. In one way this is

Figure 2

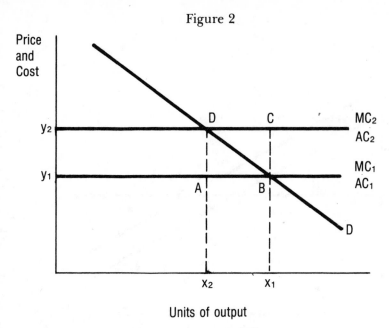

Units of output

too large since output is now only x_2, so the actual added costs incurred in production are only the area ABCF. But in another way the cost so calculated is too small because the consumers of the output are losing a consumer's surplus because of the higher price and the smaller quantity produced. The total loss in surplus is the area y_1y_2HG. The overall cost associated with the controls is therefore the area ABCF plus the area y_1y_2HG. In the situation depicted, this is more than the standard calculation of the area ABDE. But this is not necessarily the case. If demand is very elastic, the standard figure could possibly be larger than the correct figure.

The same problem also exists in a competitive situation if the output adjustment is large, although for a similarly situated demand curve and cost shift, the surplus would not be as large.

Again a standard chart may be used to illustrate the problem. Figure 2 shows cost and demand curves for a competitive industry.

For initial costs MC_1 and AC_1, price and output are y_1 and x_1. With the shift in costs to MC_2 and AC_2 price and output are y_2 and x_2. The standard method would measure total costs as the area y_1y_2CB. In a competitive industry this figure will always be too high. On the other hand, if cost is computed in a similar fashion but based on the new (after cost increase) output, the estimate will always be too low since it neglects the surplus lost (the area ABD). The correct cost is the area y_1BDy_2.

As will be discussed below, another possible (and realistic) adaptation to effluent controls is to change (reduce) the quality of the product

Price
and
Cost

Figure 3

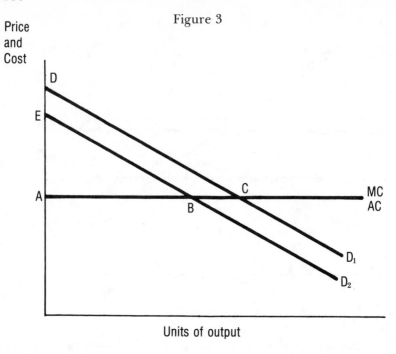

Units of output

produced. This presents a similar problem for calculating surpluses. Again referring to a competitive industry, assume that producers alter quality in such a way that their costs per unit do not rise at all. But since the quality is poorer, consumers will not be willing to pay as much per unit. The latter can be depicted by a demand curve shift. In Figure 3 the horizontal axis could relate to tons of paper, but D_1 pertains to a higher brightness than D_2.

In this particular case the standard method would register no cost at all, which would be incorrect since there is a loss of surplus equal to EBCD. Moreover, if the product were eliminated from the market entirely the loss would be ACD.

But this entire discussion of lost surplus assumes that when resource use is shifted or consumer budgets are altered, there is no gain of surplus elsewhere in the economy. One way of putting it is to think of all resource and demand shifts taking place evenly on a broad front, on the margins where there is no surplus being yielded, but this would not necessarily correspond to reality. For example, if a product is eliminated, the released resources might systematically shift to the generation and production of a substitute which would then itself be associated with a surplus. In order to calculate opportunity costs, that surplus would have to be netted out of the surplus formerly generated by the eliminated product. Thus, even leaving aside the shakiness of consumers surplus as a normative concept (which is discussed in other papers),

when cost shifts or changes in quality characteristics are large, computing costs would in principle require a general equilibrium analysis based
on detailed knowledge of production and preference functions
throughout the economy. One can only conclude that when such shifts
are large, our basis for estimating their social costs is very weak.

In the following discussion this problem is largely ignored and it is
assumed that, when all conflicting biases are considered, the standard
method gives a usable estimate of social costs. The emphasis is, rather,
on the influence of those technological options which are assumed by the
analyst. Even here, though, we will have to appeal to the opportunity
cost concept when interdependencies of the environmental media are
considered.

Estimates Focusing upon What
Can Be Done at Particular Outfalls

• *RESIDUAL LOADS GIVEN: UNIFORM REDUCTION OF DIS-
CHARGE AT ALL OUTFALLS.* By far the simplest basis for calculating
costs is to take an estimate of residual loads (since they have not yet been
measured directly, and usually we are dealing with projections anyway)
and then, neglecting the possible output effects of cost increases, assume
that some standard treatment technique is applied at each outfall. For
example, *The Economics of Clean Water* estimates the costs of secondary
treatment (or its "industrial equivalent," which is but vaguely defined) at
each outfall. During the period 1971-80, total United States expenditures by municipalities and manufacturing plants for water pollution
control programs are estimated at about $70 billion.[3] Rough estimates
have also been made for higher levels of control. For example, the
National Water Commission estimates that it would cost $470 billion to
achieve "best available technology" at all outfalls over the next ten years.[4]

While based on simple assumptions, estimates of the costs of achieving more or less uniform reduction at all outfalls cannot be dismissed as
necessarily irrelevant. The Federal Water Pollution Control Act
Amendments of 1972 in principle require uniform national effluent
standards for municipalities and for industry, by class—although one
can seriously question whether they will be achieved. In principle the
amendments do not require that residual loads be taken as given (as
most aggregate-level cost estimates have done), but the subsidy arrangements for municipalities and industries do tend to encourage
treatment and, correspondingly, to discourage the use of technologies to

---

[3]Environmental Protection Agency, *The Economics of Clean Water* (1972).
[4]"The Report of the National Water Commission" (review draft).

reduce the generation of residuals. Some of the latter are examined below.[5]

● *RESIDUAL LOADS GIVEN: DISCHARGE REDUCTION PRO-GRAMMED FOR EACH OUTFALL.* Despite the importance of non-treatment alternatives, let us stay for a moment with the assumption of given waste loads. One of the most important economic studies of water quality management was performed by the federal government in the Delaware estuary region during the early and middle 1960s.[6] Among other things, the study permitted cost comparisons to be made between strategies which impose uniform residuals reduction requirements at all outfalls and more flexible strategies aimed at achieving the same ambient standards.

The concept of ambient standards has not been introduced. In less careful discussion of water quality two distinct types of standards are often not distinguished with sufficient clarity. *Effluent* standards pertain to requirements (either by weight of materials or concentrations) set on the quality characteristics of the actual wastewater discharges. *Ambient* standards refer to quality requirements in the watercourse (usually by concentrations). As long as the focus in cost estimation is on applying specified technologies at all outfalls, no connection between the residuals discharge reductions achieved and the quality of the water in the stream is provided or needed. But a more logical procedure (on the assumption that the objective is to achieve or maintain quality characteristics in watercourses) is to start with water quality objectives and then reason back to a desirable strategy for obtaining them.

To do this, a quantitative connection must be established between what occurs at the points where the residuals are discharged and the quality of water at various other points along the watercourse. The mathematical functions making this connection are known as *transfer functions.* They form an essential tool for analyzing the cost-effectiveness of water quality management strategies whose objective is to obtain specified sets, or alternative sets, of water quality goals in the watercourse (ambient standards). Appendix I offers some further details about the nature of these transfer functions and about the mathematical models which embody them in such a way that the cost-effectiveness of alternative strategies can be analyzed.

[5]The statements about the 1972 amendments made in this paragraph are discussed in some detail in A. V. Kneese, "Congressional Performance, A Case Study of Federal Water Pollution Control Policy in the Post-World War II Period," a paper prepared for the Public Choice Society Meetings, College Park, Maryland, May 22-24, 1973. One favorable feature of the amendments in this respect is that they do require full cost reimbursement for industrial plants connected to municipal sewers. But rapid write-off provisions in the tax laws still favor end-of-the-pipe treatment.

[6]*Delaware Estuary Comprehensive Study,* (Philadelphia: Federal Water Pollution Control Administration, 1966).

Table 1 shows cost comparisons derived from such a model for the Delaware estuary area for different ambient standards and two different implementation strategies.[7] One strategy specifies uniform reduction at all outfalls; the other uses a mathematical programming procedure to minimize costs of achieving the objective sets. The latter implies substantially different levels of discharge reduction at different outfalls depending upon the cost of achieving reductions at these particular outfalls[8] and upon the effect (calculated via transfer functions) of reductions at a particular outfall on the ambient quality. For most objective sets, the cost differences are large. Several other studies have obtained analogous results, some of which are even more striking.

Table 1

Summary of Total Costs of Achieving Objective Sets 1, 2, 3, and 4. (Costs—in million 1968 dollars, present value at 3 percent discount —reflect waste-load conditions projected for 1975-80.)

Objective Set	Uniform Treatment Total Costs	Cost Minimization Total Costs
1	460	460
2	315	215
3	155	85
4	130	65

It is not the purpose of this paper to comment extensively on policy, but it is worth noting that additional studies with the Delaware model indicate that a simple system of uniform charges to each discharger could approach the least-cost solution solely because of its economic incentive effect. Results of the analysis are shown in Table 2. Note that in this table figures are annual rather than present values and, therefore, not directly comparable with the previous table. The reason the effluent charge induces a solution close to the least-cost one is basically simple. If residuals dischargers desire to minimize cost, they will reduce discharge until the cost of a unit of further reduction is just equal to the effluent charge they would have to pay on that unit if they did discharge it. If they push discharge reduction further, the incremental costs of reducing discharge for the next units will exceed the charge they would have to pay if they discharged them and their overall cost (effluent charge plus reduction cost) would rise.

[7]These are taken from the *Delaware Comprehensive Study*, ibid.

[8]This study took residuals loads as given and just considered treatment costs. Therefore, cost differences are limited to those resulting from scale economies in treatment, site characteristics, and the varying difficulty of treating difficult types of residuals.

Table 2
Cost of Treatment under Alternative Programs

| DO objective (ppm) | Least Cost | Program | |
| | | Uniform Treatment | Single Effluent Charge |
	(million dollars per year)		
2	1.6	5.0	2.4
3-4	7.0	20.0	12.0

As a result of individual cost-minimizing behavior, the highest level of discharge reduction will occur at those outfalls where the costs of making reductions are lowest. That these individual responses to the charge do not produce quite as low a cost as the actual programmed solution has to do with the nature of the particular transfer functions used. It is because some types of residuals degrade naturally in watercourses, and the contribution which a reduction in discharge makes to meeting the ambient standards consequently depends on the distance between the point of discharge and the point for which the ambient standard is specified. It is only under very specific conditions that the location of the discharger does not matter in achieving the cost-minimizing solution. A more technical explanation of these conditions is provided in Appendix II.

One deficiency of the Delaware study is that it focuses narrowly on the treatment of waterborne residuals *after* generation and does not take into account economical procedures for reducing generation itself. There is no particular reason to think, however, that the conclusions would have been changed if ways of reducing generation had been included, although they might well have been strengthened. Since possibilities for, say, changing the process or recovering by-products depend a great deal on the basic design of the plant, and since the introduction of these procedures is usually more practicable for newer plants, cost differences between individual plants might be even larger than when just treatment is taken into account.

● *RESIDUALS LOADS AFFECTED BY INTERNAL PROCESS CHANGE.* Careful, quantitative studies have amply illustrated that industrial dischargers have many opportunities to reduce the generation of residuals as well as to treat them after generation. The former range from adapting the qualitative characteristics of inputs and outputs to by-product recovery and the recycling of materials. One of the earliest studies to give close attention to these possibilities was focused on beet sugar manufacturing.[9] It was found that reduction of discharges which increase biochemical oxygen demand (BOD)—one of the primary problems of residuals from this type of enterprise—could be very greatly increased by the application of recycling and by-product recovery, and that only at the very highest level of control would "end-of-the-pipe" treatment become economical.

Another example of how the processes used influence residuals generation is found in the steel industry. When one compares the basic oxygen furnace with the open hearth, the former is found to generate 45 percent more waterborne residuals. These include degradable organics, ammonia, solids, and heat. The electric furnace, on the other hand, generates virtually no waterborne residuals.[10] The importance of process change has been observed in studies of pulp and paper manufacturing and food processing and is, no doubt, characteristic of other industries as well.

The beet sugar study illustrates another important aspect of the costs of residuals discharge reduction, i.e., how costs tend to rise rapidly (exponentially) as zero discharge is approached. This is shown in Figure 4. The curve—a marginal cost curve—shows the increase in total costs with each successive unit of reduction. For each level of reduction the least cost combination of measures for achieving that level is used in deriving the curve.

An analogous result is shown for a petroleum refinery in Figure 5. Again, it is evident that marginal costs rise rapidly as high levels of removal are approached. In this case, however, the whole range of marginal costs lies at a higher level. Thus, for a marginal cost of $0.07 the beet sugar plant can achieve more than 96 percent BOD removal. The same marginal cost permits less than 70 percent removal in the petroleum refinery, and a marginal cost of about $0.23 is required to go over 90 percent removal. These results once more suggest the inefficiencies inherent in uniform discharge reduction strategies.[11]

[9]G.O.G. Löf and A.V. Kneese, *The Economics of Water Utilization in the Beet Sugar Industry* (Johns Hopkins University Press, for Resources for the Future, 1968).

[10]The difference between the basic oxygen furnace (BOF) and open hearth is largely traceable to the greater use of hot metal in the BOF charge and hence the larger use of the coke plant, as coking is the largest source of BOD and the only source of phenols and ammonia. The electric furnace involves hardly any coking. See William J. Vaughan and Clifford S. Russell, "A Residuals Management Model of Integrated Iron and Steel Production," a paper prepared for presentation at the Steel Industry Economics Seminar sponsored by the American Iron and Steel Institute and Northern Illinois University, DeKalb, Illinois, April 1973. The different steel-making technologies also have differential effects on other types of residuals generated. The significance of this is discussed below.

[11]The information on petroleum refining is from Clifford S. Russell, *Residuals Management in Industry: A Case Study of Petroleum Refining* (Baltimore: Johns Hopkins University Press, 1973). The fact that the curve for beet sugar manufacturing is continuous and that the curve for petroleum refining is piece-wise linear is an artifact of the different methodologies used to derive them.

Figure 4

Marginal Cost of BOD Discharge Reduction
by a Beet Sugar Refinery*

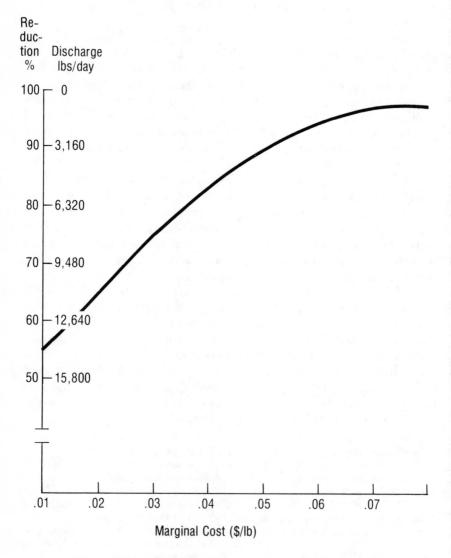

* 2700 tons of beets processed per day.

Source: Prepared by Clifford Russell on the basis of data presented in G.O.G. Löf
and A.V. Kneese, *The Economics of Water Utilization in the Beet Sugar
Industry* (Baltimore: Johns Hopkins University Press, 1968).

Figure 5

Marginal Cost of BOD Reduction in Petroleum Refining

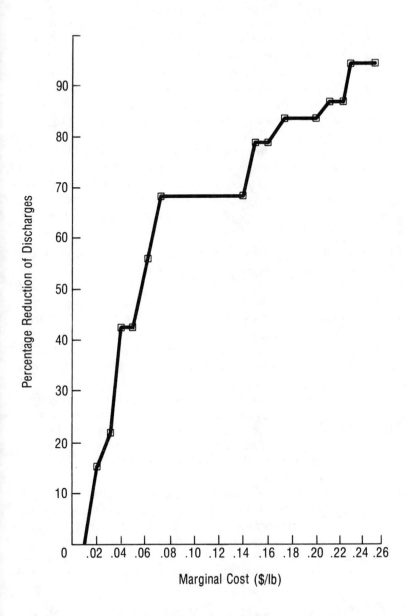

Source: Based on Figure 16 in Clifford S. Russell, *Residuals Management in Industry: A Case Study of Petroleum Refining* (Johns Hopkins University Press, 1973), p. 139.

A type of "process" change that is much less frequently analyzed than the application of materials-saving technologies (recycling or by-product recovery) is altering the qualitative characteristics of outputs. The production of certain "high quality" variants of products can have a substantial effect on the amount of residuals generated. This is strikingly illustrated by some results from a study of the pulp and paper industry. It was found that when all other specifications remained the same and the product brightness was reduced by decreasing the extent of bleaching from 80 to 25 (the brightness of unbleached kraft) dissolved solids and organic residuals were cut by over 80 percent.[12] There are many other examples where residuals generation is intimately related to product quality specifications.[13]

These results illustrate the complexity of making reasonably precise estimates of residuals reduction costs for complex industries. Types of raw materials used, internal process designs, treatment of residuals after generation, and the qualitative characteristics of the final products are all involved. Reasonable levels of precision can be achieved only by detailed studies of individual industries.

● *SPILLOVERS TO OTHER MEDIA.* Since mass is conserved, the weight of a residuals stream, once generated, cannot be reduced by treatment. In fact, mass is increased because the treatment process itself requires inputs. Treatment or, more generally, modification processes can only change the form or location of discharge. Furthermore, process changes which reduce discharges to one medium, even though they can reduce the total mass of residuals generated, often increase the flow of residuals to another medium. For example, the beet sugar plant, discussed above, with tight controls on organic matter discharge to the watercourse, discharges somewhat more oxides of sulfur to the atmosphere than a plant that does not have such tight wastewater controls. Accordingly, liquid, solid, and gaseous residuals streams are often interdependent, and the cost of modifying one will depend to a greater or lesser degree upon permissible discharges of (or effluent charges on) the other.

A striking illustration is provided once again by the steel industry. Whether or not quenching coke with the coking unit's wastewater is permitted has a major impact on the cost of reducing the discharge of waterborne residuals. If quenching with this wastewater is permitted, the removal of degradable organics remains quite low in cost, up to a level of about 85 percent discharge reduction. But if this practice is

---

[12]Blair T. Bower, George O. G. Löf, and W. M. Hearon, "Residuals Management in the Pulp and Paper Industry," *Natural Resources Journal* (1972).

[13]For a careful study on this phenomenon in the petroleum refining industry, see Russell, op. cit.

prohibited in the interest of controlling air pollution, the costs of achieving that level of removal jump by a factor of twenty-five.[14]

Interdependence of this kind becomes particularly significant when a serious effort is made to approach very low levels of residuals discharge. This fact is often neglected in cost estimation—a neglect that is fostered by the existing isolation of policy making and implementation for one medium from that for the others. The Environmental Protection Agency was created, at least in part, in recognition of these interdependencies. But the recognition remains de jure rather than de facto.

Nevertheless, in our more highly developed regions, the appropriate approach would seem to be simultaneous consideration of air, water, and solid residuals streams and the effects of various magnitudes and distributions of these streams upon the environment. For the latter type of analysis, some form of transfer functions are, once again, absolutely essential. By considering sets of ambient standards and their associated costs (including cost distribution) a representative government should be able to come up with a set of ambient standards which are coherent and reflect the preferences of the population reasonably well. The pertinent costs in such public choice-making processes certainly include the direct costs of achieving a prescribed ambient standard in one medium. But they also include the indirect costs that achieving this standard imposes on meeting ambient standards in other media, or the associated environmental degradations in those media, as well as the distribution of both types of costs.[15]

Mathematical model building which incorporates transfer functions and is aimed toward permitting an informed balancing of trade-offs is rather well advanced (see Appendix III). But policy making and the development of institutions to effectively implement such an approach lag far behind.

Non-Point-of-Discharge Technical Alternatives

ALTHOUGH THE COST ESTIMATES which normally appear in the literature pertain to discharge reductions at particular wastewater outfalls, research and practice have conclusively shown that a variety of other non-discharge-point technologies can enter efficiently into regional sys-

[14]Vaughan and Russell, op. cit.

[15]For further discussion see Allen V. Kneese and Edwin T. Haefele, "Environmental Quality and the Optimal Jurisdiction," presented at the Joint Institute on Comparative Urban and Grants Economics Conference, University of Windsor, Canada, November 2-4, 1972.

tems of water quality management.[16] Where non-point sources of waterborne residuals (agricultural runoff, for example) are important, such technologies may be the only ones that can be effectively used. They include the regulation of river flow from either surface or groundwater reservoirs, in-stream re-aeration (the mechanical introduction of air into the water to raise its dissolved oxygen content), disposal of wastewaters to the land for irrigation purposes, the temporary impoundment of wastewaters during conditions of low water quality in the stream, the diversion of wastewater to make improved use of naturally occurring assimilative capacity, and perhaps still other measures. Studies have shown that system optimization which includes these technologies as possible options can produce cost savings for given ambient quality objectives—often spectacular ones. Widespread application of this knowledge has been hindered by the absence of appropriate regional institutional arrangements, except in certain countries. The development of river basin institutions has, for example, been firmly supported by French, English, and (proposed) German laws. How effectively the institutions which have been, and are being, created will function remains to be seen. Considerable lip service has been paid to regional planning in United States laws but little has been done to implement it and virtually nothing has been done to create permanent regional agencies for continuous water quality management. If anything, our national policy has in practice hindered the construction of regional institutions. If one could realistically foresee the development of institutions capable of efficiently implementing the full range of available technologies (including discharge point and non-discharge point), anticipated costs of water quality improvement could be greatly affected.

Conclusions

THE INTENT OF THIS PAPER has been to show that the possible costs of controlling water quality vary over a wide range. Costs depend upon the levels of water quality contemplated, the strategies envisaged for implementing management programs, the range of technologies admitted, and the restrictions set upon impacts on other environmental media. So far, the most detailed empirical work has been done for particular

---

[16]See, for example, Robert K. Davis, *The Range of Choice in Water Management: A Study of Dissolved Oxygen in the Potomac Estuary* (Baltimore: Johns Hopkins University Press, 1968), and the discussion of the Ruhr situation in Allen V. Kneese and Blair T. Bower, *Managing Water Quality: Economics Technology, Institutions* (Baltimore: Johns Hopkins University Press, 1968).

treatment processes, industries, and for specific regions, but some estimates of national costs have also appeared.[17]

Detailed regional case studies have been particularly illuminating with respect to the cost-effectiveness of various implementation strategies. These studies have shown the inefficiency of uniform discharge reduction approaches, the potential of effluent charges for approximating least-cost distributions of discharge reduction at individual outfalls, the applicability and efficiency of a wide range of technologies in a regional river-basin context, and the importance of considering direct interdependencies between the various environmental media. These analyses were made possible only by combining cost estimates, economic models, and transfer functions into coherent analytical devices.

It is fair to say that policy formation at the national level in the United States, as contrasted with the major West European countries, has given little attention to cost-effectiveness—even to the extent of ignoring existing studies. This is not the place to speculate why. But it might be observed that there is a growing sense in Congress that the nation's capabilities are not limitless and that priority trade-offs can no longer by avoided. A corollary is that how far we can go toward meeting our national objectives depends upon how efficiently we go about it.

Appendix I: Transfer Functions and Economic Models for Cost Analysis

THIS APPENDIX explains the concept of transfer functions a little more rigorously than was done in the text. The starting point is a model developed many years ago to predict some of the effects of an organic residuals discharge to a flowing stream.

---

[17]National cost estimates can have a certain usefulness if the assumptions upon which they are based are clearly set out and the results are taken with a few grains of salt. Such estimates can be, and have been, used in econometric models of the U.S. economy to estimate the effect of the anticipated expenditures for pollution control on aggregative variables of interest such as employment, price levels, and the growth of GNP. (See *Economics of Clean Water,* op. cit.) Also they have been used in models of the international economy to estimate impacts upon national income and balance of payments under various unilateral and multi-lateral policy strategies. (See Ralph d'Arge and Allen V. Kneese, "Environmental Quality and International Trade," *International Organization,* University of Wisconsin Press, vol. 26, no. 2, Spring 1972, p. 419-69.) If careful attention is given to the multiple sources of uncertainty in the estimates they can often yield qualitative interpretations, e.g., international trade effects of unilateral action in the United States are likely to be insignificant for a strategy that yields an average of secondary treatment at all outfalls.

The Streeter-Phelps Model[18]

Roughly speaking, quality degrading substances discharged to watercourses may be divided into two classes—degradable and nondegradable. Sodium chloride, or ordinary table salt, is a good illustration of the latter. Its behavior in water is extremely simple since all that happens to it is dilution. Accordingly, calculating the concentrations that result when a given amount of it is discharged to a watercourse is a relatively simple matter. Degradable substances are quite different since they are transformed by biological or chemical action, and the transformation process itself often causes quality-reducing changes in the water body. The most common example is degradable organic matter which is contained in household sewage and in many industrial effluents. Oxygen-demanding material is measured in terms of pounds of biochemical oxygen demand, or BOD. Bacteria feed on the organic matter and convert it to plant nutrients—nitrates and phosphates—and carbon dioxide. In the process these aerobic bacteria use the dissolved oxygen (DO) in the water for respiration. The result is an oxygen "sag."

One way of visualizing the oxygen sag is to think of a certain amount of oxygen-demanding waste being placed in a container of well-aerated water with an appropriate bacteria culture. As the biological reactions proceed, the BOD is reduced at a rate proportional to the remaining BOD and in doing so DO is depleted. At the same time, some oxygen is restored to the water through the air-water interface. As time elapses, the oxygen-demanding waste is gradually fully oxidized and the re-aeration continues until the level of dissolved oxygen again reaches the saturation point.

One can also think of a flowing stream with uniform characteristics over a certain reach in which there is a constant input of BOD at a certain point. Distances downstream can be regarded as equivalent to passage of time in the previous illustration. Proceeding downstream, there is first a decrease in the dissolved oxygen content as oxygen is consumed in the biochemical reactions and then, still further downstream, a rise in the dissolved oxygen content to saturation level. While the following formulas are written in terms of time, they could just as well be written in terms of distance, provided appropriate equivalencies are met. These formulas are commonly known as the Streeter-Phelps equations.

Biochemical oxidation is indicated as a first-order differential equation of the form

$$\frac{dL}{dt} = -k_1 L_t,$$

[18]H. W. Streeter and F. B. Phelps, "A Study of the Pollution and Natural Purification of the Ohio River," Public Health Bulletin no. 146 (Washington, D.C.: Public Health Service, 1925).

where L_t is the unsatisfied BOD in parts per million (ppm), t is time (days), and k_1 is a rate constant which is a function of the particular characteristics of the wastewater and the water temperature. If L_a is the initial concentration of BOD in the stream and is interpreted as the constant of integration, the result of integrating the above equation is

$$L_t = L_a e^{-1t}.$$

Re-aeration is also regarded as a first-order process. It is a function of the difference between actual DO concentration and saturation concentration, as follows:

$$\frac{dC}{dt} = k_2 (C_s - C_t),$$

where C_t is the concentration (ppm) of DO at time t; C_s is the saturation concentration; and k_2 is a rate constant, primarily a function of temperature.

The two reactions are combined and the resulting equation, written in terms of the DO deficit ($D_t = C_s - C_t$), is

$$\frac{dD}{dt} = k_2 L_t - k_2 D_t. [19]$$

After substituting $L_a e^{-k_1 t}$ for L_t,

$$\frac{dD}{dt} = k_1 L_a e^{-k_1 t} - k_2 D_t.$$

This is a first-order differential equation of the general form

$$\frac{dy}{dx} + P_y = Q,$$

with $Q = k_1 L_a e^{-k_1 t}.$

The solution of this equation is

$$D_t = \frac{k_1 L_a}{k_2 - k_1} \left(e^{-k_1 t} - e^{-k_2 t} \right) + D_a e^{-k_1 t},$$

where D_a is the deficit and L_a is the BOD concentration, both at time t = 0. The time of the maximum deficit, say, (t_c), can be found by taking the derivative of the above equation with respect to time, setting it to zero, and solving for t_c. The result is

[19]See G. M. Fair, J. C. Geyer, and D. A. Okun, *Water and Wastewater Engineering*, vol. 2 (New York: Wiley, 1968), for a good explanation of the derivation.

$$t_c = \frac{1}{k_2 - k_1} \log_e \frac{k_2}{k_1} \left[1 - \frac{\left(k_2 - k_1\right) D_a}{k_1 L_a} \right]$$

At the location corresponding to this time, the maximum DO deficit is

$$D_c = \frac{k_1 L_a}{k_2} e^{-k_2 t_c}.$$

Multiple Sources and Receptors Under Steady-State Conditions

The equations just discussed transfer the number of pounds of BOD discharged per day at a particular location into concentrations of dissolved oxygen deficit (or, given the saturation level, dissolved oxygen) at other "receptor" locations downstream. The equations hold for steady-state conditions (i.e., the rate of stream flow, the BOD decay rate, the rate of re-aeration, and temperature are all constants). Even given the steady-state assumptions, this model is too simple to be useful in any moderately developed river basin. There will be multiple points of discharge as well as multiple receptors. Before the oxygen sag will have fully recovered, another discharge will enter the system, and so on down the line. Thus the stream can be thought of as a number of interconnected segments. When the differential form of the Streeter-Phelps equations are applied to these, a system of linear first-order differential equations results. But for steady-state conditions, the transfer functions, which are found by solving these equations and which relate the change in DO in segment i to an input of BOD in segment j, fortunately simplify to a set of linear relationships. In matrix notation the system can be written as follows:

$$Ax = r$$

where A is a matrix of transfer coefficients, x is a vector of BOD discharges in kilograms per day, and r is a vector of DO concentrations in milligrams per liter at various specified (receptor) locations. This form is extremely convenient because, as we will see below, it lends itself easily to incorporation within linear economic models.

The Economic Model

Assume that a watercourse consists of m homogenous segments (thirty segments were used in the Delaware estuary study discussed in the text) and c_i represents the improvement in water quality required to meet a DO target in segment i. The target vector c of m elements can be met by changes of inputs to the watercourse from various combinations of the m segments. Define another vector $x = (x_1, x_2, \ldots, x_n)$ in which the values of x refer to changes in the mass of discharges at the n discharge locations. This vector generates DO changes through the mechanism of the constant coefficients of the linear system already described.

If we let A be the (m by n) matrix of transfer coefficients, then Ax yields the vector of DO changes corresponding to x.

Now, given S (the vector of target improvements or standards), we have two restrictions, namely $Ax \geq S$ and $x \geq 0$. These are sets of linear constraints such as those found in a standard linear program. All that is needed to complete the program is an objective function. Let d' be a row vector where d_j = unit cost of x_j, j=1, . . . , n. Notice that this assumes linear cost functions.[20] The problem can now be written as a standard linear program,[21]

$$\min d'x$$
$$\text{subject to } Ax \geq S$$
$$x \geq 0$$

Of course, the transfer coefficients (a_{ij}), as already explained, relate to a steady-state condition of specified conditions of stream flow and temperature. Thus, the model turns out to be totally deterministic, and the variability of conditions is handled in this analysis by assuming extreme environmental conditions usually associated with poor water quality.

A linear programming model of the general type just described was constructed for the Delaware estuary. In addition to DO, it included other (nondegradable) types of material. Computation of the a_{ij}'s is much easier for the latter. Once done, the model provides an extremely flexible tool for the analysis of alternative policies. Some of the results are summarized and discussed in the text.

Appendix II: Cost Minimization and the Location of Discharges

THE TEXT POINTED OUT that in the Delaware case a single flat effluent charge was estimated to induce (given cost-minimizing behavior on the part of the dischargers) a pattern of residuals discharge reductions cost-

---

[20]Programs with linear constraints and nonlinear objective functions can usually be solved if the nonlinear function is not too complicated; so this condition would not necessarily have to hold.

[21]The actual programs needed to solve the problem encountered in the Delaware estuary were somewhat more complicated.

ing much less than a uniform cutback strategy. But it was further indicated that the overall system cost associated with the uniform effluent charge was not as low as the programmed solution. In this appendix the logical basis for this difference is examined. If the location of discharges did not matter vis-à-vis the achievement of the specified ambient standard, then a necessary condition for cost minimization (given rising marginal costs in the relevant range) would be that marginal costs be equalized at all outfalls. If this were not so, units of discharge reduction could be shifted from higher marginal cost locations to lower cost locations and overall system costs could be diminished in the process.

But marginal cost equalization is a necessary condition only when all the coefficients in the transfer matrix are identical. This is a more formal way of saying location does not matter. When the coefficients are not all equal, strict cost minimization requires that prices be tailored for each outfall. This explains why the solution based on a single charge only approaches rather than equals the programmed cost minimization solution. How close it gets in a particular situation is an empirical question relating to the magnitude of the a_{ij}'s. To see this, assume two industrial dischargers with the following cost functions for reducing discharge:

(1) $c_1 = f(x_1)$

(2) $c_2 = f(x_2)$

where x_1 = discharge from plant 1,

   x_2 = discharge from plant 2,

   c_1 and c_2 = costs of reducing discharges at the two plants, respectively.

Assuming segment 6 is the critical reach and recalling the meaning of the elements in the transfer matrix,

   $R_6 = a_{61}x_1 + a_{62}x_2$ (i.e. binding constraint),

where

(3) R_6 = the standard.

Form the Langrangian:

(4) $L = c_1 + c_2 + \lambda (R_6 - a_{61}x_1 - a_{62}x_2)$.

At the optimum, the necessary conditions are,

(5) $\dfrac{\partial L}{\partial x_1} = \dfrac{dc_1}{dx_1} + \lambda (-a_{61}) = 0$

(6) $\dfrac{\partial L}{\partial x_2} = \dfrac{dc_2}{dx_2} + \lambda (-a_{62}) = 0$

(7) $\dfrac{\partial L}{\partial \lambda} = R_6 - a_{61}x_1 - a_{62}x_2 = 0$

Equations (5), (6) and (7) represent 3 equations and 3 unknowns (x_1, x_2, and λ) which must be solved simultaneously. Note that equation (7) is the equality (binding) constraint $R_6 = a_{61}x_1 + a_{62}x_2$. Solving for λ,

(8) $\lambda = \dfrac{1}{a_{61}} \dfrac{dc_1}{dx_1}$ and

(9) $\lambda = \dfrac{1}{a_{62}} \dfrac{dc_2}{dx_2}$.

Lambda is the shadow price of the oxygen constraint in reach 6 (dollars per milligram per liter).

Notice that:

$$\dfrac{1}{a_{61}} \dfrac{dc_1}{dx_1} = \dfrac{1}{a_{62}} \dfrac{dc_2}{dx_2}$$

or $\dfrac{dc_1}{dx_1} = \dfrac{a_{61}}{a_{62}} \dfrac{dc_2}{dx_2}$.

Note also that $\lambda \neq 0$ unless either $\dfrac{dc_1}{dx_1}$ or $\dfrac{dc_2}{dx_2} = 0$. Because both equations (5) and (6) are equal to zero, they may be set equal to each other:

$$\dfrac{dc_1}{dx_1} = \dfrac{dc_2}{dx_2} + \lambda(a_{61} - a_{62}).^{22}$$

Hence we note that marginal costs of treatment at the optimum are not equal unless the transfer coefficients, a_{61} and a_{62}, are equal.

Appendix III: A Multiple Residuals Regional Management Model

THE TEXT INDICATED THAT where significant interlinkages exist between residuals streams, the cost of reducing one can be accurately calculated only by setting permissible discharge or ambient levels for the others, and/or taking explicit account of the value of associated environmental

[22]I am indebted to my associate Walter Spofford for developing this demonstration.

deterioration in another medium. This requires a quantitative model which not only incorporates economic evaluation of costs (in principle including the cost of environmental deterioration) and transfer functions which translate residuals discharges into environmental conditions but which also considers all residuals streams and environmental media simultaneously. One such model was developed in the Resources for the Future Environmental Quality Program, primarily by Clifford Russell and Walter Spofford. This appendix presents a skeletonized discussion of its structure.

The overall model consists of a set of three interlinked submodels depicted schematically in Figure 6 and described briefly below.[23] In developing the model it is assumed that residuals discharges are continuous and unchanging once set at a particular level and that conditions (e.g., wind speed and direction, water flow) in the atmosphere and watercourse are fixed. It will be recalled from Appendix I that the Streeter-Phelps equations have a simple solution under these conditions. The same is true of the equations characterizing the atmospheric diffusion system embodied in the model.

The essential components of the overall model follow:

• *A LINEAR PROGRAMMING MODEL* (LP) that relates inputs and outputs of the various production processes and consumption activities at specified locations within a region—including the amounts and types of residuals generated by the production and consumption of each product, the costs of transforming these residuals from one form to another (i.e., solid to gas, liquid to gas, liquid to solid) and the costs of disposing of residuals in the natural environment (e.g., transportation costs, the cost of landfill operations, etc.).

The industry components of the model permit choices among production processes, raw material input mixes, by-product production, recycling of residuals, and in-plant adjustments and improvement, all of which could reduce the total quantity or the types of residuals to be transformed by treatment processes or disposed of in the natural environment. That is, the residuals generated are not assumed fixed (or given)—either in form or in quantity. Allowance is made in the model for the reduction of the residuals originally generated. Also, provision is made for choices among transformation processes and hence among the possible forms of the residual to be disposed of in the natural environment (i.e., solid, liquid, or gas).

---

[23]A relatively full discussion of this approach is found in Clifford S. Russell and Walter O. Spofford, Jr., "A Quantitative Framework for Residuals Management Decisions," in Allen V. Kneese and Blair T. Bower, eds., *Environmental Quality Analysis: Theory and Method in the Social Sciences* (Baltimore: Johns Hopkins University Press, 1972).

Figure 6

Schematic Diagram of the Regional Residuals Management Model

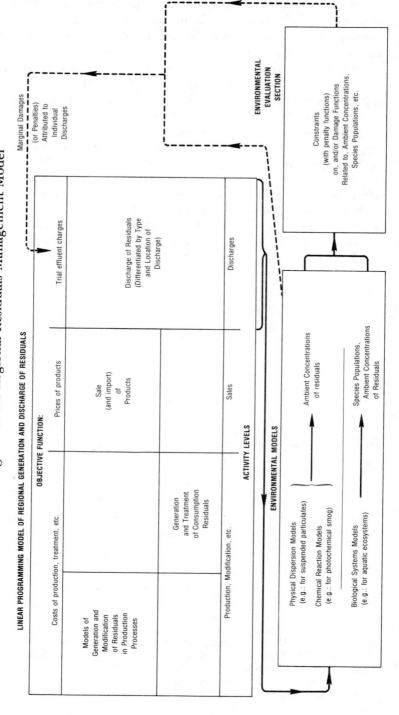

• *ENVIRONMENTAL MODELS* (atmospheric and stream dispersion) use the information which the linear programming model of production and consumption supplies regarding the amounts and types of residuals discharged into the atmosphere and the watercourses and relate it to the resulting ambient concentrations and types of residuals that are present at various receptor locations. For the purposes of this paper, it is useful to linger a bit over how these dispersion models (transfer functions) operate in the context of the overall model.

Environmental quality models (of which the Streeter-Phelps model discussed in Appendix I is one example) are used to compute the ambient concentration, at any point, i, due to a specified discharge from any source, j. Since the residuals are taken to be noninteracting and the rates of change are linear functions of concentration, the separate effects may be added to one another yielding the total effect from all sources within a specified region. This may be expressed mathematically for the kth residual and for steady-state conditions with the following set of linear relationships.[24]

Source, j

$$R_1^{(k)} = a_{11}x_1^{(k)} + a_{12}x_2^{(k)} + \ldots + a_{1n}x_n^{(k)}$$

$$R_2^{(k)} = a_{21}x_1^{(k)} + a_{22}x_2^{(k)} + \ldots + a_{2n}x_n^{(k)}$$

Receptor, i $\qquad\qquad\qquad\qquad\qquad\qquad\qquad\qquad$ (1)

$$R_m^{(k)} = a_{m1}x_1^{(k)} + a_{m2}x_2^{(k)} + \ldots + a_{mn}x_n^{(k)}$$

or in matrix notation, as $R^{(k)} = A \cdot X^{(k)}$ $\qquad\qquad\qquad$ (2)

where,

$x_j^{(k)}$, $j = 1, \ldots, n;$ $\quad =$ quantity of the kth residual discharged from the jth source during a given
$k = 1, \ldots, p$ $\qquad$ time interval.

---

[24]This set of equations is for the kth residual only. Similar equation sets would be required for other residuals. The superscript k will be omitted in discussing ambient standards with no intended loss of generality. In the actual model applied to the Delaware, the Streeter-Phelps type diffusion model is replaced by a more complex aquatic-ecosystem model which need not be discussed for our purposes here. For a more complete discussion of this, see W. O. Spofford, C. S. Russell, and R. A. Kelly, "Operational Problems in Large-Scale Residuals Management Models," a paper presented at the Conference on Economics of the Environment, sponsored by Universities-National Bureau Committee for Economic Research and by Resources for the Future, November 10-11, 1972.

a_{ij}, $i=1, \ldots, m$; $=$ unit transfer coefficient indicating the concentration at receptor location, i,
$j=1, \ldots, n$ due to a unit discharge of a given residual from source, j.

$R_i^{(k)}$, $i=1, \ldots, m$; $=$ the concentration of the kth residual at receptor location, i.

n $=$ number of sources of the kth residual

m $=$ number of receptor locations

p $=$ number of residuals in the analysis

The vector of rates of residuals discharges to the environment, x_j, $j=1, \ldots, n$, is available as an output of the LP residuals generation and discharge model and the transfer coefficients, a_{ij}, $i=1, \ldots, m$; $j=1, \ldots, n$, are computed from environmental models. The vector of ambient concentrations of a given residual at all receptor locations, R_i, $i=1, \ldots, m$, is the output of the system of equations, (1). This concentration vector, R, in turn, is used as input to the receptor-damage model (functions).

• *AMBIENT STANDARDS.* Where damage functions are not available or where one wishes to test strategies which do not incorporate them, ambient standards may be used. For this situation, equation set (1) may be modified to include ambient standards as follows:

$$a_{11}x_1 + a_{12}x_2 + \ldots + a_{1n}x_n \leqslant S_1$$
$$a_{21}x_1 + a_{22}x_2 + \ldots + a_{2n}x_n \leqslant S_2$$
$$\vdots \qquad\qquad\qquad \vdots$$
$$a_{m1}x + a_{m2}x_2 + \ldots + a_{mn}x_n \leqslant S_m$$

or in matrix notation, $A \cdot X \leqslant S$,

where S_i, $i=1, \ldots, m$ represents a vector of ambient standards. Since this is a linear constraint set, it can be transferred in its entirety to the LP model.

• *DAMAGE FUNCTIONS.* When damage functions are available and are to be used, it is possible to compute marginal damages for all sources at the locations of residuals discharge. It is assumed that the functions are continuous and have continuous first derivatives. For this case, the damages at each location, due to the kth residual, $D_i^{(k)}$, i=1, . . . , m, are assumed to be functions only of the corresponding ambient residual concentrations, $R_i^{(k)}$, i=1, . . . , m, or, expressed mathematically,[25]

$$D_i^{(k)} = f(R_i^{(k)}) \qquad i = 1, . . ., m \qquad (3)$$

The form of the assumed damage function is depicted in Figure 7 below.

The purpose of this exercise is to compute the total marginal damages corresponding to each residual source. For the kth residual, the total damages to all receptors and/or uses in the region are given by:

$$D_T^{(k)} = D_1^{(k)} + D_2^{(k)} + . . . + D_m^{(k)} = \sum_{i=1}^{m} D_i^{(k)}.$$

Figure 7

Assumed Shape of Damage Function for the kth Residual *

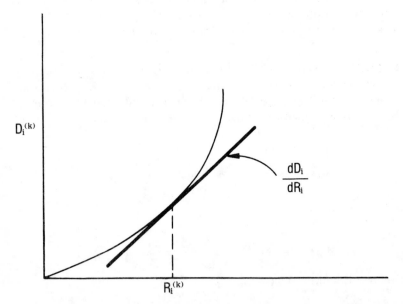

*The shape of this function is assumed to be monotonically increasing. The actual path of the curve would depend upon the specific residual, and the type and numbers of receptors present.

[25]Damages which are a function of several residuals can also be handled conveniently using this scheme. As we shall see, partial derivatives rather than total derivatives could be used in the evaluation of marginal damages.

As shown by equation (3), the damage, D_i, is a function of the ambient concentration, R_i. Therefore, the following expressions for marginal damages associated with discharges of the kth residual obtain:[26]

$$\frac{\partial D_T}{\partial x_1} = \frac{dD_1}{dR_1} \cdot \frac{\partial R_1}{\partial x_1} + \frac{dD_2}{dR_2} \cdot \frac{\partial R_2}{\partial x_1} + \cdots + \frac{dD_m}{dR_m} \cdot \frac{\partial R_m}{\partial x_1}$$

$$\frac{\partial D_T}{\partial x_2} = \frac{dD_1}{dR_1} \cdot \frac{\partial R_1}{\partial x_2} + \frac{dD_2}{dR_2} \cdot \frac{\partial R_2}{\partial x_2} + \cdots + \frac{dD_m}{dR_m} \cdot \frac{\partial R_m}{\partial x_2}$$

$$\frac{\partial D_T}{\partial x_n} = \frac{dD_1}{dR_1} \cdot \frac{\partial R_1}{\partial x_n} + \frac{dD_2}{dR_2} \cdot \frac{\partial R_2}{\partial s_n} + \cdots + \frac{dD_m}{dR_m} \cdot \frac{\partial R_m}{\partial x_n}$$

However, from equation set (1), $\dfrac{\partial R_1}{\partial x_1} = a_{11}, \dfrac{\partial R_2}{\partial x_1} = a_{21}, \dfrac{\partial R_m}{\partial x_1} = a_{m1},$ and so on, and the above equation set reduces to:

$$\frac{\partial D_T}{\partial x_1} = a_{11} \frac{dD_1}{dR_1} + a_{21} \frac{dD_2}{dR_2} + \cdots + a_{m1} \frac{dD_m}{dR_m}$$

$$\frac{\partial D_T}{\partial x_2} = a_{12} \frac{dD_1}{dR_1} + a_{22} \frac{dD_2}{dR_2} + \cdots + a_{m2} \frac{dD_m}{dR_m}$$

$$\frac{\partial D_T}{\partial x_n} = a_{1n} \frac{dD_1}{dR_1} + a_{2n} \frac{dD_2}{dR_2} + \cdots + a_{mn} \frac{dD_m}{dR_m}$$

or in matrix notation as, $\dfrac{\partial D_T}{\partial x} = A' \cdot \dfrac{dD}{dR}$.

The marginal damages are evaluated using this equation set, and the resulting vector

$$\frac{\partial D_T}{\partial x_j}, \quad j=1, \ldots, n,$$

---

[26]Here again, the superscript k has been omitted for purposes of exposition.

is returned to the residuals generation and discharge linear programming model as prices on residuals discharged to the environment.

The procedure outlined in the last few pages is summarized below:

(a) Given a residual discharge set, x_j, $j=1, \ldots, n$, and a matrix of
transfer coefficients, a_{ij}, $i = 1, \ldots, m$; $j=1, \ldots, n$, solve equation set
(1) for R_i, $i=1, \ldots, m$.

(b) Evaluate the slopes of the various damage functions, $\dfrac{dD_j}{dR_i}$, $j=1$,

$\ldots, m$, $i=1, \ldots, m$, at the points R_i, computed in step (a) above.

(c) Using this result, evaluate the marginal damages associated with
residuals being discharged to the natural environment.

(d) Use the marginal damages, computed in step (c), as prices on
residuals discharged in the residuals generation and discharge linear
programming model, and resolve the linear programming model for a
new set of residuals discharges, x_j, $j=1, \ldots, n$.[27]

A model of this general form is being applied to a regional analysis
of the Delaware estuary region. It has also been adopted to examine
some aspects of collective choice-making with respect to environmental
quality and other public goods. One objective is to determine what form
of collective choice-making (representative government) process would
serve best to reflect the preferences of the affected population for environmental quality, considering explicitly the trade-offs between media,
vis-à-vis other goods and services.[28]

DISCUSSION

DISCUSSANT: *Robert K. Davis*

With the 1972 amendments in mind, I would question whether transfer functions really have any role to play, since effluent limits are
defined on a technological basis. That is, they are defined by what
can be achieved rather than by what can be derived from ambient
standards through the use of transfer functions. I guess what I am
asking is: Do transfer functions still have a role to play in exploring
cost surfaces, if we accept the concept of the 1972 amendments?

---

[27]For technical reasons, a charge or "penalty function" approach is used in the actual
computations, even when the objective is to meet a set of ambient standards. The reasons
for this need not detain us here but are discussed in Spofford, Russell, and Kelly, op. cit.

[28]Clifford S. Russell, Walter O. Spofford, Jr., and Edwin T. Haefele, "Environmental
Quality Management in Metropolitan Areas," paper prepared for the International
Economic Association's Conference on Urbanization and the Environment, Copenhagen,
Denmark, June 19-24, 1972.

REPLY: *Allen V. Kneese*

I think that despite the 1972 amendments there is still a clear role for transfer functions, the consideration of alternative strategies, and so on. Furthermore, I consider the 1972 amendments—and this is a personal bias—a transitory phase. I think it will take us a few years to find out that they do not work, especially when energy and attention are being centered on other environmental media. I think we will be forced to do what the English have recently done and what the Germans are moving toward—the analysis of environmental problems in the regional context in which they exist. In that kind of situation transfer functions are critical devices for the analysis of alternatives.

Environmental Policy Costs: Definition, Measurement, and Conjecture

RALPH C. D'ARGE

ALFRED NORTH WHITEHEAD ONCE REMARKED THAT "definition is the half-way house toward measurement." To discuss the quantification of public policy costs, one must define them operationally, yet I have been unable to develop a definition which is sufficiently encompassing. Do public policy costs include all of the social costs of designing and implementing a new institution aimed at achieving a public objective; transaction costs in operating quasi-markets for public goods; opportunity costs for private as opposed to public use of common property resources; or only the social costs of implementing alternative policies within an already rigidly mandated institutional structure? The domain of the policy cost "set" appears to be as large as the imagination of the analyst will allow. Implementing the United States Constitution has and will continue to induce policy costs!

What I wish to do is provide a very limiting definition of policy costs so, to paraphrase Whitehead, we may proceed at least halfway to the "halfway house." Policy costs are defined here as the costs of information, enforcement, and administration to achieve a predefined public objective, such as water quality improvement. The objective may be predefined in terms of a water quality standard that is multidimensional and invariant with respect to shadow price, the minimization of the sum of *all* costs including damages, or multiple-objective in character. However, this analysis will emphasize the efficiency criterion of cost minimization.

I will assume the existence of some constitutional body (a congress or state legislature) with powers to define objectives, design environmental management institutions to achieve these objectives, and evaluate social costs associated with alternative designs.[1] The policy costs of the constitutional body in designing such institutions are beyond the scope of this discussion. It is also assumed that the geographical jurisdiction of the constitutional body is of such scope that it has complete (or at least predominant) authority for the particular water quality problem. Finally, I assume that the constitutional body is interested in maximum efficiency, that it wants to establish an institution which will achieve a given level of goal fulfillment at minimum cost. Policy costs in this context are then quite simply the design, initiation, and operating costs of institutions mandated to manage water quality.

The constitutional body must decide on two fundamental considerations besides general objectives: (a) the mandate identifying the scope, objectives, and relevant processes of the water quality management authority, and (b) the magnitude of its initial and expected future budget. The remainder of this paper—using a simple deterministic and, in some brief instances, a probabilistic model—will be concerned with typifying the considerations which should enter the decision on objectives, mandate, and budget. The measuring of policy costs is then examined within the context of this model. Two fundmental problems are omitted at the onset. First, we assume that what is "optimal" will vary according to the objectives of the constitutional body, and that this body is willing and able to define the Pareto-efficient point subject to its own mandate and constraints. Second, by so doing we are implying that distributional consequences are either: (a) not of concern to the constitutional body, or (b) are resolved via lump-sum transfers *or other means* that are not instrumental in defining the structure, mandate, or practices of the water quality control agency. These assumptions insure at least a partial conceptual clarity in the results, despite a certain lack of political relevance.

The measurement of policy costs is examined with reference to "ideal" measures emerging from the model. The model and its underlying assumptions are examined in light of the "degree of directness" of control measures for water quality. It is postulated that the degree of directness increases with a decline in the number of behavioral or technological assumptions interposed between control agency and private decisions. In the absence of these assumptions on behavioral or technological linkages, controls (such as penalties and rules) become "absolutely" direct. The paper concludes with a brief discussion of some of the conditions which must be mandated for a water quality management authority to be viable and efficient.

---

[1] The potential powers of existing constitutional bodies have been clearly established by acts of Congress in 1965 and 1972 for water, and in 1969 for air. The translation of potential powers into viable solutions will be a major concern of this paper. However, existing legislation will not be scrutinized except insofar as these mandates may provide illustrative examples for the conceptual models which are discussed.

Basic Model

L<small>ET</small> $\overline{Z}_{it}$ <small>DENOTE A VECTOR</small> of various characteristics of water quality (i.e., dissolved oxygen, lack of dissolved solids or salinity, etc.) at location i during time interval t *which are observed and known with absolute certainty.* Then we can identify the traditional damage function and control cost function that are location and time specific as:

$$D_{it} = D_{it} \ (\overline{Z}_{it})$$
$$C_{it} = C_{it} \ (\overline{Z}_{it}) \tag{1}$$

With zero policy costs, from the standpoint of static efficiency, the control agency can simply minimize the sum of D_{it} and C_{it} and select the optimum level of water quality characteristics. We may wish to complicate the model by introducing technological (or transfer) functions relating emissions to water quality characteristics. Letting $\overline{X}_{it}$ denote a vector of actual emissions of waterborne residuals at site i during time interval t, and $f(\cdot)$ denote a deterministic (and instantaneous) "transfer function" of emissions to quality measures, then the simple model in (1) can be rewritten as follows:

$$D_{it} = D_{it} \ (\overline{Z}_{it})$$
$$C_{it} = C_{it} \ | (\overline{X}_{it})$$
$$\overline{Z}_{it} = f_{it} \ (\overline{X}_{it}) \tag{2}$$

If there are locational or time dependencies, then additional functional relationships must be added to express these interdependencies and the minimization is over time and space as well.

Explicit Policy
Cost Functions

A <small>LOGICAL DEFICIENCY WITH THIS SIMPLE MODEL</small> is that if policy costs (as defined earlier) were truly zero for all parties, there would not be a need for a management agency of *any* type. Private negotiation would instantaneously achieve minimum costs. Thus, to make this simple model more realistic, we must at least add costs associated with information and perhaps enforcement as well. A second difficulty is that control costs are incurred in most instances prior to observations on actual emissions. Thus, these costs are based on anticipated rather than actual emissions whereas damages result totally from actual emissions. It appears that this type of dichotomy should also be explicitly recognized and included along with policy costs.

Information costs are encountered for obtaining knowledge on every aspect of the water quality problem from design criteria and damages to costs of enforcement. Here we shall examine only the costs of obtaining accurate measures of $\overline{Z}_{it}$ and $\overline{X}_{it}$, and presume that all other costs of information (such as information on damages) are zero.[2] Defining new vectors $\hat{Z}_{it}$ and $\hat{X}_{it}$ as the best estimates of characteristics of water quality and emissions, a hypothesized cost-of-information equation can be specified as follows:

$$CI_{it} = CI_{it} (\hat{Z}_{it} - \overline{Z}_{it}, \hat{X}_{it} - \overline{X}_{it}) \qquad (3)$$

which indicates that information costs are determined solely by the difference between estimated emissions and actual emissions, and between estimated ambient water quality and actual water quality. Such a relation can be thought of as a general specification of the classical statistical problem of sampling where there is a cost per sample, where additional samples reduce the degree of estimated variation in the mean of the population, and where the mean is taken as the best estimate. It can be also thought of as analogous to the more modern problem of finding the optimal stopping point, where the goal is to continue sampling or searching until the expected benefits of continuing such a search become less than the a priori known costs of sampling. CI_{it} then measures the least-cost method of undertaking such sampling experiments.

Enforcement costs are the costs of requiring a particular emitter to comply with a predetermined emissions standard or to pay for emissions once the optimal level of emissions (in a static or dynamic sense) is determined. Such costs include the costs of courts, taxing or subsidy mechanisms, and administrative costs totally related to compliance. Quite arbitrarily, they do not include monitoring costs which are subsumed under information costs. In order to include the obvious dependence between enforcement costs and monitoring effects in a rudimentary manner, I make the assumption that enforcement costs are inversely related to information costs but positively related to the difference between estimated and actual measures of X_{it} and Z_{it}.[3]

---

[2]The structure of damage relationships and their magnitudes may be the most uncertain or require the most expenditure on information for accurate estimation. Many economists have asserted such a hypothesis in recent years. In this case, the cost-of-information relationship specified in (3) would need to include the parameters or coefficients of the damage function as well as the differences between estimated and actual quality characteristics. In order not to complicate the model unduly, specification of the cost of information as being partially determined by the accuracy of the damage relationship is not included. If the magnitude of damages is positively related to the error in estimation of water quality characteristics, then the impact on policy decisions of errors in damage estimation will be similar to errors in measuring quality characteristics. In other cases, the model needs to be respecified to take explicitly account of informational costs incurred in estimating damage relationships.

[3]In other words, the greater the expenditure on monitoring, the lower the cost of enforcing a particular quality standard. With less than perfect monitoring, we should anticipate that a greater degree of latitude would exist in court or other legal disputes. Also, the greater the difference between predicted and observed emissions, the more difficult it would be, *ceteris paribus*, to provide an adequate level of enforcement. For

The proposed enforcement cost relationship is as follows:

$$CE_{it} = CE_{it} (CI_{it}, \hat{X}_{it} - \overline{X}_{it}, \hat{Z}_{it} - \overline{Z}_{it}, D_{it} + C_{it}) \qquad (4)$$

where $D_{it} + C_{it}$ denotes the sum of control costs and damages, a sum which reflects the severity of the resulting emissions standard and thereby financial obligations arising directly through effluent charges or, indirectly, through severity of standards. As this sum rises, firms will be more prone to make expenditures toward thwarting enforcement regulations through legal or nonlegal means.[4] In either case, enforcement costs could be anticipated to rise. To reflect this observation, we include the sum of damage and control costs in the enforcement-cost specification.

Administrative costs are defined to be simply the overhead costs associated with the information and enforcement efforts and therefore are considered to be a function of these costs:

$$CA_{it} = CA_{it} (CE_{it}, CI_{it}) \qquad (5)$$

In this context, the costs of administration per se are little more than "cost-plus" components of information and enforcement costs. This is an extremely naive assumption. We should expect such costs to be highly dependent on the fundamental nature of the agency—whether it is highly aggressive and regulative, or primarily information oriented—on its mandate and budget, and on the underlying philosophical structure. Perhaps, as an analogue to Parkinson's Law, the time interval during which the agency has existed should also be introduced. For now, however, we shall presume that administrative costs are analogous to "overhead" costs in the private sector where strict dependence on the size of budget is often postulated for the costs of management.

The basic model, suitably modified by (3) through (5), if known explicitly, could conceptually be used to derive the optimal level of emissions and water quality levels where the optimum is defined as the minimized sum of the damage, control, *and* policy costs.

In addition to the cost and damage relationships, to adequately solve the minimization problem, information must be available on the cost of making errors in estimation, if such costs are not already embedded within the damage and control cost functions. For example, separate control cost relationships might be specified for water pollution control on the basis of: (1) estimated waste loads, and (2) emergency procedures for large deviations of actual from estimated waste loads. Also, informa-

---

example, if an agency could provide a "perfect" prediction of an emissions profile, the strength of its case in debating a violation would be much stronger than if its predictive capability were low or zero for a particular level of monitoring.

[4]Such expenditures should be explicitly included in any complete behavioral model of environmental regulation. Here they are totally subsumed within the costs of enforcement.

tion must be available or be obtained on the *mechanism* leading to errors in measurement or enforcement. For example, can errors in estimation be reduced by more than one type of sampling process? Is the sampling process the only mechanism for reducing errors in estimating waste load emissions? To analyze these questions within a tractable framework, we turn to some simple examples.

A Simple Example

IN ORDER TO ILLUSTRATE THE COMPLEXITIES in unraveling the context in which policy costs should be examined, some extremely simplifying assumptions are introduced into the above model. The first assumption is that information is obtained through a simple sampling process where samples on water quality and emissions are made and where the variances associated with their true (and assumed normal) distributions are known in advance. Given elementary sampling principles we can easily divide the general information-cost specification into two parts. First, if total costs per sample are constant for estimating both $\overline{Z}_{it}$ and $\overline{X}_{it}$ (this implies no *fixed* costs associated with the sampling process), (3) becomes:

$$CI = CI_Z N_Z + CI_X N_X \tag{6}$$

where N_Z and N_X are the number of sample measurements of $\overline{Z}_{it}$ and $\overline{X}_{it}$, respectively, identified here for convenience as scalars rather than vectors. The CI_Z and CI_X coefficients denote the per unit cost of sampling $\overline{Z}_{it}$ and $\overline{X}_{it}$. In addition to this equation, we must introduce two equations that act as constraints on the minimization by relating sample size to the accuracy of the measures of $\overline{Z}_{it}$ and $\overline{X}_{it}$. These are:

$$x_{it} = \hat{X}_{it} - \overline{X}_{it} = \alpha \frac{\sigma_X}{\sqrt{N_X}}$$

$$z_{it} = \hat{Z}_{it} - \overline{Z}_{it} = \beta \frac{\sigma_Z}{\sqrt{N_Z}} \tag{7}$$

where α and β denote "t" statistics appropriate for the preselected confidence intervals of the estimates of $\overline{X}_{it}$ and $\overline{Z}_{it}$. Thus, the more general problem of identifying and measuring information costs is subsumed into the specific problem of minimizing the sampling costs, along with damage and control costs, subject to a preselected degree of accuracy of the estimates.

With this simplification, the entire problem can be established as a straightforward minimization problem as follows:

$$\text{Minimize } L = D_{it} + C_{it} + CE_{it} + CI_{it} + CA_{it} \qquad (8)$$

$$\text{subject to: } x_{it} = \alpha \, \frac{\sigma_X}{\sqrt{N_X}}$$

$$z_{it} = \beta \, \frac{\sigma_Z}{\sqrt{N_Z}}$$

$$\overline{Z}_{it} = f(\overline{X}_{it}) \text{ or } \hat{Z}_{it} = f(\hat{X}_{it})$$

with the transfer function $f(\cdot)$ completely determined but subject to measurement errors in estimates of $\overline{Z}_{it}$ and $\overline{X}_{it}$. The problem now confronting the control agency (or its parent) is to simultaneously select $\hat{Z}_{it}$, x_{it}, N_Z and N_X to minimize costs. The first-order conditions for such a problem can be easily derived but do not yield interpretations without further simplification.

In order to make the problem tractable, we shall presume CE and CA are zero and concentrate on a problem with a simple combination of damage, control, and information costs and with hypothetical functions for each of these costs. In addition, we shall concentrate only on the emissions estimate, $\hat{X}_{it}$, and not on the transfer of emissions into an estimated index, $\hat{Z}_{it}$, of ambient water quality.[5] With explicit functional forms of equations, we are able to illustrate some of the fundamental differences between zero information-cost problems and those with positive information costs.

In List 1, a hypothetical problem is posed with nonlinear damages but with linear information and control costs related to the emissions variable X where $\overline{X}$ is the true value, while $\hat{X}$ is the best estimate of that value. Thus, three decision variables (two that are independent) emerge for this problem: (1) the number of samples on emissions, denoted as N; (2) the estimate of $\overline{X}$ denoted by $\hat{X}$; and (3) the estimated deviation of $\hat{X}$ from its true value, denoted by x. The deviation is therefore equal to $\hat{X}-\overline{X}$.

We presume the goal of the control agency is to minimize the sum of damage, control, and information costs. It is also assumed that a con-

[5]For many water quality management problems, the uncertainty with regard to the *identification* of the transfer function in time and space is undoubtedly greater than other informational uncertainties. In fact, many of the sampling costs may be devoted to more precisely identifying the transfer function. We shall make the simplifying assumption here that some underlying deterministic function (such as the Streeter-Phelps oxygen sag equation) is valid but that uncertainty is introduced via lack of knowledge on emissions and water quality in the watercourse. Thus, the estimates of water quality and emissions are constrained by a deterministic transfer function (presumed to be valid) such that there must be a precise correspondence between the two vectors.

214

HYPOTHETICAL PROBLEM
List 1

$$\text{CI} = 1 \cdot \text{N} \tag{a}$$

$$\text{D} = (\hat{\text{X}} + \text{x})^2 \tag{b}$$

$$\text{C} = 500 - 30\hat{\text{X}} \tag{c}$$

$$\overline{\text{X}} = \hat{\text{X}} + \text{x} \tag{d}$$

$$\text{x} = \left(\frac{\alpha\sigma}{\sqrt{\text{N}}} \right) \text{ with } \alpha \simeq \text{``t''} \tag{e}$$

$$\sigma = 40, \quad \alpha = 2 \tag{f}$$

$$\text{Min L} = \text{CI} + \text{D} + \text{C} + \lambda \left(\frac{\alpha\sigma}{\sqrt{\text{N}}} - \text{x} \right) \tag{g}$$

$$\frac{\partial \text{L}}{\partial \hat{\text{X}}} = 2(\hat{\text{X}} + \text{x}) - 30 \geqq 0 \tag{h}$$

$$\frac{\partial \text{L}}{\partial \text{x}} = 2(\hat{\text{X}} + \text{x}) + \lambda \geqq 0 \tag{i}$$

$$\frac{\partial \text{L}}{\partial \text{N}} = 1 - \lambda \left(\frac{40}{\text{N}\sqrt{\text{N}}} \right) \geqq 0 \tag{j}$$

$$\frac{\partial \text{L}}{\partial \lambda} = \left(\frac{80}{\sqrt{\text{N}}} \right) - \text{x} \geqq 0 \tag{k}$$

$$\text{N}^* \simeq 113 \tag{l}$$

$$\hat{\text{X}}^* = 7.48 \tag{m}$$

$$\text{x}^* = 7.52 \tag{n}$$

$$\text{Min L} = \text{D} + \text{C} = \overline{\text{X}}^2 + 500 - 30\overline{\text{X}} \tag{o}$$

$$\frac{\partial \text{L}}{\partial \overline{\text{X}}} = 2\overline{\text{X}} - 30 \geqq 0 \tag{p}$$

$$\overline{\text{X}} = 15 \tag{q}$$

*denotes optimum level with slide rule accuracy

straint on the accuracy of the estimates, x, is imposed where the level of the constraint depends exclusively on the "t" statistic (or confidence interval) given for x specified as α, and on a known variance of the distribution of $\overline{X}$ denoted as σ^2. A final set of assumptions relate to damage being a function of only what actually occurs, i.e., $\overline{X}$, while control costs are determined exclusively on the best estimate of emissions, $\hat{X}$. Such a set of assumptions appears reasonable in that most controls (in the sense of treatment plants, low-flow augmentation, process changes, etc.) must be developed and placed in operation on the basis of expected emissions (and variability) prior to the time when actual emissions occur. Alternatively, damages result from the actual level of emissions.

With these assumptions and the explicit functional forms enumerated in List 1, (a) through (f), the first-order conditions for a minimization can be specified along with numerical results. These are tabulated in (g) through (n) in List 1.

Contrasting (m) and (q) in List 1, we observe a reduction in the optimum level of $\overline{X}$ when information costs are introduced which should be expected a priori. Information costs introduce uncertainties that tend to make it pay to reduce emissions below the zero information cost optimum derived in (o) through (q). However, the best estimate of $\hat{X}$ and the error x *sum* to the optimum $\overline{X}$. This result is obvious in that there are no penalties for error except the one imposed by the constraint on accuracy. Note that even with a large variance in the distribution of $\overline{X}$ (i.e., with σ relatively large), the magnitude of information costs acts to reduce the number of samples taken through the relation shown in (j).

This particular hypothetical problem involving information, damage, and control costs simultaneously, selects the optimal standard (or marginal damage level) and determines the number of samples to ascertain the appropriate level of information. If these activities are undertaken separately there will generally be too high an emissions rate initially established and too much information obtained (in response to the higher emissions rate).

To expand this extremely simple numerical example a little further, a specification for enforcement costs will be introduced. Note that our previous conjecture was that enforcement costs are determined by the amount of information obtained (and the amount of information obtained is proportional to the information costs incurred); by the variability of the difference between the estimated and the actual emissions (and the difference between the estimated and actual quality measures); and by the magnitude of control plus damage costs. To simplify, we shall assume here that enforcement cost is negatively related to information cost but positively related to the sum of control and damage costs. The numerical equation assumed for enforcement costs is:

$$CE = -.1\ CI + .2\ (D + C) \tag{9}$$

HYPOTHETICAL PROBLEM
List 2

$$\text{Min } L = CI + D + C + CE + \lambda\left(\frac{\alpha\sigma}{\sqrt{N}} - x\right) \tag{a}$$

$$\text{Min } L = 600 - 36\hat{X} + .9N + 1.2(\hat{X} + x)^2$$
$$+ \lambda\left(\frac{80}{\sqrt{N}} - x\right) \tag{b}$$

$$\frac{\partial L}{\partial \hat{X}} = -36 + 2.4(\hat{X} + x) \geq 0 \tag{c}$$

$$\frac{\partial L}{\partial x} = 2.4(\hat{X} + x) - \lambda \geq 0 \tag{d}$$

$$\frac{\partial L}{\partial N} = .9 - \lambda\left(\frac{40}{N\sqrt{N}}\right) \geq 0 \tag{e}$$

$$\frac{\partial L}{\partial \lambda} = \frac{80}{\sqrt{N}} - x \geq 0 \tag{f}$$

$$N^* = 131 \tag{g}$$

$$\hat{X}^* = 8.01 \tag{h}$$

$$x^* = 6.99 \tag{i}$$

Using the numerical equations specified in List 1, (9) can be simplified to:

$$CE = 100 - .1 N - 6 \hat{X} + .2 (\hat{X} + x)^2 \tag{10}$$

Now, by minimizing the sum of C, D, CI, and CE an optimal estimate of $\hat{X}$ can be obtained. The derivation is provided in List 2.

The introduction of enforcement costs has increased the number of samples taken (i.e., it pays to be more accurate) and slightly increased the best estimate of $\overline{X}$ but has reduced the error estimate.[6]

This particular model is perhaps the most simple that can be constructed in that it includes specific accuracy in the form of a constraint but

---

[6]An important implicaton of even this most simple model is that applying the marginal damage functon as a tax on the polluter will generally not yield an efficient result unless there is "perfect" information on the emissions.

does not explicitly include the distribution of $\overline{X}$ or how the best estimate, $\hat{X}$ is affected by sample size. It also does not include additional damage costs which may result from errors which induce greater than "normal" damages. The reasons for error have not been explicitly included in the definition and structure of these simple examples. Conceptually, it could arise because of measurement problems, faulty operation of pollution control equipment, lack of knowledge of transfer functions, or other means. For an adequate analysis, the reasons for such error should be explicitly included in the model.

Quantifying Policy Costs

IN PREVIOUS SECTIONS, we have identified a very simple and incomplete model of water quality management where policy costs are explicitly introduced. Only very simple dependencies among the various policy costs were examined. Additionally, a dependence per se between damages, control costs, and policy costs was not studied except via common variables in functions. But, even with the "mild" dependencies introduced, it is obvious that policy cost functions require a great deal of information for estimation, a case of the information cost being necessary to estimate information costs! Ideally, we should desire to estimate the following function delineating minimum policy costs defined here as $CP_{it}*$.[7]

Thus:

$$CP_{it}* = CE_{it}* + CI_{it}* + CA_{it}*$$
$$CP_{it}* = g(D_{it}*, C_{it}*, CI_{it}*, CA_{it}*, CE_{it}*) \tag{11}$$

where the function $g(\cdot)$ is derived by solving the first-order conditions of the more general model or the simple model equations (h) through (k) in List 1. For the hypothetical problem given in List 1, the $g(\cdot)$ function can be readily derived and interpreted.

$$CP* = CI*$$
$$CI* = 1 \cdot N*$$
$$CI* = \frac{\alpha\sigma}{\sqrt{N*}} \ (\lambda*) = \frac{\alpha\sigma}{\sqrt{N*}} \left(\frac{\partial D*}{\partial \hat{X}*}\right) \tag{12}$$

[7]Superscript asterisk denotes the value of the variable is at an optimum.

It is clear from (12) that the optimal information cost is positively related to both the variance of $\overline{X}$ and the prespecified confidence interval for x and inversely related to the optimal number of samples. That is, if the optimal number of samples increased by a slight amount, with no other changes, then the optimal information cost would decrease by definition. Of course, optimal costs of information vary directly with optimal sample size and cost per emissions sample, and on marginal damages at the optimum, $\partial D^*/\partial \hat{X}^*$. To construct the cost of information function for this simple problem requires complete information on marginal damage plus anticipated errors in estimation obtained through prior information on $\overline{X}$. We could further complicate the problem by including sampling on the damage and control cost *functions* in the cost-of-information function as well. But such complications require us to examine even more precisely identified problems to obtain answers.[8]

In conclusion, this very simple example provides several lessons on measuring policy costs. First, the optimum level of information costs cannot be separated from exact knowledge on: (1) the damage function, (2) the control cost function and, most importantly, (3) the variation between actual emissions and estimated emissions derived from samples. Second, the optimal level of damages and control costs cannot be determined without reasonably precise estimates of information costs and of the underlying variability (characteristics of the statistical distribution) of the estimators for emissions. And yet, the optimal level of information costs needs to be estimated so that a standard or base of measurement for comparing actual policy costs can be developed.

While these conclusions are certainly negative as regards the establishment of general qualitative principles for measuring policy costs, they clearly indicate what is needed: experimental efforts where not only are effluent charges, standards, and other controls examined, but also where they are studied simultaneously with the relevant information costs that have been measured. Such experimental efforts might lead to some applicable rules of thumb for policy cost measurement with a generality beyond the case-by-case basis. Rather than end on this rather dismal note as regards purely conceptual models, I shall attempt a more qualitative and less rigorously deductive discussion of policy costs in the context of types of controls and control agencies.

In recent years, much attention has been paid to the question of when charges, subsidies, and standards will yield necessary and sufficient conditions for an optimum adjustment with zero policy costs.

---

[8]This was demonstrated at least partially in a paper by Porter where a joint information decision was made on a set of variables *and* functions connecting them. See: W.R. Porter, "The Value of Information in Making Public Decisions," *Intermountain Economic Review* (August 1971).

There also has been much theoretical analysis about the possibilities for private negotiation to resolve externalities, and about the general effect of transactions costs on resource allocation. What appears to be missing is a partial or complete taxonomic analysis of the relative efficiency of different environmental planning systems which would include not only the type of controls to be selected (e.g., taxes or zoning ordinances) but also the type and structure of the planning organization responsible.[9]

My purpose here is to present briefly some simple ideas on the nature of interdependence and efficiency of control strategies and on the types of institutions used for control.

Control Strategies
and Degree of Directness

THE FIRST DISTINCTION TO BE CONSIDERED is the one between direct and indirect controls or management strategies. Direct controls are those which are applied at or to the source of the water quality problem and which threaten sufficient penalties to make avoidance extremely costly. The continuous monitoring and closing down of a particular factory which habitually violates emission standards is an example of such a direct control. The construction and operation of treatment facilities strategically located along watercourses is another example of direct control. Direct controls leave little or no latitude for private decision making. There is only a single causative link between infraction and agency action. The price of not complying is established at a level high enough to insure universal compliance.

Indirect controls are defined here as management strategies with at least two links of expected causation between problem source and application of control. Examples include Pigovian taxes which operate on two behavioral postulates. The first is that if the polluting firm is taxed for waste discharge, it will attempt—within the bounds of efficiency—to avoid this charge by altering production, adopting waste-controlling technologies, relocating, or by some other legal means.[10] The second

---

[9]Recently, several researchers have begun to analyze alternative types of environmental control agencies or management systems and their relationship to the efficiency and other performance properties of management strategies or controls. See B.T. Bower and W.R.D. Sewell, *Selecting Strategies for Air Quality Management*, Resource Paper no. 1, Policy Research and Coordination Branch, Department of Energy, Mines and Resources, Ottawa, Canada (1971); T.D. Crocker, "On Air Pollution Control Instruments," Working Paper Series no. 8, *Program in Environmental Economics*, University of California, Riverside (August, 1971); and L.E. Craine, "Institutions for Managing Lakes and Bays," *Natural Resources Journal* 2, no. 3 (July 1971).

[10]Whether actions taken are legal or illegal may depend on how direct the control is. If the tax or charge is viewed as oppressive then the firm may resist it illegally.

postulate is that if the firm reduces waste discharge by a certain amount, societal losses will also be reduced.

As indicated above, direct and indirect controls are distinguished by the number of their technological or behavioral links. However, it appears that most economic controls to date have been mixtures of both direct and indirect controls. For example, the law requires motorists to stop for a red light, but if it is not obeyed and the violator is caught, he is penalized. However, the penalty is usually not high enough to command universal compliance. Zoning laws can also combine direct and indirect control strategies by containing both penalty and variance provisions, the latter based on the behavioral assumption that most affected parties will comply and not request variances.

Behavioral linkages or relations, even in what seem the most simple control strategies, can require a substantial amount of research and background information before the control strategies in which they are embodied can be satisfactorily implemented. A gasoline tax to reduce automotive emissions seems relatively simple, but the assumption that —even after one allows for differences in gas consumption between vehicles of different weights and makes—there is a fixed technological link between miles driven and pollutants emitted is weaker than one might think. It has been carefully documented that, apart from mileage driven, the ways in which different people drive (i.e., acceleration, stopping, downshifting, etc.) also have markedly different impacts on exhaust emissions of reactive hydrocarbons, carbon monoxide, and oxides of nitrogen. In analyzing the relationship between gasoline taxes and vehicle emissions, one must consequently include the possibility that a higher gasoline tax, aimed at encouraging vehicle operators to drive shorter distances at slower speeds to reduce gasoline consumption, might, instead, increase total exhaust emissions. Such a perverse type of emissions impact could also be anticipated for waterborne emissions. A standard which is based on mean monthly flow rather than on daily variations might reduce information costs but increase the probability of perverse incentives.

A second difficulty may arise with the behaviorial relationship between gasoline taxes and automotive emissions. Increased gasoline prices may stimulate the purchase of automobiles with better gasoline mileage, i.e., vehicles with smaller engine displacement but higher compression ratios. There is some evidence that engine size and the emission rates of certain pollutants are not positively correlated, but might even be inversely correlated. To cite an extreme example, in testing 1971 automobile models, the federal government found that a vehicle with 79 cubic inches had emission rates per mile of hydrocarbons and carbon monoxide at least 50 percent higher than a vehicle with 472 cubic inches.[11] Other data collected in the same tests also indicated a slight negative correlation between cubic inch displacement and emission rates of hydrocarbons and carbon monoxide. Thus, the linkage between engine size and emission rates is not necessarily positive and in consequence, the substitution of vehicles with smaller engines for gasoline economy in response to a gasoline tax may have no effect or may even have a negative effect on automotive emission rates.

Our example of the gasoline tax (a very indirect control) clearly points out (even though far removed from waterborne emissions) that indirect controls require a substantial amount of information and knowledge of behavior prior to their implementation—less than controls with more direct linkages. However, it should be noted that even very direct controls may embody high information costs through the need for accurate monitoring and information on the technical relationship between reduced waterborne emissions and an improvement in water quality. If one views the direct and indirect control dichotomy as a continuum starting with absolutely enforced controls on a known source and proceeding to more behavioral linkages in indirect controls, it is defensible to hold that information requirements (and thereby costs) generally increase as more indirect control strategies are employed (assuming each control achieves the same impact). For example, a desired pattern of water quality might be achieved through direct emissions standards by zone, or by a more indirect control such as zone emissions taxes. But the assessment of taxes on waterborne emission patterns requires substantial a priori behavioral research and perhaps significant on-site experimentation particularly to be able to contrast policy costs of such taxes with the policy costs of standards to achieve a given objective.

Alternatively, if direct controls become overly oppressive and lead to massive resistance, legal or otherwise, the costs of enforcement and monitoring will be extremely high. At the very least, resources will be wasted in trying to implement the controls. At worst, the attempt could induce a serious misallocation of resources, from the standpoint of the existing statutory rationale for allocating goods, services, and resources. Figure 1 depicts the effects of enforcement costs and information costs as the degree of indirectness increases. An optimum in directness may exist which balances the incremental costs of enforcement *and* information. Such a comparison, however, presumes that information costs are known before the control is attempted, which is usually not the case, as was demonstrated earlier. While a diagram like Figure 1 is relatively easy to deduce theoretically, we have no empirically based knowledge about the shape or dimensions of such a curve, or even about what factors most

[11]See *The Federal Register*, vol. 36, no. 70, Washington, D.C. (April 10, 1971).

222

Figure 1

Hypothesized Relation Between Total Cost of Controls
and Directness of Controls

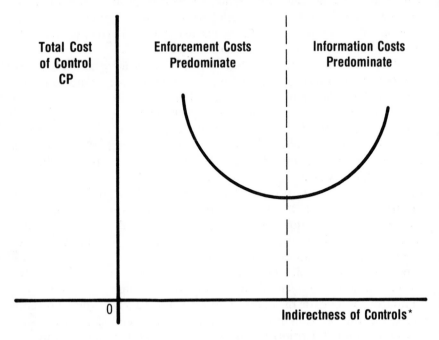

Total Cost of Control CP **Enforcement Costs Predominate** **Information Costs Predominate**

0

Indirectness of Controls*

*Along the horizontal axis are presumed to be sets of different controls with the most direct to the left and most indirect (greater behavioral relation) to the right.

strongly affect the cost of various control strategies in water quality management.

It is important, in determining where controls should be applied, and to what degree, that those subject to the regulatory effort should know how many and what kinds of alternative controls are available, and how much they cost. Thus, to make the buyer pay a purchase or excise tax on steel from a plant where emissions from the coking or quenching processes are damaging the water is a highly indirect method of control, since it assumes that purchasers will buy less steel, that the plant's production will be cut accordingly, and that emissions will thereby, indirectly, be reduced. The steel purchasers obviously have fewer methods of control, i.e., to reduce demand or possibly to negotiate with the steel producers. But, the steel manufacturer can resort to precipitators, use sealants on coke oven doors, relocate or redesign the coke ovens, or institute almost any conceivable process change which is less costly than paying an effluent charge.

Given this brief discussion on direct and indirect controls, can some assertions be made? I think so. Controls that are more direct may be favored over less direct controls because the linkages between cause and effect are fewer in number. State water quality regulations (not explicit pricing or emissions standards systems) for the use of public watercourses are examples of existing indirect controls with relatively many behavioral linkages. How behavioral relationships influence the actions of the public agency in relating water quality measures to emissions is uncertain. There is also uncertainty surrounding the transfer functions relating emissions to water quality measures. Quality standards are generally established independently of the level of damages, control costs, mandate, or budget of the regulatory agency. In consequence, a very indirect control is implied with a large number of behavioral and technological "transfer" functions and assumptions thereby included.

A less indirect control than Pigovian taxes or charges contains, by definition, fewer behavioral linkages and thereby implies fewer uncertainties. Thus, information costs tend to be lower for controls which are more direct. However, enforcement and related monitoring costs may become prohibitively high if controls are too direct and oppressive.

Control Agency Structure and Directness of Controls

OUR PREVIOUS DISCUSSION EMPHASIZED the type and efficiency of different controls by examining the degree of the control's directness. There is a problem, however. Once the control agency is established through authorization, funding, and listing of responsibilities, constraints may be placed on it which preclude analyzing certain types of controls. For example, water management agencies in the United States are typically authorized to consider issues only within precise geographical, hydrological, or institutional boundaries. Again, when these agencies are formed, they may or may not be given the power of taxation, condemnation, and arbitration. The effect of such restrictions is, more often than not, to determine in very rigid terms what types of controls an agency will consider as feasible or within its legal jurisdiction.[12]

The initial structure of the agency may largely determine the background and prior training of agency personnel. The types of personnel in a control agency developed purely for the purpose of monitoring water

---

[12]A well-documented case of agency restrictions and their effects on the possibility of selecting efficient controls was completed by R.K. Davis on the U.S. Army Corps of Engineers' efforts to improve water quality in the Potomac River using low-flow augmentation. See R.K. Davis, *The Range of Choice in Water Management: A Study of Dissolved Oxygen in the Potomac Estuary* (Baltimore: Johns Hopkins University Press, 1969).

quality are likely to be quite different from those in an agency with more inclusive environmental protection or control functions. And with such personnel differences, it is unlikely that the two agencies will agree on the feasibility of the same set of controls.

These considerations lead me to the conclusion that those elements which define the type of control agency—its initial structure, function, and mandate and the degree of directness in the type of controls it will consider—are not separable. The effect of inseparability is to require, from the standpoint of an "ideal" set of controls and control agencies, simultaneous decisions on the function and type of control agency as well as the type and directness of controls. Otherwise, a control agency may be developed which is incompatible with the least-cost types of control measures. However, this conclusion, while simple, may not be practicable since it is not likely that a governing body would have the requisite information to make all such decisions at the same time. In fact, in most instances, the control agency would also be responsible for gathering and processing information on alternative types of control strategies. It can perhaps be taken as a datum that a control agency will be developed before all or even a small number of the environmental control issues or control strategies have been identified. In that case, to evaluate alternative types of control agencies, we must develop a set of criteria which at least partially reflect the range of potential problems and controls that might be anticipated.

The Environmental Planning Process

BEFORE PROCEDING TO THE CRITERIA THEMSELVES, we turn to a tentative (and certainly incomplete) list of seven phases which characterize the process of environmental and land-use planning regardless of the institutional structure of the agency or management strategy:

1. Perception of a particular water quality problem
2. Initial information process (by public and private actors) on environmental and social consequences
3. Tentative set of management strategies set forth and studied with public and private actor participation
4. Additional information acquired on specific costs and impacts
5. Selection and implementation of management and/or control strategies for resolving the water quality problem
6. Additional information acquired on effects of control strategies in resolving problems
7. Revision of management strategies to yield greater efficiency of control or to achieve other criteria

Given such a framework for an environmental decision making process, the next step is to develop explicit criteria for evaluating and compar-

ing alternative types of control agencies. I will briefly discuss seven potential criteria, although others may be important in particular planning
situations.

The first criterion is efficiency in terms of cost. The measuring rod
for comparison is quite simple for this criterion; namely, how much are
the policy costs of operating a particular agency of type A compared with,
say, an agency of type B, where A and B designate mutually exclusive
differences in mandate, geographical responsibility, jurisdiction over
types and number of waste residuals, enforcement capability, and any
other differences which pose a necessary choice. Of course, to compare
adequately agencies of type A and B requires that both be able to accomplish identical objectives in terms of water quality management. If the two
cannot accomplish the same tasks and are therefore not comparable on
the basis of policy cost efficiency, then additional criteria on priorities of
solving particular water management problems must be introduced so
that other criteria can be used in choosing between alternative types of
control agencies.

A second criterion is the reliability of the control agency in resolving
management issues. This criterion is perhaps highly related to jurisdiction and the extent of power given to the agency under its original
mandate. A third criterion is the extent to which changes in the existing
social structure (or the existing set of social and cultural values) must
take place in order to accommodate the particular type of control
agency. A fourth criterion might be the speed with which the control
agency can complete phases 2 through 7. Speed in instituting controls is
especially important whenever even slight delays may induce irreversible
impacts in a particular water-related food chain or other ecological cycle.
Related to this is the speed with which a particular type of control agency
can alter management strategies once they have been imposed. For example, if the agency is so mandated that court suits are allowed to be
initiated each time there is a change in effluent standards, it is doubtful
that the agency will be efficient. A fifth criterion is simplicity, and the
degree to which the public understands the control agency and how it
operates. An agency with a very complex mandate or set of jurisdictional
powers may have difficulty in responding to the public's perception of a
water pollution problem because of a diffusion of responsibility within
the bureaucracy. A sixth criterion is the extent of information costs
encountered in implementing management strategies. A purely public
control agency may find it difficult to obtain residuals information from
emitters unless the emitters themselves are directly involved in agency
decision making. This is one of the differences between river commissions in the United States, which are highly public, and equivalent organizations (*Genossenschaften*) in West Germany where emitters have explicit representation on governing boards. Whether information costs in

the United States are actually higher has not yet been adequately documented, but it seems likely that they would be.

A seventh criterion is the flexibility of the control agency in responding to new types of water quality problems. A particular type of agency may have relatively low information costs, efficiency in control, simplicity, and reliability. It may induce few changes in social structure but be totally unresponsive in analyzing new problems. How one can build this type of flexibility into control agencies remains to be studied, but I suspect that flexibility increases, the more diversified or interdisciplinary the planning staff of the control agency is. Diversity, however, may preclude excellence in designing specific management strategies, so there is a potential price to pay for it.

The list of possible criteria for evaluating different types of control agencies is undoubtedly very large and the list of seven very briefly described here is meant to be no more than suggestive. What, hopefully, has been established is that the structure, mandate, and jurisdiction of control agencies will determine their value in resolving water quality problems. The mandate and jurisdiction of agencies will also determine the types of management strategies, and the latter, in turn, will partially dictate the mandate and jurisdiction of control agencies and thus their efficiency, information costs, flexibility, and other outputs.

I would like to suggest several tentative hypotheses about the relationship between policy costs, directness of controls, and agency structure and mandate without giving supporting evidence. To simplify, I would suggest two organization types for control agencies. Type A has an engineering orientation, is single-purpose in scope, and emphasizes monitoring as a major function. Typically, this type would rigorously enforce standards, regulations, or zoning ordinances, which were politically determined outside the agency. Type B has a broader interdisciplinary orientation and is multipurpose in scope and in providing solutions, while monitoring and enforcement are only subsidiary functions. Type B might be characterized by the use of a wider spectrum of controls, including indirect controls of the tax-subsidy type. Perhaps both types can be incorporated in a single super-agency, but I am doubtful.

The type A agency, given its single-purpose orientation (to increase the average level of dissolved oxygen in a river, or to reduce photochemical smog in the air), would undoubtedly be more efficient in achieving a particular level of control at least cost using direct controls. The type B agency, while not as efficient in applying direct controls, might be more efficient in applying indirect controls or in devising multiple control strategies to achieve the same water quality improvement as an agency of type A. Such a hypothetical case is depicted in Figure 2.

227

Figure 2

Type of Control Agency and Policy Costs

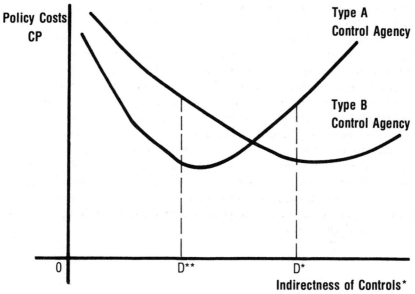

*Along the horizontal axis are presumed to be sets of different controls with the most direct to the left and the most indirect to the right.

A set of rather direct controls indicated by D** in Figure 2 can be implemented at less cost by an agency of type A. Such an agency can undoubtedly achieve lower monitoring and administrative costs with specific jurisdictional powers to enforce controls. Alternatively, when the control strategy is relatively indirect, the agency with broader jurisdictional powers and ability to apply indirect controls as a part of its mandate could be expected to be more efficient as regards policy costs. This case is depicted by the point D* indicating a set of relatively indirect controls with the type B agency hypothetically having lower costs than type A. Of course, there will be many alternative configurations of agency structure between A and B, and these configurations may be very different in orientation. But what is important to note is that, depending on the structure (or directness) of control strategies, different types of control agencies might be more or less efficient than others in achieving implementation.

Conclusions

To summarize my points of view, decisions on resolving water quality management problems ideally would require careful, simultaneous choice of both control institutions and control strategies. Yet, we have little research on what constitutes an efficient institutional control system. For a control agency to adequately consider the use of extremely indirect but perhaps more efficient control strategies, a much more comprehensive mandate and jurisdiction must be designed for it than is presently in vogue in the United States. In addition, since some water quality problems cannot be adequately anticipated, criteria other than policy cost efficiency should be considered in choosing between alternative structures for the water quality management agency.

Until we have a general theory of design for water quality management institutions, policy costs cannot be adequately evaluated in opportunity cost terms. What I have attempted here is to offer a rather cursory review of factors that may be important to that theory of design in terms of assessing policy costs.

DISCUSSION

Discussant: *Henry S. Rowen*

It is clear that some cost elements which d'Arge distinguishes between would be extremely difficult to measure. Consider the number of damaged constituents and the variety of costs that might be involved. I doubt the wisdom of trying to make all of these cost categories commensurate. This does not seem useful, unless one can say something about the distribution of the costs. This, of course, is central to one of the later sections of the paper on the design of instruments and agencies, since who bears the cost will have a major effect on agency design.

Second, with respect to his focus on direct versus indirect incentives, I think that a point has been omitted as to why direct regulations might be preferred—namely that direct regulations are conceivably more equitable to people, more evenhanded, than is the case with fee systems. In addition, I am not sure whether I agree with the point that indirect controls require much more information than direct controls do. To come up with direct controls that are efficient, requires a great deal of information. With respect to production functions or processes, you may not need as much information for a system of indirect controls, because many more matters are left to individual decision makers to decide for themselves. Moreover, standards are often set without this information being in hand, allowing for the possibility of amending the law, so I am not sure that the information requirements for an indirect control system are as great as assumed.

DISCUSSANT: *Fred H. Abel*

While I think this is a good paper, it seems to me that enforcement and administration costs are a significant part of any implementation strategy and that something should be added about the magnitude of the differences in these costs when you move from a standards approach to an effluent approach. In what direction do these magnitudes move? Do enforcement costs go up as you move from effluent charges to standards? As standards go up, do the information costs go up or down? I do not think the earlier comments got to these questions. Nor did they mention that no matter how indirect the links are, the overhead costs are large only if you insist you have to know everything about all the links—the cost may even be zero if you monitor the end result independent of all the links. We need to know the cost magnitudes, and whether they vary greatly between alternatives, or whether they do not. If they do not, they may be irrelevant, and if that is true, then it should be stated.

DISCUSSANT: *Allen V. Kneese*

In exploring the question of enforcement costs, it might be interesting to determine whether there are alternative designs of incentives, or perhaps other regulatory instruments that can be used to shift the burden of information from the private to the public sector, or vice versa. In particular, I was thinking of an example concerning gasoline taxes. One can imagine a situation in which there is a one-time test of emissions from automobiles. On the basis of that test a surcharge is placed on the gasoline bought by that automobile and a sticker is affixed to the windshield. The automobile owner is permitted to have his car tuned and tested in order to provide evidence that his emissions are lower, and have the sticker changed. It will revert to the old sticker in the course of six months unless he provides further verification. This type of regulatory instrument would tend to push most of the enforcement and informational costs into the private sector, which in some cases might be quite desirable.

Estimates of Industrial Waterborne Residuals Control Costs: A Review of Concepts, Methodology, and Empirical Results

STEVE H. HANKE and IVARS GUTMANIS

THIS PAPER FOCUSES on a series of studies that have attempted to estimate the costs of industrial water pollution abatement and that have been conducted or sponsored in the United States by the Environmental Protection Agency. These studies are critically examined and recommendations for their improvement are made. Social costs which include damages associated with various levels of residual discharge, however, are not reviewed. Water pollution abatement costs resulting from control of pollutants discharged by households, by agricultural activities, by trade and by service sectors, are not included in this review. Nor have we presented here an analysis of water pollution abatement cost estimates prepared for various limited geographic areas in the United States, such as specific river basins. Exclusion of such cost studies from this review has a limited, if any, affect on our findings, because most of the costing methodologies used in those studies are very similar, or even identical, to cost estimates reviewed here. This, in turn, indicates that the national water pollution control cost estimation procedures discussed here are readily applicable to smaller geographic areas such as states, counties, or river basins, as indeed is the case.

We begin by describing a conceptual framework that has been successfully used, largely by researchers associated with Resources for the Future, to estimate industrial water pollution abatement costs. Within this framework one can effectively deal with the major factors that must be considered when analyzing such costs. We discuss these factors and use them as criteria in evaluating the cost studies under review. In addition to our description and review of the basic methodologies and results of EPA cost studies, we discuss the implications that can be drawn from these studies with reference to public planning and decision making.

Estimation of Industrial Water Pollution Abatement Costs: Conceptual and Methodological Framework

Conceptual Framework

The basic concepts, the methods of analysis, and the empirical cost information usually used to determine the costs of industrial water pollution control programs limit comprehensive investigations of the real costs to society associated with water pollution abatement. They are, therefore, of little assistance in formulating public policies in which one is attempting to balance incremental social costs and benefits. As a result of these highly selective and restrictive cost estimation practices, most past and present control cost estimates represent only a subset of possible estimates. There is mounting evidence that the estimates prepared do not represent the least costly way in which industry can comply with the various standards governing the discharges of residuals, since end-of-pipe treatment methods are usually the only abatement alternatives considered.[1] Other methods available to industry for the reduction of residuals, such as changes in production processes, have frequently proved to be less costly in reducing the residuals associated with a given level of industrial production than end-of-pipe treatment. [2]

Most studies of abatement costs are also misleading for purposes of guiding public policy because they usually focus on reducing the discharges of one residual into one environmental medium. This approach

---

[1]It should be pointed out that the exclusive reference to end-of-pipe treatment as the only method for the reduction of residuals is a rather recent phenomenon. Sanitary engineers have recognized, even though they often failed to systematically examine, the various ways in which residuals could be reduced. For an excellent early treatment of alternatives to end-of-pipe treatment, see the following works by John C. Geyer: *Textile Waste Treatment and Recovery* (Washington: The Textile Foundation, 1936), and "The Effects of Industrial Wastes on Sewage Plant Operation," *Sewage Works Journal* 9, no. 4 (July 1937): 625-34. See, for example, Blair T. Bower, "Studies of Residuals Management in Industry," paper presented at the meeting of Universities-National Bureau of Economic Research and Resources for the Future, Chicago, Ill., November 10-11, 1972.

[2]Leslie Ayres and Ivars Gutmanis, "A Model for Strategic Allocation of Water Pollution Abatement Funds," International Research and Technology, Inc. (August 1971).

ignores the laws of conservation of energy and mass.[3] For example, studies that concentrate on end-of-pipe treatment fail to acknowledge that only the form of the residual or the medium to which it is discharged has been altered by treatment, and thus ignore the possible trade-offs that exist between types of residuals and various receiving media.

In an attempt to deal with alternative approaches to the reduction of waste discharges, researchers at Resources for the Future have developed and made operational a framework for the study of residuals management and industrial water pollution abatement costs.[4] The objectives of their efforts include the determination of:

- The factors which influence residuals generation in an industry and that industry's quantitative responses to variations in those factors.

- The range of options available in an industry to respond to increasingly stringent constraints on the discharge of residuals to the environment, i.e., constraints on the use of common property resources as inputs to the production process.

- Models of residuals generation for different industries for use in analyzing regional residuals management.

- The extent to which the biological, physical, technological, and economic interrelationships among the types and states of residuals require that all residuals be considered simultaneously in determining the optimal residuals management strategy for an industrial plant.

- The proportion of total production costs represented by net residuals management costs, taking all impacts upon other costs into consideration. These costs are then analyzed under increasingly stringent constraints on residuals discharges and in relation to different sets of other assumptions affecting costs (i.e., fuel, raw material, and other input variables; the technology of production; and product output specifications).

Since our review concentrates on this last point, the meaning of net residuals management cost must be more fully explained. Firms use

[3]For a theoretical discussion of this topic and its relationship to residuals management see: R.U. Ayres, and A. V. Kneese, "Production Consumption and Externalities," *American Economic Review* 59, no. 3 (June 1969): 282-97. For an applied treatment of the subject see: Regional Plan Association, *Waste Management,* a Report of the Second Regional Plan, (New York 1968).

[4]See Clifford S. Russell and Walter O. Spofford, Jr., "A Quantitative Framework for Residuals Management Decisions," in Allen V. Kneese and Blair T. Bower, eds., *Environmental Quality Analysis: Research Studies in the Social Sciences* (Baltimore: Johns Hopkins University Press, 1971); and George O. Löf and Allen V. Kneese, *The Economics of Water Utilization in the Beet Sugar Industry* (Baltimore: Johns Hopkins University Press, 1968).

inputs to produce both products and nonproduct materials. Nonproduct materials can either be discharged into environmental media or recovered. If firms maximize profits, they will engage in materials recovery until the marginal benefits of this activity are equal to their marginal costs. Nonproduct materials that are not recovered (those not produced in the range from zero to the profit maximization level of recovery) but are discharged into environmental media are defined as residuals. Therefore, the quantity of residuals generated is equal to the total nonproduct materials formed in production minus those that are economically recovered.

In calculating net residuals management costs, one should not include all of the costs associated with the modification of nonproduct materials but only the net costs associated with handling the residuals portion of the nonproduct materials. The *gross* residual costs of residual management are equal to the difference between (1) total production costs, including the costs of modifying residuals to the required level, and (2) total production costs at the level where the profit from nonproduct recovery is highest. To compute the *net* residuals management costs, any benefits from modifying nonproduct materials beyond the profit maximization level must be subtracted.

The conceptual framework for properly studying residuals management is presented in simple terms. Where there are no legal constraints placed on residuals discharges, residuals generation is expressed as follows:[5]

R_{git} = f (RM, PP, PO), where

R_{git} = quantity of residual, i, generated, g, per unit time, t;
RM = type of, and hence characteristics of, raw material inputs;
PP = technology of production process, including technology of materials and energy recovery and technology of by-product production; and
PO = product output specification.

Although there are other variables (e.g., operating rates, plant layout, and the cost of in-plant water recirculation) which affect residuals generation,[6] the above expression is generally representative of the major factors which, in the absence of legal constraints, determine residuals generation.

The residuals that are actually discharged into environmental media are a function of the above variables plus legal constraints on the choice of plant technology available for residuals modification. Hence:

---

[5]This expression was used by Blair T. Bower in "Studies of Residuals Management in Industry," op. cit.

[6]W. Wesley Eckenfelder, Jr., *Water Quality Engineering for Practicing Engineers* (New York: Barnes and Noble, 1970).

R_{dit} = f (RM, PP, PO, EC, TR), where

R_{dit} = quantity of residual, i, discharged, d, per unit time, t;
RM, PP, PO are the same as above;
EC = environmental controls imposed on discharge of liquid, gaseous, solid, and energy residuals (heat and noise), i.e., standards, charges, etc.; and
TR = technology of residuals modification.

In addition to the above factors, the relative prices of factor inputs (including environmental inputs) and various exogenous variables (such as technical change and governmental policies) influence residual discharges.

These factors are accounted for in Figure 1 which was constructed by Clifford S. Russell. This model illustrates the misleading nature of cost studies based upon fixed-cost coefficients that relate costs to the level of residual reduction by end-of-pipe treatment in municipal waste treatment plants.

Waterborne industrial residuals often possess characteristics which differ from domestic wastes and therefore may require abatement practices considerably different from those used by municipal treatment plants. In some cases (as with "red mud" resulting from alumina reduction), the required abatement is difficult to achieve by standard waste treatment procedures; instead, highly specialized (and costly) pollution control methods must be used. The chemical and physical properties of industrial wastes not only differ from domestic wastes, but also differ markedly from one industry to another and even between manufacturing plants in the same industry (these differences may be caused by the use of different production technology subprocesses and different raw materials).

Furthermore, the concentration and volume of residuals generated by industrial processes may show considerable variation. Peak concentrations of some of the wastes may require special handling (such as equalization basins) which can affect industrial waste treatment costs.[7]

Industrial waste is often treated according to the residual characteristics of separate manufacturing processes. In some industries, segregation, rather than collection, of industrial waste streams may be the best way to minimize net residuals management costs. Under such conditions, each residual stream receives only that treatment which is appropriate for its volume and constituents, allowing uncontaminated wastewater streams to be discharged directly or recycled. Because residual

---

[7]For data on the temporal variability of residuals discharges in the canned and preserved fruits and vegetables industry see National Canners Association, "Comments on the Draft Development Document for Effluent Limitations Guidelines and Standards of Performance: Canned and Preserved Fruits and Vegetables Industry," mimeographed.

Figure 1

A Proposed Model of Industrial Residuals Generation and Discharge

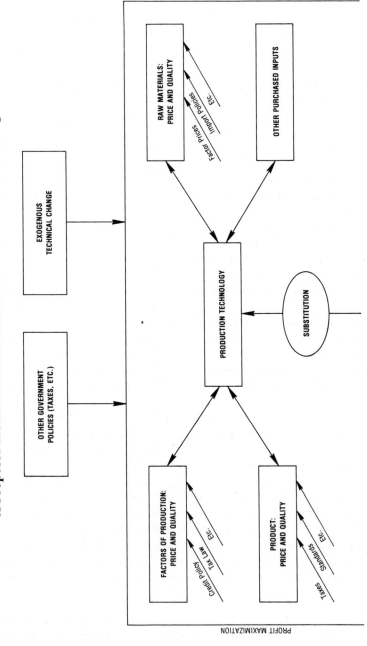

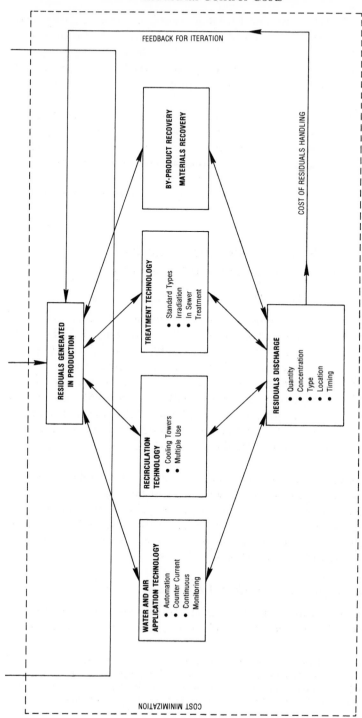

Source: C.S. Russell, "Models for Investigation of Industrial Response to Residuals Management Actions," *Swedish Journal of Economics* 73, 1: p. 137.

streams may be segregated and treated according to waste characteristics, some processes become integral parts of the manufacturing operation rather than waste treatment per se.[8] This not only argues against the use of aggregate data on end-of-pipe treatment, but also points out the difficulty of separating and determining net residuals management costs.

Factors which may influence the costs of industrial waterborne residual control and abatement costs are presented in Table 1. The factors are listed in the sequence in which they are related to the decisions which must be made to implement the industrial residual control processes. Since each decision is associated with specific levels of expenditures, cost estimating procedures should be closely related to each of the specific factors listed.

Factors Affecting Industrial Waterborne Residual Control Costs

Residual Control Standards

Among the factors which determine industrial water pollution abatement costs, the level of waste control (i.e., the limit placed on discharge of residuals) is perhaps most important,[9] since the cost of reducing discharges generally rises rapidly as the proportion of residuals removed is increased. This applies to all wastes and is illustrated in Table 2 in terms of biochemical oxygen demand (BOD) removal for four industries. These cost data indicate the rapid rise in the marginal costs of residuals treatment as the proportion of waste removed is increased. Also refer to Figure 2 for a graphical representation of this phenomenon.

To develop a rational industrial wastewater control policy, the costs associated with various levels of control must be calculated and compared with the benefits of control. At present, this point is often overlooked. Cost studies usually do not analyze the costs associated with alternative levels of control. Moreover, the control costs are rarely, if ever, compared with the benefits of control.

Production, Input, and Output Alternatives

The costs of controlling industrial residual discharges may be altered at the tail-pipe end of a production process. However, effective

---

[8]Herbert F. Lund, *Industrial Pollution Control Handbook* (New York: McGraw-Hill); and McGauhey, *Engineering Management of Water Quality* (New York: McGraw-Hill, 1968).

[9]For problems related to control standards see: William B. Halladay "Industrial Waste Treatment—Fact and Fiction," *Journal of Water Pollution Control Federation* 42, no. 11 (November 1970).

Table 1

FACTORS AFFECTING INDUSTRIAL WATERBORNE RESIDUAL
CONTROL COSTS

FACTOR	NOTES
1.0 *Quantity of Residual Discharged Per Unit of Time* (Control Standards)	As a rule prescribed by federal or local authorities. Are principal determinants of waste reduction costs.
2.0 *Production Processes and/or Inputs and Outputs* 2.1 Alternative Technologies 2.2 Raw Materials 2.3 End Products 2.4 Materials Recovery 2.5 By-Product Production	Usually difficult to change after production activities have begun.
3.0 *Institutional Arrangements for End-of-Pipe Treatment* 3.1 Industrial Treatment Plants 3.2 Municipal Treatment Plants 3.3 Regional Treatment Plants	Depend on the capability of the governmental unit to treat industrial wastes which may be affected by: 1) Chemical composition 2) Volume of flow 3) Local regulations Cost savings possible because of economies of scale in larger treatment plants.
4.0 *Other Control Alternatives** 4.1 Dispersion 4.2 Dilution 4.3 Detention 4.4 Diversion 4.5 Environmental Treatment	These depend on the industry, the chemical and physical characteristics of wastes and the local climatological, geological, and related conditions, all of which may have considerable impact on control costs.
5.0 *Intermedia Transfer*	
6.0 *Engineering and Other Economic Variables* 6.1 Volume and Composition of Waste Flow 6.2 Choice of Specific Processes Used in Waste Control 6.3 Availability of Land 6.4 Fuel Costs 6.5 Water Costs 6.6 Energy Costs 6.7 Level of Technology Employed in Production (age of plant and equipment)	The first two of these factors are difficult to modify, although both have considerable impact on control costs.

*Given the 1972 Clean Water Act Amendments, not all of these may be legal abatement procedures.

Table 2

INCREMENTAL COSTS OF BOD REMOVAL IN FOUR SELECTED
PLANTS

Type of Plant	Cost (in cents) of Removal of One Pound of BOD Waste One Additional Percentage Point Above		
	30 Percent BOD Waste Removal	90 Percent BOD Waste Removal	95 Percent BOD Waste Removal
Malt Production[a]	3	30	62
Meat Processing[b]	6	60	90
Ice Cream and Frozen Dessert Manufacture[c]	4	30	50
Shoe Leather Tannery[d]	12	45	75

Notes: (a) Annual Capacity 3,000,000 bushels of barley steeped.
   (b) Annual Capacity 210,000,000 pounds, live weight.
   (c) Annual Capacity 800,000 gallons of milk processed.
   (d) Annual Capacity of about 270,000 hides.
Source: Ivars Gutmanis, *The Generation and Costs of Air, Water, and Solid Waste Pollution: 1970-2000.* (Background analysis for The Brookings Institution, May 1972.)

cost reduction can also be achieved by selecting and/or modifying the production processes per se, by changing the raw materials used, and by changing the end products that are produced. Further alterations in control costs may be achieved by recycling or by recovery and reuse of certain residuals generated in the production process.

Since some of these changes may significantly affect the cost of industrial pollution abatement, estimates of the affects of these changes must be made. The impact on treatment costs of the following alternatives should therefore be analyzed:[10]

1. alternatives in technology (kraft vis-à-vis sulfite process in pulp production)
2. alternatives in raw materials and fuels (different sulfur content in coal)
3. alternatives in end products produced (different finishing processes in steel products)
4. materials recovery (recovery of mercury used in production of caustic soda)
5. by-product recovery (recovery of blood and related by-products in slaughterhouses)

[10]Robert U. Ayres and Ivars Gutmanis, "Methodology: Technological Change, Pollution and Treatment Cost Coefficients," in Ronald G. Ridker, ed., *Resources and Environmental Consequences of Population Growth in the United States* (Washington, D.C.: GPO, 1972).

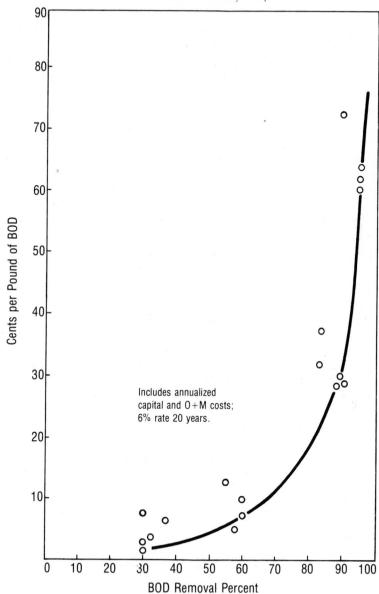

Figure 2

Incremental Cost of Removing BOD:
Malt Plant Annual Capacity - 3,000,000
Bushels of Barley Steeped

Includes annualized
capital and O+M costs;
6% rate 20 years.

Cents per Pound of BOD

BOD Removal Percent

Source: Ivars Gutmanis, *The Generation and Costs of Air, Water, and Solid Waste Pollution: 1970-2000*. (Background analysis for The Brookings Institution, May 1972)

242

Changes in Technology

Technological changes in production processes can greatly influence the generation of residuals.[11]

Faced with increasingly stringent residual control standards, some industries cannot use end-of-pipe abatement because of technological or cost factors. In these situations, many industries have developed production technologies which produce fewer residuals. For example, the older mercury-cell production process for caustic soda results in significant mercury wastes; the newer diaphragm technology does away with mercury residuals entirely.[12]

A more complex example of the impact of alternative technologies can be seen in steel processing. Steel can be produced either by the prevailing ingot casting technology or by continuous casting. The use of continuous casting technology has increased rapidly since its introduction in the early 1960s and by 1985 is expected to surpass the ingot casting process. When estimating pollution control costs for the steel industry, it is important to realize that continuous casting generates less waste lubricating oils and produces superior metal surfaces which require less scraping, sand blasting, or treating with acids and other corrosive chemicals. Consequently, continuous casting results in fewer acid wastes and surface scale particles, and less process wastewater, all of which reduce the pollution abatement costs associated with given levels of steel production.[13]

The impact of alternate technologies on control costs can also be seen in the pulp and paper manufacturing industry. In wood pulp production, the sulfate or kraft process has increasingly dominated the sulfite process and other methods. The kraft process brings about a major reduction of waterborne dissolved solids (DS) (see Table 3), which are difficult and costly to treat. Moreover, it results in large reductions in the generation of solid wastes, sulfur dioxide, and the wastewater load. However, it increases particulates and creates odorous sulfides and chlorine gas.[14]

[11]See for example: "Dry Paper Process Has Economic Benefits," *Chemical Week* (August 23, 1971), p. 37; "Hydrometallurgy: Copper's Solution for Pollution?" *Chemical Week* (May 17, 1972), p. 27; see also R. U. Ayres, op. cit.

[12]Ivars Gutmanis, *Control of Mercury Pollution: Some Economic Considerations*, Washington, D.C.: prepared by International Research and Technology Corporation for the Environmental Directorate of the Organization for Economic Cooperation and Development (1971).

[13]Federal Water Pollution Control Administration, *The Cost of Clean Water*, vol. 3, "Industrial Waste Profile No. 1: Blast Furnaces and Steel Mills" (Washington, D.C.: GPO, 1967).

[14]Ivars Gutmanis and Adele Shapanka, *Economic Costs Associated with the Environmental Quality Alternatives in the United States, 1970, 1980 and 1986; An Input-Output Analysis*, a research report prepared for the Interagency Economic Growth Project, Council of Economic Advisors, Office of Management and Budget, and the U.S. Department of Commerce and Labor (1972).

Table 3

POLLUTANTS FROM THE KRAFT AND MAGNEFITE (SULFITE)
PROCESSES IN TISSUE PAPER PRODUCTION
(in pounds per ton of tissues)

Residuals	Kraft Technology	Sulfite Technology
Airborne		
Chlorine	1.8	1.1
Sulfur dioxide	5.6	26.0
Sulfides and hydrogen	26.0	0
Particulates	62.0	28.0
Waterborne		
Dissolved solids	507.0	931.0
Suspended solids	118.5	113.1
Solid Wastes		
Organic and inorganic solids	70.0	163.0

Note: Brightness scale of tissue is 80-81.
Source: B.T. Bower, G.O.G. Löf, and W.M. Hearson, "Residuals Management in the Pulp and Paper Industry," *Natural Resources Journal* 11, no. 4.

From this we can easily see the difficulties of designing national pollution control policies. From a wastewater management point of view, the kraft technology is preferable to the sulfite process, since waterborne residuals and wastewater control costs are significantly lower. However, with respect to air pollution management, the sulfite process is preferable. To develop policies for residuals control in the tissue paper industry, one must take a broad view of residuals management. One must calculate the benefits and costs of various types of control for various types of residuals and various environmental media.

The production of paper provides another example of the effect that different production technologies can have on the generation of residuals and on residual control costs. Currently either the dominant wet-forming process or the dry-forming process can be used in production. The dry-forming process uses wood pulp but does not require water as a carrier, whereas large quantities of water are required for the wet-forming process. Therefore, dry-forming can eliminate or considerably reduce the high costs of treating the large volume of process wastewater used in the wet-forming process.[15]

As we have indicated, comprehensive estimates of industrial waterborne residual control costs must take into account the effects which may

[15]Private communication from Blair T. Bower, Resources for the Future (1973).

be realized by the use of available alternative production technologies. But these estimates must also recognize the potential for changes in the pertinent technologies over time in order to determine future abatement costs consistent with production technologies which will then be possible.

As with the assessment of entire production technologies, it is also possible to identify residual generating subprocesses, and to evaluate alternatives for that particular activity which will generate fewer residuals or result in reduced pollution abatement costs. Because there is a relatively wide latitude for selectivity, subprocess changes can be directed specifically toward easing specific pollution loads and minimizing specific waste control expenditures.

Several examples illustrate the possibilities for such reduction. In the canned and frozen foods industry, a shift from water-conveying to dry-conveying suction systems reduces residual water flows as well as BOD and DS; similar results, with considerably reduced residual water loads, can be obtained from dry caustic or cryogenic peeling, and by blanching with hot air instead of steam or hot water.[16] However, these new subprocesses do tend to be more energy intensive. A careful evaluation of trade-offs and their associated benefits and costs must be conducted if rational policies are to be made.

In the production of plastic materials and polyvinyl chloride (PVC) resins, switching to the bulk method of producing PVC resin can considerably reduce wastewater—BOD, suspended solids (SS), and DS.[17] Since the use of each of these (or similar) production subprocesses will result in a reduction in water pollution control costs, estimation of pollution control costs must include such alternatives for production subprocesses, and, as in the case of changing technologies, projected future abatement costs must also reflect the future use of such alternatives.

Changes in the Materials Used

The type and especially the quality of material inputs used in production often have a pronounced affect on the quantity of residuals generated and therefore residual treatment costs. In fruit-processing operations, the increasing conformity of fruits (inputs) in size and ripeness has substantially decreased wastage. The improved size uniformity of harvested pineapples, for example, has reduced fruit wastage by approximately 40 percent over the last 30 years.[18] Examples of changes in the materials used which adversely affect the environment in general

---

[16]"New Tomato Peeling Process," *American Tomato Yearbook* (1970), p. 43.

[17]Robert U. Ayres, et al., *A Materials-Process-Products Model for Chlorinated Plastics Industry* (second Quarterly Report), prepared by International Research and Technology Corporation for National Science Foundation (Washington, D.C.: December 1971).

[18]Personal communication from Blair T. Bower, Resources for the Future (1973).

and water pollution control costs in particular can be found in many studies of mineral supply and availability.[19] These reports project a considerable decline in the quality of ores to be used, which will increase residuals generated per unit of ore processed as well as the costs of pollution abatement. Finally, the projected increase in the use of coal, in place of oil and gas, provides another example of increasing residuals generation and abatement costs. The effect of such changes in the use of raw materials on pollution abatement costs must be considered in efforts to estimate the abatement costs for any industrial sector.

Changes in End Products Produced

The characteristics of end products also have a substantial effect on residuals generation and on the costs of their control. In food manufacturing, for example, there has been a trend toward an increasing number and variety of end products. Two decades ago, the only peach product that was canned was halved peaches. Today peach products also include peach pie pieces, peach concentrate, peach nectar, and peach irregulars. As a result, the yield has risen from 40 cases per ton of peaches to over 55 cases in the last 20 years.[20] The result has been substantial decreases in BOD and SS per ton of raw material processed, with a corresponding decrease in control costs for this part of the industry.[21]

Similar increases in the variety and mix of end products have resulted in higher yields and declining residuals per unit of input processed for most other food products. Except for small volumes of manure and waterborne residuals, large modern slaughterhouses discharge very few residuals. However, large quantities of residuals are generated in older and smaller slaughterhouses, where end products consist only of traditional meat cuts. In the older plants, BOD generation is about 70 percent higher, dissolved solids about 60 percent higher, and solid wastes about 40 percent higher.[22]

The opposite trend, where changes in the end products result in increased residual loads and control costs, can be illustrated by the trend toward paper products with higher brightness and pastel colors. Due to the greater degree of pulp bleaching required for high brightness paper, larger quantities of residuals are generated. For pastel-colored tissues, the generation of dissolved solids is double what it would be if the tissue had its usual white color. By comparison, eco-tissue in a brown color, the color of the kraft process pulp without bleaching, generates

[19]U.S. Bureau of Mines, *Mineral Facts and Problems* (Washington, D.C.: GPO, 1970).

[20]Personal communication from Blair T. Bower, Resources for the Future (1973).

[21]National Canners Association, Western Research Laboratory, *Liquid Wastes from Canning and Freezing Fruits and Vegetables,* 1971.

[22]Federal Water Pollution Control Agency, *The Cost of Clean Water,* vol. 3, "Industrial Waste Profile no. 8: Meat Products" (Washington, D.C.: GPO, 1967.)

Table 4

NET RESIDUALS MANAGEMENT COSTS PER TON OF OUTPUT,
INTEGRATED KRAFT MILL PRODUCING 500 TONS PER DAY OF
TISSUE PAPER

Bleached Tissue Paper P.C. Ti 4:

Type of Treatment[a]	LEVEL OF DISCHARGE STANDARDS		
	I[b]	II[c]	III[d]
Gaseous residuals modification, $/ton	0.16	0.59	1.66
Liquid residuals modification, $/ton	3.07	4.09	6.76
Solid residuals disposal, $/ton	0.38	0.38	0.38
Total, $/ton	3.61	5.06	8.80

Unbleached Tissue Paper P.C. Ti 31:

Type of Treatment[a]	LEVEL OF DISCHARGE STANDARDS		
	I[b]	II[c]	III[d]
Gaseous residuals modification, $/ton	0.12	0.46	0.83
Liquid residuals modification, $/ton	0.10	0.40	0.72
Solid residuals disposal, $/ton	0.33	0.33	0.33
Total, $/ton	0.55	1.19	1.88

(a) Costs are in 1970 dollars and are based on estimates of operating labor, maintenance labor and supplies, power and material requirements, 12.5 percent annual charge on estimated capital investment, and are credited with chemical recoveries at typical market prices. Operation 350 days per year was assumed.

The costs of any secondary solid residuals generated in liquid and gaseous residuals modification (i.e., sludge) are included in the liquid and gaseous residuals modification costs.

(b) LEVEL I indicates a discharge of 30 lbs. of particulate per ton of tissue paper.

(c) LEVEL II indicates a discharge of 8 lbs. of particulate per ton of tissue paper.

(d) LEVEL III indicates a discharge of 4 lbs. of particulate per ton of tissue paper.

Source: Blair T. Bower, "Studies of Residual Management in Industry," a paper presented at the Conference on Economics of the Environment, sponsored by the Universities-National Bureau Committee for Economic Research and by Resources for the Future, November 10-11, 1972.

---

only 10 percent as much dissolved solids as the colored tissues. Eco-tissues may also be produced entirely from wastepaper. The differences in the net residuals cost associated with bleached and unbleached tissue paper have been calculated by Blair T. Bower and are displayed in Table 4. The costs of residuals discharge control are roughly five times as high for the bleached tissue paper. Bleaching is now used in over 50 percent of the wood pulp industry and, assuming continuation of present trends, will rise to 60 to 70 percent in 1986, creating higher pollutant loads and resulting in much higher abatement costs.[23]

Materials Recovery and By-Product Recovery

The recycling of residuals depends on the interaction between prices in raw material markets, costs of recovery, new technology, and environmental regulations. It is reasonable to expect that recycling will increase as cumulative use continues to press on virgin sources of raw material inputs and as new technology evolves which allows for the reduction of costs via recycling processes.[24]

Data for the glass industry, for example, suggests that if the use of purchased cullet (glass scrap) as raw material could be increased from the current rate of 4.5 percent to 20 or 30 percent, the cost of water pollution abatement in this sector would be reduced by about one-fifth.[25] Similar waste treatment cost reductions, resulting from increased use of salvaged materials are also reported in textile manufacture, in paper production, and in the primary nonferrous metals sectors.[26]

In addition to the use of the materials recovery methods to reduce pollution abatement costs, these expenditures can be reduced further by applying by-product recovery processes. For example, in the case of slaughterhouse wastes, recovery and utilization of blood, hooves, horns, and other non-meat products considerably diminishes the waterborne waste loads. Similarly, recovery of wool grease in woolen mills results in such economic by-products as lanolin, as well as in reduced waste loads and in lower costs of pollution abatement.

Institutional Arrangements for End-of-Pipe Treatment

The costs of industrial waterborne waste pollution control may be reduced if the industrial wastes are treated by municipal or regional treatment plants. Such a joint treatment approach has been in effect for some time and is expected to increase.

The treatment of industrial wastes along with municipal wastewater offers economies of scale. Further, it is claimed that combining an area's industrial and municipal wastes into common collection and treatment centers improves the probability of more highly qualified professional operation and management.[27]

---

[23]Blair T. Bower, "Studies of Residual Management in Industry," op. cit.

[24]"Another process has been developed for recovering hydrochloric acid and iron oxide from steel mill pickling liquor," *Chemical and Engineering News* (May 18, 1970), p. 32.

[25]Personal communication from Arsen Darney, EPA (1972).

[26]Ibid.

[27]Personal communication from John C. Geyer, Johns Hopkins University (1973).

However, there are some problems with this mutually advantageous relationship. Current fees charged by municipal treatment plants to handle industrial wastes appear to be inadequate to cover actual treatment costs. While comprehensive data on municipal charges are not available, a survey of information on fees charged indicates they cover less than one-half of the actual treatment cost.[28] These discrepancies are the result of past laws and regulations which allowed about 50 percent of municipal treatment plant costs to be absorbed by the federal government as subsidy payments.

Developing an appropriate financing formula for industrial waste treatment by municipalities, however, leads to further complexities, especially when considered and planned within a context of regional water management. Furthermore, there are limitations on the types and amounts of industrial residuals that can be accepted in municipal waste treatment facilities. Present facilities may lack the size and ability to handle large volumes of complex industrial wastes. Finally, in some cases it has been more economical for large industrial waste treatment plants to accept and treat municipal wastes. Such an approach may be efficient in localities where municipal water use is small in relationship to the treatment needs of a large industrial plant nearby.

Other Control Alternatives

Emphasis on pollution abatement cost estimation is centered on the end-of-pipe treatment of residuals. Only limited attention is given to other abatement alternatives. As mentioned earlier, these alternatives include: changes in production technologies, production subprocesses, raw materials, and end products, as well as the recovery of residuals and the production of by-products. However, once the level of residuals is determined, several other alternatives for the control of residual discharges might be applicable. They must be analyzed to determine which provides the best cost-benefit solution. Among those alternatives are:

- *Dispersion*—the distribution of waterborne residuals discharge over a large area or into a larger volume of water in nature
- *Dilution*—the *artificial augmentation* of the volume of the receptor used to assimilate waterborne residuals
- *Detention*—the temporary hold-up or storage of waterborne pollutants for later, more gradual release, or release at a more advantageous time
- *Diversion*—the transportation of waterborne residuals to another location for treatment and/or discharge
- *Environmental treatment*—the treatment of the surface waters to remove residuals or diminish their harmful effect
- *Desensitization*—rendering the potential waterborne residuals harmless through defensive efforts of those affected by the pollution

[28]Ivars Gutmanis, *The Generation and Costs of Air, Water, and Solid Waste Pollution: 1970-2000.* (Background analysis for The Brookings Institution, 1972).

These alternatives are not equally effective in the control of all residuals nor are all permitted under current legislation. For example, control of toxic, persistent chemicals such as mercury is ineffective using the dispersion, detention, or diversion strategies. Similarly, control of phosphate emissions is accomplished in a cost-effective manner by product modification, not by the abatement strategies listed above.[29]

Unique costs are associated with each pollution abatement alternative. To determine the cost-effective way to achieve desired levels of residual reduction, it is necessary to compare the costs of all pertinent control alternatives for any given source of pollution.

Intermedia Transfer

Intermedia transfer of residuals generated from industrial activities is becoming an increasingly important aspect of residuals control. A number of variations are possible (see Figure 3), each of which is related to unique costs which need to be investigated. For example, particulates from metal surface treatment operations in steel mills may be controlled by using either the conventional bag-house method or a cylone-washer. If the former is used, the waste medium is air; in the latter case the particulates are transferred from air to a water medium. Similarly, residuals present in sludge from any secondary treatment process may be abated by the use of landfill, or they may be further controlled by using incineration which results in a transfer from a land medium to that of air. The pollutants from the incineration operations may be subject to further intermedia transfer. For example, if these pollutants are abated by use of a Venturi scrubber, an intermedia transfer from air to water takes place.[30]

Thus, since the waste medium is a significant determinant of the waste-disposal costs, intermedia transfers need to be considered.

Engineering and Other Economic Variables

Among the engineering variables which affect pollution abatement costs is the scale or size of treatment plants, measured in terms of residual flow or the flow of the residual carrier. For the waterborne residuals, the principal abatement cost determinants (assuming a pre-

---

[29]*Phosphates in Detergents and the Eutrophication of America's Waters,* Twenty-Third Report by the Committee on Government Operations, April 14, 1970, and, *Lake Erie Report: A Plan for Water Pollution Control,* U.S. Dept. of Interior, Federal Water Pollution Control Administration, Great Lakes Region (August 1968), p. 56. J. R., Vallentyne, "Phosphorus and the Control of Eutrophication," *Canadian Research and Development* (May/June 1970), pp. 36-49.

[30]For a more detailed statement on intermedia transfer, albeit one limited to a single industry sector, see: *The Economics of Clean Water,* vol. 3, "Inorganic Chemicals Industry Profile," FWPCA (March 1970).

Figure 3

Potential Variations in Intermedia Transfer of Waterborne Residuals

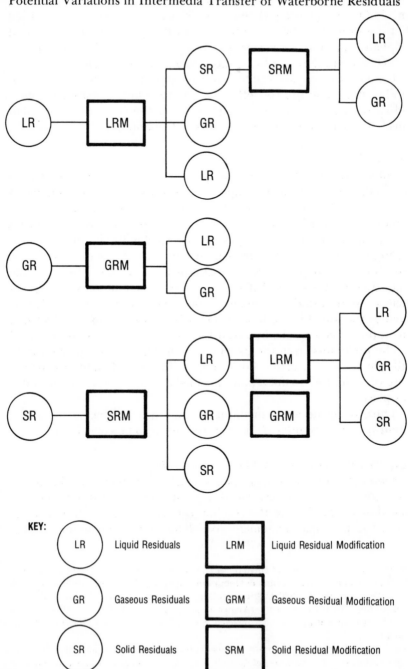

Figure 4

Capital Cost, Operating and Maintenance Cost, and Debt Service
vs. Design Capacity for Primary Treatment Plants

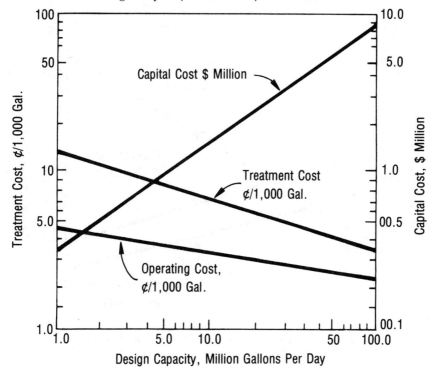

scribed level of waste removal efficiency) are the volume of wastewater
flow or hydraulic load and the proportion of residuals removed.[31] While
total costs increase for treatment plants designed to handle larger hydraulic loadings, the per unit costs of wastes treated become lower. Thus,
economics of scale characterize the investment and operation costs of
waste treatment plants. Figures 4 and 5, which relate the capital and
operating costs to hydraulic loadings, illustrate these economies, which
should be considered in studies of residual control costs.

Another determinant of treatment costs is the choice of a particular
waste treatment subprocess.[32] While the specific subprocesses used for
waste treatment and their sequence are determined, to a considerable
degree, by the type, kind, and concentration of the residual (as in the

[31]Robert Smith, "Cost of Conventional and Advanced Treatment of Wastewater,"
Journal of Water Pollution Control Federation 40, no. 9, (September 1968).

[32]For presentation of alternative waste treatment subprocesses see *The Economics of
Clean Water,* vol. 3, "Inorganic Chemicals Industry Profile," FWPCA (March 1970).

Figure 5

Capital Cost and Treatment Cost vs. Design
Capacity for the Granular Carbon Absorption Process

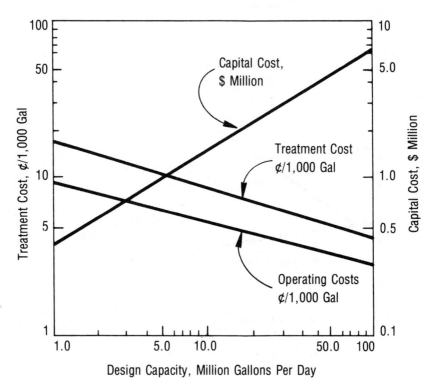

Design Capacity, Million Gallons Per Day

case where a pretreatment operation is required), a considerable number of alternate subprocesses and treatment sequences are available to accomplish the same waste removal efficiency, as shown in Figure 6. The analysis of these subprocesses is required, if proper abatement costs are to be determined.

The third factor in determination of treatment costs pertains to the age and technology of the production process which emits the residuals.[33] As a general rule, residual control in older industrial plants or in facilities which employ older types of production technology is much more expensive than in more modern establishments.

[33]See, for example, Federal Water Pollution Control Administration. "Industrial Waste Profile no. 4, Textile Mill Products," *The Cost of Clean Water* (Washington: GPO, 1967).

Figure 6

Wastewater Treatment Sequence and Processes Substitution Diagram

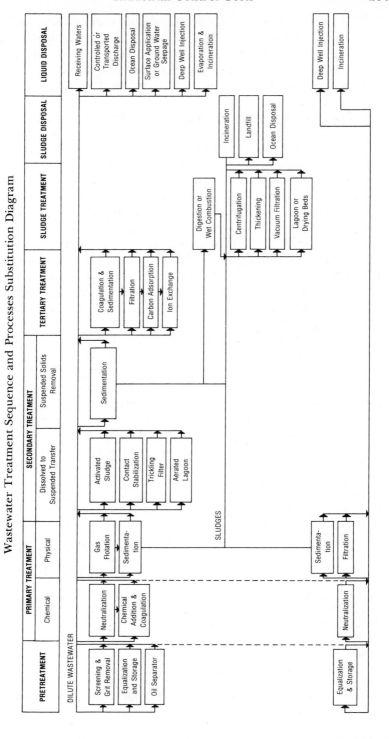

The fourth factor is related to changing the product mix.[34] Large changes in the end product mix and/or increasing the number of end products may—although this is not always the case—increase the wastewater volume and the treatment costs. For example, more complex petroleum refineries require considerably larger wastewater flow, as indicated in Figure 7.

The fifth factor determining control costs is the timing of the control equipment installation. While comprehensive data on this issue are not available, partial information from a number of case studies indicates that for a manufacturing establishment with a certain magnitude of output and residual removal requirements, the capital costs of an identical treatment facility, with the same output and the same removal requirements, may vary by a factor of seven, depending on whether the waste treatment facility is installed at the time of the plant's construction or added later.[35] The cost of adding a treatment facility to an existing establishment depends on such factors as the capital requirements for the facility, space availability within the production plant, and the cost and availability of adjoining land. Other economic factors which play an obvious and important role in control cost calculations are the costs of fuels, energy, and water.

In this section, we have presented a methodology which should be used when systematically evaluating industrial waterborne residual control costs. The major factors that determine the level of residual control costs have also been discussed. The proper determination of the cost-effective way to control residuals, even though a complex undertaking, should account for these factors. If costs are not properly calculated, rational pollution control policies cannot be determined, because the real costs cannot be appropriately balanced against the benefits of control.

In the next section of this paper, we survey the major studies that have been conducted or sponsored by EPA. Our critique of these studies will be based upon the criteria presented in this section.

The Costs of Industrial Water Pollution Control

Increasingly comprehensive environmental legislation has evolved in the past twenty years. During this period, the Federal Water Pollution Control Act of 1956 and the subsequent enforcement provisions for

---

[34]See W.L. Nelson, "Clean Water Needs of Refineries," *Oil and Gas Journal* 61, no. 3 (1963).

[35]Ivars Gutmanis, "Costs of Waste Control Facilities in Ohio," Internal Memorandum, P.H.S.D.A.P. (1963).

Figure 7

Relationship Between Refinery Complexity
and Clean-Water Needs

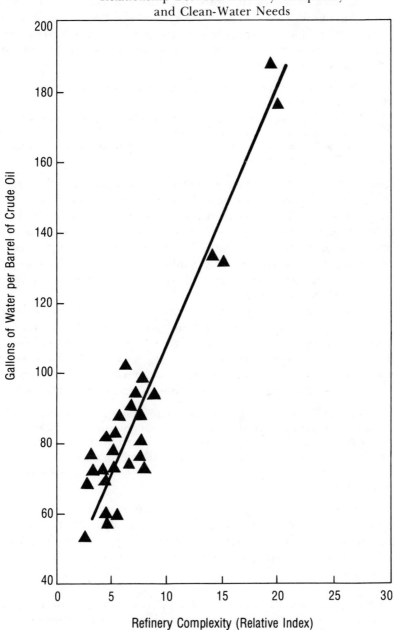

Source: W.L. Nelson, "Clean Water Needs of Refineries," *Oil and Gas Journal* 61,
no. 3, Table 1 (1963): 80.

water quality standards have established the legal basis for industrial water pollution abatement expenditures.

Public recognition of the costs which may be required to control industrial pollution was explicitly manifested in the 1965 amendments to the act of 1956. The 1965 amendments authorized each state to establish water quality standards and in Section 16(a) directed the Secretary of the Interior to conduct three studies:

  • Study the cost of carrying out the Federal Water Pollution Control Act, as amended
  • Study the economic impact on affected units of government of the cost of installing waste treatment facilities
  • Study the national needs for and the cost of treating municipal, industrial, and other effluents

These studies were to cover the five-year period beginning July 1, 1968, and were to be updated each year thereafter. Most of the federal government water pollution abatement cost estimates are the result of this legislation. Early studies reported substantial costs for water pollution control and generated considerable discussion and controversy. This, in turn, brought about other water pollution control cost studies, undertaken by affected industries and by private organizations interested in the quality of the environment.

The purpose of this section is to criticize the methodology of some of these past cost estimates undertaken both by government and nongovernment groups.[36] This is done in light of the conceptual and analytical framework presented in the preceeding section. The discussion of these studies is presented in three parts, each representing a methodology used to obtain industrial water pollution control expenditures:

  • Cost estimates derived from unit cost (operating, maintenance, and replacement) for a typical industrial water pollution treatment plant
  • Cost estimates derived from estimated expenditures for control of residuals per unit of wastewater flow (i.e., hydraulic loading)
  • Costs derived from a survey of past and proposed expenditures for residual treatment

The cost studies reviewed are subjected to two types of critical appraisal. The first is concerned with the specific analytical techniques and data sources employed. The second evaluates the general methodologies used in terms of the conceptual cost framework presented in the previous section.

---

[36]The basic characteristics of each of the studies reviewed are catalogued in a separate Appendix which may be obtained directly from the authors. The characteristics that best describe the studies under review include: the industry studied, time period, abatement alternatives considered, levels of treatment specified, economies of scale, general cost estimating methodology, and cost components specified.

Unit Costs for a Typical Industrial
Water Pollution Treatment Plant

One of the best examples of this abatement cost estimation methodology is presented in EPA's initial attempt to estimate national costs of industrial residual control.[37]

The costs were estimated, assuming secondary treatment or *equivalent* residual removal levels equal to 85 percent reduction of BOD and settlable solids, for the 1968 to 1973 period, for eleven sectors of industry. The estimates included the following three components:

- Costs of meeting current (1968 to 1969) needs for industrial residual abatement
- Costs for meeting industrial residual control needs for the 1968 to 1973 period resulting from industrial growth during the period
- Costs resulting from replacement of industrial residual control equipment due to depreciation (1968 to 1973)

The methodology employed (referred to in the EPA reports as "census projections") consisted of the following seven tasks:

- Determine the total number of establishments in each of the eleven industrial sectors in 1968 and project these to 1973
- Determine for each sector the number of industrial establishments which have industrial residual treatment facilities in 1968
- Subtract the number of establishments with residual treatment facilities from the total number of establishments to determine the number of establishments which require residual treatment facilities in 1968 and in each future year
- Determine capital expenditures for a typical industrial water pollution treatment plant
- Multiply the capital expenditures of a typical industrial water pollution treatment facility by the number of establishments during the period 1968 to 1973 without such facilities to determine total annual national capital costs for residual control in this period
- Determine the annual operation and maintenance costs for the 1968 to 1973 period, assuming these to be equal to a certain percentage of capital replacement values
- Determine replacement costs of residual control equipment (1968 to 1973) due to depreciation and based upon a fixed percent of invested capital

The data required for the first two tasks were obtained from the *1963 Census of Manufacturers*.[38] Capital costs for a typical industrial

---

[37]See *The Cost of Clean Water*, vol. 1, *Summary Report*, and vol. 2, *Detailed Analysis*, prepared by the Federal Water Pollution Control Administration (Washington: GPO, January 10, 1968).

[38]U.S. Bureau of the Census, *1963 Census of Manufacturers*, vols. 1 and 2, (Washington: GPO, 1964).

water pollution treatment plant were estimated by EPA staff, as was the percentage of capital replacement value which represents the annual operating and maintenance costs. This cost estimation technique contains a number of significant shortcomings.

The first of these is that capital cost estimates are based upon 85 percent reduction of BOD and SS without any reference to other residuals. The second is the explicit assumption that in the "primary metals," "inorganic chemistry" and "machinery" sectors the primary treatment is equal to the secondary treatment.[39] The third major source of error is the assumption that it is possible to estimate capital costs for a typical average industrial waterborne residual abatement facility. As described in the previous section, capital costs for abatement facilities differ markedly from one establishment to another because they depend on the volume of hydraulic loading (economics of scale), use of particular residual control processes, the mix of end products, the availability of land for treatment plants, and other variables.

A fourth cause of error concerns the operation and maintenance cost components. As was pointed out in the previous section, the calculation of operating and maintenance costs cannot be based upon a single percentage of capital costs since the latter do not take into account economies of scale, whereas operating and maintenance costs do reflect such economies.

A fifth source of error is that all estimates are based on data from plants using water at the rate of 20 billion gallons or more per year. In many manufacturing sectors, the inclusion of only those plants with annual water withdrawals of 20 billion gallons or more may indeed cover 97 percent of all water withdrawals as reported by the Bureau of the Census but this cannot be assumed for all sectors. In the "food and kindred products" sector, for example, because of the prevalence of many very small manufacturing enterprises, estimates made on the basis of plants with annual water withdrawals of 20 billion or more gallons may significantly underestimate the magnitudes of residuals.

Given the large number of cost variables not taken into account by this methodology it is doubtful if a truly representative typical industrial residual treatment plant can be specified which would yield average operating, maintenance, and capital costs that were universally valid.

In applying the "census projections" methodology, EPA has forgotten a basic fact which was pointed out as early as 1937 by John C. Geyer, who said, "If there is an absolute rule governing these problems [industrial residual control] it is the rule that each is different and must be solved individually."[40]

[39]The Cost of Clean Water, vol. 2, p. 95 fn.
[40]John C. Geyer, "The Effect of Industrial Wastes on Sewage Plant Operation," Sewage Works Journal 9, no. 4 (July 1937): 625.

It should be noted that only single values for abatement costs were reported. Ranges should be reported that are based upon various reasonable assumptions regarding the parameters contained in the estimation model. Due to the extreme degree of aggregation and the lack of parameter specification, this type of sensitivity analysis cannot be conducted using the "typical treatment plant" methodology.

In addition to these specific criticisms, cost estimates derived through use of the typical treatment plant cost methodology are deficient in providing information on perhaps less costly alternatives. Only estimates of end-of-pipe treatment costs can be obtained using this methodology. Alternative pollution abatement cost estimates resulting from changes in production technologies, from application of in-plant modifications to the residual streams, from changes in the mix of end products or raw material inputs, and from intermedia transfers are not incorporated into this methodology. Even if this method yielded accurate estimates of treatment costs, the appropriate net residuals management costs could not be determined by its sole use.

Estimated Expenditures to Control Residuals per Unit of Wastewater Flow

Most of the industrial waterborne residual cost estimates undertaken by government and nongovernment groups have used a "cost per unit of wastewater flow" methodology.

While the details of this method vary from one study to another the basic methodology is essentially the same and consists of the following five tasks:

   • For each industry sector determine the volume of wastewater generated per unit of time
   • Differentiate total wastewater volumes by size of hydraulic loadings per unit of time
   • Determine the required residual control processes for each industry sector to achieve the desired level of control
   • Determine cost functions for capital and operation and maintenance expenditures, relating these to the volumes of wastewater flow per unit of time
   • Multiply the appropriate cost estimates—derived from the cost functions—by the corresponding wastewater volumes to determine total capital and operating and maintenance costs for each industry sector

This general methodology—although it has serious shortcomings, due to the fact that its use is limited to estimating end-of-pipe treatment costs—does have an advantage over the "cost of typical treatment plant" method. When using the "cost per unit of wastewater flow" method, it is possible to disaggregate estimates of costs, so that estimates are not solely confined to those associated with a "typical treatment plant." The cost variations associated with the treatment of residuals for various types of industry, and firms within given industries, vary greatly and are dependent on many factors. Hydraulic load is one important factor. The effects of hydraulic loading on the costs of various types of industrial treatment methods are captured by the proper use of the "cost per unit of wastewater flow" method, but they are *not* by the "typical treatment plant" method. This is because loading effects are aggregated and averaged for the latter method and they are not for the former.

The initial attempt to apply the "wastewater flow" methodology to estimations of national industrial water pollution abatement costs was undertaken in 1968 to supplement estimates obtained using the "typical treatment plant" method (i.e., "census projections"). The costs were estimated for the same eleven sectors of industry for the 1968 to 1973 period, with the results reported in volume 3 of *The Cost of Clean Water*, which comprises a total of ten reports, or "Industrial Waste Profiles."[41]

All ten reports discuss the prevailing practices of the end-of-pipe treatment in each industry and provide, albeit on a limited basis, information on other waste abatement alternatives, particularly on the role played by changes in production technologies and by in-plant modifications. No information is provided on the changes in end-products or intermedia transfers.

The principal handicap with most of these studies, however, is the paucity or outright absence of empirical cost information. That is, while these reports provide useful insight into the waterborne pollution abatement practices by the ten industry sectors, only a few of them provide estimates of industrial waterborne residual control costs, and these are calculated on the basis of end-of-pipe treatment only.

Since the initial attempt, in 1968, to use the "cost per unit flow" method, EPA has continued to rely on it. Improvements in the methodology have been made. Accordingly, in the remainder of this section, we concentrate on the 1972 and 1973 cost estimates, since they represent the most refined attempts to use the "cost per unit flow" method.

---

[41]*The Cost of Clean Water*, vol. 3, prepared by a consulting firm for the Federal Water Pollution Control Administration (Washington: GPO, 1967): "Industrial Waste Profile No. 1: Blast Furnaces and Steel Mills" (September 28, 1967); "Industrial Waste Profile No. 2: Motor Vehicles and Parts" (November 24, 1967); "Industrial Waste Profile No. 3: Paper Mills, Except Buildings" (June 30, 1967); "Industrial Waste Profile No. 4: Textile Mill Products" (June 20, 1967); "Industrial Waste Profile No. 5: Petroleum Refining" (September 1967); "Industrial Waste Profile No. 6: Canned and Frozen Fruits and Vegetables" (September 1967); "Industrial Waste Profile No. 7: Leather Tanning and Finishing" (September 1967); "Industrial Profile No. 8: Meat Products" (September 1967); Industrial Waste Profile No. 9: Dairies" (September 1967); and "Industrial Waste Profile No. 10: Plastic Materials and Resins" (October 1967).

The fundamental improvements in the "cost per unit flow" 1972 and 1973 calculations, as compared to the previous studies, consist of more reliable estimates of cost functions and more comprehensive and explicit recognition of all end-of-pipe cost components. However, as with the previous "cost per unit flow" estimates, the 1972 and 1973 costs are calculated assuming end-of-pipe waste abatement. Only gross costs are estimated and net residual treatment costs remain absent. This precludes meaningful analyses of public policies. The cost estimates for 1972 and 1973 are summarized in Table 5.

On the basis of a comparison of the aggregate cost data for the twenty-nine industries involved, capital outlays for end-of-pipe control devices as reported in the 1973 report are estimated to be a third less than those estimated in the *Economics of Clean Water* report prepared in 1972. This occurs despite the fact that the estimates reported in 1973 reflect expenditures required to meet the more stringent water quality standards provided in the 1972 Clean Water Amendments.

A number of factors may account for these differences, although it is not likely that the differences result from variations in the "cost per unit flow" estimates per se. On the contrary, the cost functions used to calculate "cost per unit flow" appear to be almost identical in the 1972 and 1973 reports. The real cause for the large differences in estimates results from the number of manufacturing establishments included and from the size distribution of these establishments as measured in terms of wastewater volumes. Based on these differences in estimates, it is clear that, in addition to the unit cost of treatment, one must be careful in making comparisons of cost figures when the industry coverage and resulting volume of residual flows can differ markedly among studies.

Industrial Water Pollution Cost Estimates Derived from Surveys

The use of survey methodology to estimate industrial water pollution control costs has not received wide application in the past. Among the best known of such surveys is the McGraw-Hill survey of pollution control expenditures by industry undertaken annually since 1967.[42] Unfortunately these surveys do not differentiate expenditure data for air pollution control, water pollution control, and other forms of environmental protection, making comparisons between different studies of water pollution control expenditures impossible.

An example of the application of a more disaggregated survey methodology is the one used by the Conference Board[43] in 1971 to

[42]*McGraw-Hill Survey of Business Plans for Plants and Equipment* (New York: McGraw-Hill, 1972).

[43]*The Economics of Clean Water,* vol. 3, "Industry Expenditures for Water Pollution Abatement," conducted by the Conference Board (New York: 1972).

Table 5

COMPARISON OF CAPITAL COST ESTIMATES FOR WATER
POLLUTION CONTROL BY INDUSTRY, 1977
(millions of 1971 dollars)

Industry	SIC	1972 Report	1973 Report	Percentage Change
Meat Products	201	192	416	+117
Dairy Products	202	142	211	+ 49
Canned, Frozen Preserved Foods	203	421	210	− 50
Wet Corn Milling	2046	35	23	− 34
Sugar Refining	2062	348	46	− 87
Beet Sugar	2063	116	31	− 73
Beverages	208	174	182	+ 5
Weaving & Finishing Mills, Wool	2231	73	63	− 14
Textile Finishing, Except Wool	226	111	122	+ 10
Pulp Mills	2611	936	142	− 85
Paper Mills	2621	740	486	− 34
Paperboard Mills	2631	586	232	− 60
Building Papers	266	26	33	+ 27
Chemicals	28	2,882	1,015	− 65
Alkalies & Chlorines	2812	16	9	− 44
Inorganic Pigments	2816	21	16	− 24
Paints & Allied Products	2851	13	46	+254
Industrial Inorganic Chemicals	2819	195	59	− 70
Organic Chemicals	2813,15, 18,87	2,637	885	− 66
Petroleum Refining	2911	1,530	245	− 84
Leather Tanning & Finishing	3111	133	69	− 48
Cement, Hydraulic	3241	36	55	+ 53
Blast Furnaces and Steel Mills	3312	2,320	1,901	− 18
Steel Rolling & Finishing	331X	36	32	− 11
Primary Aluminum	3334	225	54	− 76
Fabricated Metals	34	260	1,607	+518
Machinery (except electrical)	35	146	559	+283
Electric Machinery	36	236	358	+ 52
Motor Vehicles	371	196	221	+ 13
Total		11,900	8,313	− 30

Sources: *The Economics of Clean Water*, vols. 1 and 2, Environmental Protection
Agency (Washington: GPO, 1972); and Manuel L. Helzner and Rita
McBrayer, *Estimating Water Pollution Control Costs for Selected Manufacturing Industries, in the United States, 1973 to 1977* (Washington, D.C.: National Planning Association, 1973).

estimate industrial water pollution abatement expenditures. The purpose of this survey was to determine industry expenditures for water pollution control in the 1965 to 1970 period. The survey covered establishments in seven major industry categories which accounted for 92.3 percent of the water used and discharged in 1968 by all of the manufacturing industries included in the *1967 Census of Manufacturers* report.

The responses covered four of the five product classifications which are responsible for major quantities of wastewater discharge: paper, paperboard, organic chemicals, and petroleum. Information on a fifth major product classification, steel, was subject to limitations because of an inadequate response.

The major conceptual difficulty with these as well as other surveys in this area is the lack of a proper definition of what constitutes water pollution abatement costs in industry. Expenditure data do not represent net residuals management costs for various levels of control, and by themselves are not useful in formulating rational control policies. As noted earlier, benefits in the form of revenues received or production costs reduced due to various forms of reclamation can be realized by making expenditures for "waste control." These benefits must be subtracted from expenditure data to yield net costs—which are the appropriate data to compare with benefits (damages foregone) if efficient control policies are to be made.

With the exception of the detailed industry studies conducted by Resources for the Future, a study developed by the National Canners Association (NCA) and the American Food Institute provides one of the few methodologically sound approaches to the estimation of net residuals management costs. A questionnaire survey of the canned, frozen, pickled, and dehydrated fruits and vegetables processing industry was administered in 1972. The preliminary results are summarized in a progress report issued in August of 1973.[44]

The survey took the following factors into account in the determination of residual management costs at various levels of control:

- Costs associated with treatment lagoons, aeration, or other systems for reducing the strength of the wastewater
- Costs associated with disposal by irrigation, including pre-irrigation treatment costs
- Costs associated with disposal to municipal sewage plants, including pretreatment costs
- In-plant changes to reduce wastewater flows or the generation of BOD or SS

[44]*Liquid Wastes and the Economic Impacts of Pollution Control: Fruit and Vegetable Processing Industry* (National Canners Association, August 1973).

As the above mentioned factors indicate, this study departs from other surveys and non-RFF studies in that the NCA coverage of factors that affect residuals management costs is much greater.

The basic method that was used to determine the costs of the first three items was the "cost per unit flow" method. Cost curves for treatment were calculated by plotting the volume of wastewater against the percent removal of BOD and SS for various-sized treatment facilities. Economies of scale characterized these data. Also, costs rose at an accelerating rate as the level of BOD and SS removed was increased, regardless of the size of the treatment facility.

Aggregate treatment costs (operating, maintenance, and replacement) for a specified level of removal for company-operated facilities were determined by estimating the wastewater flow and multiplying by the cost per unit of BOD and SS removed by a given sized treatment operation. These were then summed, to yield annual costs of $11 million in 1972 and an estimate of $24 million (in 1972 dollars) for annual costs in 1975. Capital costs of $58 million are expected to be $127 million in 1975.

Irrigation costs, including pretreatment, were calculated in a manner similar to that for treatment. Again, economies of scale were found. Annual operating and maintenance costs for irrigation disposal were $13 million in 1972 and were estimated to increase to $20 million annually (in 1972 dollars) in 1975. Capital costs of $79 million were expected to reach $122 million in 1975. Municipal disposal and pretreatment costs, like company-operated treatment and irrigation, were characterized by economies of scale. The operating and maintenance costs were estimated at $10 million annually and were expected to increase to $24 million (in 1972 dollars) annually in 1975.

One of the most striking aspects of the survey was the data on in-plant abatement costs. The reported in-plant changes reduced wastewater flows, BOD, and SS by relatively small percentages and could not be relied upon as the sole means of satisfying control regulations. However, data indicate that these alternatives should be used in conjunction with other forms of control, if cost-effectiveness is to be achieved.

Economies of scale were also evident for in-plant process changes. For example, the capital costs of new peeling methods, which reduce BOD and SS generation, increased 2.5-fold with a 10-fold increase in plant size. Cooling towers to reduce waste flow, had capital costs that increased 3-fold with a 10-fold increase in plant size. The average costs of in-plant abatement under various conditions are displayed in Table 6. Of particular interest is the negative entry in the first column. This represents a situation in which the benefits from the investment in in-plant abatement outweigh the costs. It illustrates the danger of gathering data only on gross abatement expenditures, instead of calculating net

Table 6

AVERAGE COSTS OF IN-PLANT POLLUTION ABATEMENT
(Dollars per Ton of Raw Product)

% discharge flow reduced		0-10		More than 10	
% BOD + SS reduced		0-10	More than 10	0-10	More than 10
Plant size, 1000 tons per year	1-19	(-.08)	.92	.41	.82
	20-49	.14	.60	.28	.24
	50-99	.11	.06	.12	.27
	100-169	.08	.02	.04	.17
	170+	.02	.42	.10	.22

Source: *Liquid Wastes and Economic Impacts of Pollution Control: Fruit and Vegetable Processing Industry,* National Canners Association (August, 1973).

figures. The industry costs of $6 million annually for in-plant abatement were expected to be $16 million annually by 1975.

The National Canners Association's survey is one of the more innovative studies that we have reviewed. It more adequately deals with the factors that determine residuals management costs, and presents an approach that could be more widely applied.

Concluding Observations

IN CONSIDERING THE PRACTICAL ASPECTS of making efficient pollution control policies, most observers have concluded that estimating damages from pollution is much more difficult than estimating the costs of controlling them. While this conclusion is probably correct, it has led many to believe that appropriate incremental cost figures for residuals management are readily available. This is clearly not the case. The control cost studies that have been conducted or supported by EPA have been conducted ex post, i.e., after pollution control policy has been made. The studies are not based on a methodology that can be used ex ante to guide policy makers in determining trade-offs between costs and benefits and in arriving at informed judgments as to appropriate levels of residuals discharges.

The framework of analysis that we outlined in the first section of this paper and the factors that affect net residuals management costs have generally not been incorporated in cost studies conducted or sponsored by EPA. The following represent some of the major shortcomings of these studies:

- Gross costs or expenditures, rather than net residuals management costs, are presented. These are of little value in policy making and over-state the net resources that are foregone in the control of residuals
- End-of-pipe treatment has been considered the most efficient and in most cases the only means of controlling residuals
- There has been a grossly inadequate consideration of the product output and input variables, and of their influence upon costs
- The role of in-plant process changes that affect recovery and by-product production—which, in turn, affect the discharge of residuals—has not been given adequate attention
- The importance of the single plant firm vs. the multiple plant firm has not been treated. Since multiple-plant firms have more options for materials recovery and by-product production, their control costs may be lower than those for single-plant firms
- The studies have not dealt with the administrative and bureaucratic costs of implementing various types and levels of control. These costs are real and must be incorporated into the estimate of net residuals manage-ment costs if efficient levels of control are to be set

A proper methodology for answering the trade-off questions in formulating residual discharge strategies is embodied in the industry studies that have been conducted at Resources for the Future. These studies properly analyze the relevant factors that influence net residuals management costs and avoid the shortcomings listed above. If EPA, or its predecessor, had conducted industry studies years ago when the cost issue first arose, it would now have a backlog of studies and information upon which it could more efficiently base decisions.

However, this "industry study" approach has not been taken. We recommend that EPA adopt it and help industry trade associations, pri-vate groups, etc., to develop a general method of cost analysis similar to that used by Resources for the Future. This is a realistic approach. Factors that affect costs could then be incorporated in studies, and the costs of conducting these studies would not be unreasonable. For exam-ple, the total costs of a relatively simply industry study, such as the RFF sugar beet study,[45] would probably be in the $30,000 to $50,000 range.[46]

A more complicated and thorough study[47] would probably incur a total cost of $100,000 (if completed in two or three years) and $150,000 (if completed in one and a half years).[48]

In conjunction with the RFF-type studies, it is important to recog-nize two other elements that should be incorporated in the study of

---

[45]George O. G. Löf and Allen V. Kneese, *The Economics of Water Utilization in the Beet Sugar Industry* (Baltimore: Johns Hopkins University Press, 1968).

[46]Personal communication from Blair T. Bower, Resources for the Future (1973).

[47]Clifford S. Russell, *Residuals Management in Industry: A Case Study of Petroleum Refining* (Baltimore: Johns Hopkins University Press, 1973).

[48]Personal communication from Blair T. Bower, Resources for the Future (1973).

Table 7

MATRIX OF VARIABLES FOR ANALYSIS OF COSTS TO INDUSTRY OF REDUCING DISCHARGE OF RESIDUALS

LEVEL OF TECHNOLOGY	PRODUCT MIX			
	Same		Changed	
	Raw Material		Raw Material	
	Same	Changed	Same	Changed
Present				
RM Processing				
P Processes				
Near-Future				
RM Processing				
P Processes				
Future				
RM Processing				
P Processes				

RM = raw material
P = production

Source: Personal communication from Blair T. Bower, Resources for the Future (1973).

control costs: (1) existing technology and future trends must be carefully assessed, and (2) the present status and trends in various industry characteristics, such as size, product mix, etc. must be assessed. These two factors are displayed in Table 7.

The approach that we recommend is not without problems, however.[49] These include: lack of data on residual generation; lack of data describing the physical transformation and flow of materials; lack of data on costs of factor inputs, process units, and residual modification measures; lack of cost data on the effects of short-run variability in residuals generation; and difficulty in obtaining data on the technology and total costs of producing a given product without pollution control regulations. This information is essential for determing the level of residuals generated and the net costs attributed solely to residuals management.

[49]Blair T. Bower, "Studies of Residuals Management in Industry," op. cit.

Since the major difficulty encountered in conducting industry studies is a lack of data, we hope that a basic EPA policy change with reference to the study of residual management costs will be made in the near future. If it is not, we will further delay the date when rational pollution control policies can be made.

Part Three

THE FUTURE AND UNCERTAINTY

Valuing Long-Run Ecological Consequences and Irreversibilities

ANTHONY C. FISHER and JOHN V. KRUTILLA

Introduction and Statement of Main Results

IN THIS PAPER WE CONSIDER the nature and the implications, for economic theory and policy, of investments in water (and other) resources that may involve, in addition to conventionally treated costs, adverse effects on the physical environment that are difficult, or even impossible, to reverse—as with pollution from nondegradable contaminants like mercury. The approach is primarily theoretical, with, however, a number of illustrative examples presented along the way, especially at the outset, to motivate the discussion. Generally, our remarks apply to any economic activity that may lead to destruction of the biotic or abiotic resource base for the production of some alternative goods or services, such as recreational amenities.

Note that there is a "dual" to this problem, which can be formulated as investments in abatement of environmental disruption, such as the accumulation of nondegradable contaminants, that will become impossible to reverse unless the investment is made within some (assumed known) period of time. In this case, "not investing" is irreversible, in the sense that if the investment in abatement is not made within this period, it will not be possible to make it and to capture the benefits later on. This version of the problem is perhaps more relevant to an agency with an interest in abatement programs, although, as we shall see, the differences between the two versions are not substantive but merely differences in labeling.

The task of this paper, then, will be first to indicate the nature of the long-run or irreversible ecological consequences with which we are concerned (i.e., how they differ from the consequences of ordinary economic decisions) and then, in the context of some very simple models, to derive some implications for policy with respect to investments

271

which can either give rise to these effects or, alternatively, avert them. In the course of the discussion, a number of related issues are addressed: the choice of welfare criteria, the choice of discount rate, and the role of uncertainty.

However the problem is formulated, our main conclusion is that a conservative policy with respect to irreversible modification of the environment is indicated. One reason for this conclusion is that a shift in the vantage point from which an investment plan is evaluated—say from one generation to the next—can result in a change in the "optimal" plan. The optimal plan, however, cannot be implemented if it is in a direction which has been foreclosed by earlier activity. Moreover, even keeping the perspective of the present, suppose the costs and benefits of alternative uses of the environment are not known with certainty. For a society averse to risk, there will be value in retaining an option to use the environment in a way that could otherwise be foreclosed, which is another reason to hold back from an irreversible decision. For a society neutral toward risk, this same "option value" will exist if information about the costs and benefits is developed in an early period and can be used to improve the investment decision in a later period. A third reason for a conservative policy is that—even assuming perfect certainty about the costs and benefits of alternative actions, and again keeping the perspective of the present—an activity which yields positive net returns in the short run and negative returns thereafter, but which cannot be "shut down," should perhaps not be undertaken in the first place. Following the discussion of the nature of irreversibility in the following section, all of these matters are considered in detail, and the assertions are demonstrated in the context of simple but appropriate models.

Reversible and Irreversible Consequences

IN ECONOMICS, one can use differences between production and investment decisions to make a fairly easy distinction between reversible and irreversible consequences of decisions. Consider the problem of a farmer in deciding whether to allocate his stock of land to the production of, say, carrots or of beets. After taking into account such relevant factors as the anticipated future prices of associated inputs and the products he wishes to market, he will exercise his best judgment and choose the crop that appears in foresight to be most profitable. Assuming that this does not imply expensive, long-lived equipment that can be used only for growing carrots or only for growing beets, he can review his decision at the beginning of the next planting period. If his judgment regarding the relative profitability of the two alternatives was faulty and he has good reason to suspect that a change in production plans is called for, he can adjust his plans after only one crop production cycle.

If the decision maker's problem is not choosing between two different crops, or different mixes of products in a production batch—either of which is possible with the land, plant, and equipment in his possession —but rather choosing between alternatives which involve different land, a different type of plant, and different equipment, the consequences of the problem are of longer duration. Assume that an individual wishes to consider two possible extractive pursuits, one involving truck gardening and the other ranching. A choice here will involve the commitment of resources to more or less specialized land, plant, and equipment. A faulty decision, with benefit of hindsight, could not be as readily adjusted to the changes that would be desired in some future period. So long as his receipts covered his variable production costs, the decision maker would be "locked into" his decision, at least over the period required to depreciate his specialized equipment and other specialized resources associated with the enterprise.[1]

From the farmer's point of view, the consequences of his decision might not be trivial, but from the standpoint of society at large, the particular decision impinges marginally on the total production of beef, trucks, and specialized capital equipment. The consequences are neither ephemeral nor trivial for the decision maker, but to society at large, the irreversible consequences of the particular decision are of no great moment. This is probably true of all such marginal investment decisions.

There are, however, long-run effects of individual actions that may be anything but trivial for society as a whole. Consider, for example, the consequences of developing some parts of central Florida for agricultural use.[2] To lower the water table and make the land suitable for crops, a great deal of water in the Everglades would have to be drained off by diversion canals leading eventually to the ocean or the Gulf. A decision to reserve the Everglades for nature appreciation and related recreation, or scientific interests, is not an immutable decision, since no *technical* constraint would prevent a reversal of the decision.

If the land in central Florida is to be dedicated to agricultural production, on the other hand, unless special efforts are mounted to preserve the flow of water through the Everglades, the consequences stemming from this decision would have more permanent and therefore more significant implications. Draining off water would diminish the number and sizes of the pools necessary for preserving aquatic and

[1]This, of course, would not be true if there were a ready market for the specialized resources which could no longer be used profitably in the chosen pursuit. More likely, however, an unprofitable venture results in a loss of the capital value of resources associated with such an enterprise.

[2]We are indebted to Dr. Robert Kelly of Resources for the Future for this example.

amphibious animal life, thus seriously restricting populations of fish, frogs, and especially alligators. In addition, the lowering of the water table allows the chemical oxidation of the surficial sediments, greatly changing the nature of the soils, and possibly preventing for all time the re-creation of a plant association similar to the one which now exists. Finally, a dry Everglades would allow large fires to blaze, additionally changing the ecological relationships. A decision to restore the original Everglade ecosystem after water allocative decisions had been made would not be technically possible, especially if localized or widespread extinction of animal life had occurred, or if the soils had oxidized appreciably. The consequences of the decision to drain the Everglades, in terms of the foregone opportunity returns associated with aesthetic and scientific uses of the Everglades, would be experienced in perpetuity. There are no technical means of restoring the original character of the Everglades.

The matter of irreversibility seems simple enough, yet it is elusive for some.[3] It may be well to devote some additional discussion to this matter at the outset.

Investment in specialized plant and equipment, for example, represents an irreversible commitment of capital to an undertaking, and, for an individual making the decision, it is not to be taken lightly. But from society's viewpoint, it is somewhat like the irreversibility which attends the death of any single member of a particular population. There does not appear to be overwhelming concern for the demise of a member of a species, provided that reproductive capability is retained within the population. The risk of losing the last viable mating pair of a species —hence the genetic constituents essential to survival of that species —however, is a matter of much greater moment. Reverting to the example of the investment decision, for society the loss of capital value through a misjudged investment, is not unlike the loss of a member of a given, viable population. The loss of the scientific-technological information necessary to reproduce the capital goods for which there may be a future demand, would be a matter of greater gravity, and may be likened to the loss of genetic information for a threatened species. This is the basis for society's concern for dying arts and crafts. Their extinction reduces the cultural diversity of a society just as the loss of species reduces biological diversity, and while the two may have different implications for system stability (cultural or ecological homeostasis) they both represent reductions in the options available, and thus illustrate a central postulate of welfare economics; i.e., the expansion of choice represents a welfare gain; the reduction of options, a welfare loss.

---

[3]See, for example, the discussion in Cummings and Norton (1974).

Not all environmental modifications need to be absolutely irreversible to provide a basis for societal decisions more deliberate than those that make up the bulk of our problem decisions. That is, we recognize that the case of absolute irreversibility is a special one. But this does not imply that decisions entailing long-run welfare losses that may eventually be retrieved have no need for special consideration. As Weisbrod (1964) has noted, if restoration to the original state of affairs is excessively costly —whether in terms of the resources that must be allocated to restore that state of affairs or in terms of the duration over which opportunity returns are foregone—a case exists for explicitly recognizing the option value of preserving the original state of affairs when future demand for its services is uncertain.

Problems associated with restoration are considered below. There are two important considerations that require attention. One involves the time during which the adverse consequences of an irreversible decision must be suffered. The other involves the fidelity or authenticity of the reproduction or restoration.

Duration of Welfare Loss

Consider the change in a natural ecosystem as a result of man's production activities. If man eliminates the habitat essential for a given species, restoration of the species is impossible—or at least incomplete without the organisms dependent on the undisturbed environment. But, even if the survival of a species is not at issue, restoration is not simply a remedy for redressing the impact of an inappropriate decision that disturbed the original ecological environment.

The introduction of persistent pesticides, certain heavy metals, and nutrients into lakes and rivers gives rise to effects apparent long after the cessation of inputs to these bodies of water. The time span of the effects could be many human generations for large lakes, or only a few months to a few years for rivers and streams. DDT provides a good example of a long-term effect. Once introduced into a lake, this pesticide will collect in the sediments. The sediments will remain an appreciable source of DDT because of benthic organisms which ingest the sedimentary material, pick up the pesticide, and in turn are eaten by fish, which in turn concentrate the DDT to high levels. When the fish dies, the pesticide is again released to the sediments or taken up by other organisms. To transfer water out of the lake serves very poorly in removing the DDT because of the minute amounts dissolved or suspended in the water. The only effective way to prevent the pesticide from being circulated through the ecosystem is to bury it deeply enough under other sediments to make it essentially unavailable to benthic organisms. With sedimentation rates measured in millimeters per year, it can take more than a century to eliminate the deleterious biological effects of the DDT in the top twenty centimeters of sediment.

Biologically active nutrients in lakes present similar problems by increasing certain types of productivity in ways that tend to be self-maintaining. Large concentrations of nitrogen and phosphorus lead to large numbers of algae in surface waters. The algae, in turn, provide organic matter which can decay in the bottom waters and deplete the available oxygen. With the loss of oxygen, the mechanisms which normally strip the water column of nutrients become inactive, thus perpetuating the high concentrations. Lake Washington has shown that over decades there is a gradual return to the original conditions after the source of nutrients has ceased, but that certain events have occurred which cannot be technologically changed: the lessened volume of lake caused by the increased rate of sedimentation due to plankton growth is probably the most important of these. The advanced eutrophication of Lake Erie is a similar and perhaps more publicized example.

On the other hand, streams and rivers are affected over relatively short time spans because the sediments are moved downstream during high flows, so that surface waters are purged of a large portion of the pesticides, nutrients, or other potentially damaging materials. It should be apparent, however, that since most rivers eventually reach the oceans, the short-term effects are only local phenomena.

Artificial Restoration and the Significance of Authenticity

A question can be raised, however, whether it is not possible to mobilize the scientific-technological knowledge that also accumulates with time, to short-circuit the time element in restoration. This is doubtless a possibility in many cases involving rather ordinary land and water scapes, particularly in the more rapid restoration areas of subhumid and humid climatic zones. But when we consider the extraordinary natural environments that are prized for their scientific research materials or their unusual scenic or natural features as part of the nation's ecological heritage, the problem takes on a different dimension. If the objective is simply to restore some type of outdoor recreational facility in place of the original, it is doubtless possible to replicate in some particulars the original features that would satisfy the bulk of the demands of those seeking outdoor recreation. But, there is a legitimate question whether undisturbed natural environments should be used in connection with industrial activities on the supposition that they can be restored eventually to provide, with the most painstaking application of modern scientific-technological resources, replicas of the original that would satisfy the interests of only the less discriminating observer. The matter turns on the importance of authenticity as an attribute of the recreational experience, quite apart from the matter of preserving relevant research materials for advancing knowledge in the life and earth sciences.

The matter of authenticity as an attribute in the demand for undisturbed natural areas may be likened to the demand for authenticity in the visual arts. For the casual visitor to an art museum, the difference between an original work by one of the masters and an exact forgery by one of his protégés or a contemporary artist, is undetectable and may well satisfy his tastes. But to a connoisseur, the mere suspicion of a forgery, although so expert that even art critics differ in their opinions about its authenticity, will drastically reduce the market value of the object, as many museum curators have been embarrassed to discover. The question then turns on what is the clientele, or market, that a particular amenity resource is to satisfy?

The outdoor recreation market is a vast and complicated structure of numerous submarkets. A developed campground in an attractive roadside location may be all that the "typical" family car-camper desires. But there are others, and of growing numbers,[4] who seek solitude and primeval settings for the gratification of their recreational tastes. This clientele group represents a distinct submarket among outdoor recreation enthusiasts. To members of this puristic submarket, no less so than to the life and earth scientist, an artificial replica no matter how "exact" is as unsatisfying a remedy for a disturbed natural environment as is the expert art forgery to a connoisseur.

Why even expert forgeries are unsatisfying to the purist is not a matter of economic knowledge, nor perhaps of scientific knowledge generally. There may be an aura about the works of creative genius that the work of the most gifted imitator cannot provide. There may be even a cult comprised of those who revere the works of nature in a sense similar to that in which the Zen-Buddhists revere it—which, in turn, may be similar to the reverence that many primitive societies confer on nature in their religious observances. To those who number among the purists, preservation of the constituents of the biosphere in precisely the way it has evolved, undisturbed by industrial man, is a matter of great significance in a profound personal sense. Such feelings, in fact, have been captured in the works of Wordsworth, Emerson, and Thoreau in very moving fashion. Whatever the reasons may be, whether mystical or religious, they are felt with great intensity. For analytical purposes this translates into a

[4]Wilderness recreation, and recreation in undeveloped natural areas is the most rapidly growing outdoor recreational activity. This has been at the rate of approximately 10 percent per year over the past several decades without evidence of slackening. It seems to be a result of changing tastes in addition to elasticity of demand with respect to income and educational achievement. See George Stankey (1972) for a discussion of this phenomenon.

highly inelastic demand for the "originals." These purists currently comprise a significant market that appears to be recruiting members rapidly as the income, educational, and urban composition of American society changes (Stankey,1972). Moreover, this is a market for which refinements in restorative technology will do little by way of recreating "undisturbed" natural environments. Accordingly, the argument for technical irreversibilities[5] is a powerful one where the clientele group places a high value on the attribute of authenticity in the amenity services yielded by given natural environment resources.

Modeling
Irreversibility

LET US BEGIN WITH WHAT SEEMS TO US the simplest possible formal statement of the problem, which is, however, sufficient to derive an important implication for investment behavior under irreversibility.

Definitions and Assumptions

b_t = t^{th} period benefits, net of internal costs
c_t^e = t^{th} period external environmental costs
$\alpha_t = (1 + i)^{-(t - 1)}$, the tth period discount factor
i   = the social discount rate, assumed appropriately chosen and unchanging
t   = 1, 2
V   = present value of net returns

Investment, and associated environmental disruption, are irreversible.

As noted in the introduction, we shall have more to say about discounting and the choice of welfare criteria generally, but for now assume that we wish to maximize the appropriately discounted present value of an irreversible investment that can be made in either the first period, the second period, or not at all, *net* of its effect on the environment. Immediately following the analysis of this problem, we consider the "dual": investment in abatement of environmental disruption which will become irreversible if the abatement is not undertaken in time.

The simplest outcome, in which present value V is maximized by not undertaking the investment in either period, occurs if $c_1^e > b_1$ and $\alpha_2 c_2^e > \alpha_2 b_2$. If $c_1^e > b_1$, $\alpha_2 c_2^e < \alpha_2 b_2$, value is maximized by not investing

---

[5]Again, the irreversibility may be relative rather than absolute, and its degree may be measured in terms of costs and time required for authentic restoration.

in period 1, then investing in period 2, with $V_{max} = \alpha_2(b_2 - c_2^e)$. If $c_1^e < b_1$ $\alpha_2 c_2^e < \alpha_2 b_2$, $V_{max} = (b_1 - c_1^e) + \alpha_2(b_2 - c_2^e)$.

Thus far, the investment decision has been entirely conventional in the sense that the constraint on reversibility has played no role. Now suppose $c_1^e < b_1$, $\alpha_2 c_2^e > \alpha_2 b_2$. If investment were reversible, V_{max} would equal $(b_1 - c_1^e)$; i.e., the polluting activity would be run only for the first period, when benefits exceed costs, and would be shut down for the second, when costs exceed benefits. Since this solution is, however, not feasible (by assumption), value is maximized either by refraining from investment in both periods, thereby foregoing net returns of $(b_1 - c_1^e)$ in the first period, or by investing in the first period, and realizing a loss of $\alpha_2(c_2^e - b_2)$ in the second. Suppose $(b_1 - c_1^e) < \alpha_2(c_2^e - b_2)$. Then it is clear that value is maximized by not investing, even though this involved an opportunity loss of first period returns $(b_1 - c_1^e)$.

Now let us consider the problem of investment to avert irreversible degradation of an environmental asset like Lake Erie.

Definitions and Assumptions

$b_t = t^{th}$ period benefits from the investment, in the form of increased capacity of the asset to provide amenity services

$c_t = t^{th}$ period costs of the investment, borne ultimately by consumers in the form of higher product prices, or by taxpayers

$t = 1, 2$

Other symbols are defined as before, and there is again a constraint on reversibility, but in this case, *not* making the investment in the first period is irreversible, since the amenity (and perhaps also health) benefits are irretrievably lost for the second period as well as the first, if the investment is not made in the first. As before, the interesting outcome is the one in which $c_1 > b_1$, $\alpha_2 c_2 < \alpha_2 b_2$, and $(c_1 - b_1) < \alpha_2(b_2 - c_2)$. Then $V_{max} = (b_1 - c_1) + \alpha_2(b_2 - c_2)$, although $(b_1 - c_1) < 0$, or in other words, value is maximized by undertaking the cleanup in the first period even though costs exceed benefits, since it is not feasible to do so only in the second period, when benefits exceed costs.

It should be clear that these "perverse" outcomes are likely to be empirically significant. The destruction of a valuable resource like Lake Erie because of the accumulation of residuals that will result in the absence of some investment in abatement within, say, the next five years (our "first period"), is likely to be far more costly in terms of the resulting damage than the required investment, even though all costs are discounted back to the present. Admittedly the relative magnitudes of these costs are a matter of empirical determination, but it is our impression that there are important situations (Lake Erie may be one) in which a relatively modest investment may forestall much more serious and

even catastrophic consequences—which are irreversible if the investment is not made in time.

Discounting, Uncertainty, and Information

EVEN IN THE DETERMINEDLY SIMPLE FORMULATION of the preceding section, the optimal policy for investment in pollution control is affected by
the discount factor, α_t, as well as by the costs and benefits of the investment. It may be appropriate therefore to consider at some length the
choice of a social discount rate. In so doing we shall review briefly the
vast literature in this area. After discussing the question of how uncertainty affects the discount rate we shall take up the broader question of
how it affects optimal pollution control. Under this heading some results
are obtained concerning risk, information, and the existence of option
value.

On the Social Discount Rate

Two broad approaches to determining a social discount rate may be
distinguished, although specific positions differ further in detail. The
first we discuss might be termed the "private opportunity rate approach." It is prominently identified with theoretical arguments by Hirshleifer (1965, 1966), Mishan (1967), and Baumol (1968), among others,
and an example of its use can be found in Krutilla and Eckstein (1958).
According to this view, the source of funds for any public project is
ultimately the private sector, so that net returns to the public project
ought to be discounted by the private opportunity rate, i.e., by the rate
of return on investment in the private sector. In a world with less than
perfect capital markets, however, this raises a question: which rate, on
which investment, is relevant for determining a social discount rate? One
imperfection noted by Baumol is that the tax on corporate income drives
a wedge between the rate of time preference and the rate of return on
investment.* Thus the rate of time preference, as measured, say, by the
yield on government bonds, will be below the rate of return on private
investment, with the result that a single rate of interest, equilibrating
rates of preference and productivity in the Fisherian framework and
representing "the" private opportunity rate, cannot be attained. Other
issues, in particular the valuation of foregone consumption, are touched

---

*The rate of time preference is the interest rate necessary to make a person indiffer-
ent between consuming in one period or postponing his consumption (i.e., saving) for the
next period. The rate of return on investment measures the return which the investment
yields per period. In a world of perfect capital markets these two rates would be equal.
[Editor's Note.]

on in the exchanges between proponents of the opportunity rate approach. Any differences, however, should not obscure the important common element, a finding of normative significance in some private rate or combination of rates.

On the other hand, the "social time preference approach," associated prominently with Marglin (1963), notes the dependence of some individuals' utility on future consumption by others, and deduces from this dependence a difference between social time preference and an aggregation of individual preferences as reflected in private capital markets. Consumption by future generations is treated as a public good. In providing that public good, present individuals collectively agree to some amount of current capital formation beyond that which would be undertaken by each acting in isolation. The bond rate, in this view, could not then serve even as an appropriate measure of pure (social) time preference. Rather, some explicit political decision making procedure, perhaps democratic voting as suggested by Marglin, would be required to determine the overall rates of growth, investment, and discount in a society. In this connection we might note the difficulties in designing a theoretical mechanism by which a social consensus would emerge from the welter of individual values (see Arrow [1951] and others).

At this point an interesting question arises. Suppose we accept the principle of a social discount rate given by social time preference, on the grounds both that the private opportunity measure is ambiguous, and that future consumption has attributes of an externality and a public good for present consumers and voters. The question is, how is this rate to be made operational? The obvious answer, as noted by most students of the problem, is that the government through appropriate monetary and fiscal policy ought to drive down interest rates throughout the economy to the level of the predetermined social rate. Clearly this would be required to avoid waste in shifting resources from higher yield (private) to lower yield (public) projects, as would be implied by discounting the net benefits solely of the public projects at the lower social rate.

If, however, this is not a feasible policy alternative, as several people—notably Eckstein (1958), Steiner (1959), Marglin in a later work (1963), and Feldstein (1964)—have suggested, and if it is therefore not possible to attain the "first-best" solution, then it would be necessary to resort to some "second-best" solution, which consists of discounting returns in the private sector by the lower social rate as well. The resulting increase in the present value of these returns should then be reflected in the criteria for public investment—for example, in a minimal requirement of a present value greater than some positive number. Although this procedure does avoid the inefficiency of a two-rate system, it should be recognized, as Hirshleifer (1961) has pointed out, that it favors alternatives with a "higher futurity of yield."

Apart from second-best solutions, there is a reason why public and private discount rates might differ, even in theory. This involves the relationship of the attitude toward risk to the choice of a discount rate. Baumol argues that risk, like the corporation income tax, causes the rate of time preference to diverge from the rate of return, inducing firms to invest only to the point at which expected returns are higher than they would be in the absence of these distortions, and that the transfer of resources from private to public sectors therefore involves a correspondingly higher opportunity cost. An implicit assumption in this part of Baumol's analysis appears to be that risk-bearing as a factor of production is just as costly for a public project as for a private one. It is precisely this assumption that is challenged by Arrow and Lind (1970), who prove that if the net returns (and therefore also the risk) from a single investment are sufficiently spread among individuals—which could happen if the project were publicly undertaken—the cost of risk-bearing, the aggregate risk premium, approaches zero. It follows, then, the public investments, unlike private ones, should be evaluated at a riskless rate.

This result is certainly correct, but it does depend on at least two conditions that may not hold in a real situation. The first, as Sandmo (1972) has recently observed, is that the returns to the Arrow-Lind investment are not correlated with other income. The second condition is, somewhat paradoxically, that the investment should not produce a public good output, such as the avoidance of environmental disruption. That is, returns to a public investment may be appropriately discounted at the riskless rate, but only if the investment produces a marketable (private good) output, such as electric power or irrigation water. To demonstrate the validity of this assertion it is necessary to consider the Arrow-Lind theorem in a bit more detail. In doing so, we shall also find it helpful to explore more generally how uncertainty affects the evaluation of benefits and costs, and therefore optimal investment policy.

Before proceeding in this fashion, let us present our understanding of the social discounting issue. It is clear, first of all, that to evaluate investment projects undertaken by the government or subject to government regulation, a discount rate is required. Also, there seems to be fairly general agreement that a single (social) discount rate ought somehow to be applied to all the investment opportunities in an economy, to avoid reallocation of investment resources from higher to lower yielding alternatives. An important exception to this principle *may* arise however where conditions are such that the cost of risk associated with a project is lower, or even vanishes, if it is undertaken in the public sector. Although there is agreement on the efficiency of a uniform rate (abstracting from the risk considerations just noted), there is no consensus on what the rate should be, or even how it should be determined. Sliding over the differences within each category, it is possible to distinguish two broad approaches: (1) the market opportunity rate approach, and (2) the social time preference approach.

Uncertainty, Information, and Option Value

Weisbrod (1964) first suggested that where the demand for a publicly provided good or service is uncertain, there may be benefits to the individual consumer from retaining the option to consume (or "option value") in addition to the conventional consumer's surplus. More recently Cicchetti and Freeman (1971) have shown that, where there is uncertainty in either demand or supply, Weisbrod's option value will exist and be positive for a risk-averse individual. Suppose a public agency has under review an (otherwise profitable) activity which will result in irreversible destruction of an environmental resource, thereby foreclosing the option to consume the amenity services of the resource in the future. What Weisbrod and Cicchetti and Freeman are saying is that the activity causes an opportunity loss to the individual in addition to the loss of his expected consumer's surplus from the amenity services. The opportunity loss is in fact equivalent to the cost of risk-bearing (Cicchetti and Freeman), and though the option value models are not explicitly dynamic, it is clear that the risk cost can be entered in a multiperiod cost-benefit calculation as an adjustment to the discount rate.

The Arrow-Lind conclusion, on the other hand, is that, even with risk-averse behavior by *individuals,* the spreading of costs, benefits, and the associated risk across very many individuals renders the cost of risk-bearing negligible to each individual as well as to the aggregate, and therefore indicates that the investment should be evaluated at the riskless rate. The seeming contradiction here is resolved by noting that, if the investment produces a public good, then making it available to increasingly many does not reduce benefits, costs, or risk. It seems clear that environmental disruption is a (negative) public good in this sense and, accordingly, that the option value loss, if any, should be taken into account in evaluating an investment with (negative) environmental spillovers. To put this conclusion in the framework of the preceding discussion, we may note that in principle there is some adjustment of the discount rate which will have the same effect on the net present value of the investment as the indicated adjustment of benefits has for loss of option value.

Where option value exists, then, it seems there is an additional incentive to avoid an activity with irreversible adverse effects on the environment. Note that it has not been established that such an activity should never take place. Rather, the point is simply that irreversibility coupled with risk aversion results in a lowering of the present value of the activity. Though this does make it less likely to show a positive return, such an outcome is not ruled out.

Since there is however some controversy (see Sandmo) concerning the assumption of risk aversion required to obtain these results, it would obviously be desirable if they could be established without the assumption, in particular with the weaker assumption of risk neutrality. If option value is defined as a premium for risk-bearing, then strictly speaking this is not possible. But assuming a neutral attitude toward risk, the existence of something like option value—"quasi-option value," which similarly reduces the value of an activity having irreversible adverse environmental spillovers—has most recently been established by Arrow and Fisher (1974), though in a somewhat different context. In the remainder of this section a part of the Arrow-Fisher analysis is adapted to demonstrate the importance, under irreversibility, of a new element introduced by these authors: the improvement of information, in the form of future expectations determined by past realizations.

Definitions and Assumptions

b_t = t^{th} period benefits from an investment project
c_t = t^{th} period (internal) capital costs of the project
c_t^e = t^{th} period external environmental costs of the project
S_t = fraction of project completed in period t
t = 1, 2
V = present value of net returns

Investment, and associated environmental disruption, are irreversible; b_2 and c_2^e are conditional on b_1 and c_1^e; i.e., are known at the start of the second period on the basis of information (the values of b_1 and c_1^e) accumulated over the first period.

To avoid clutter, time discounting as in the earlier models has not been explicitly represented, so that second period benefits and costs may be interpreted as present values. Also for simplicity we assume uncertainty only about the benefits and the environmental costs of the activity, not about the internal capital costs.

Suppose $b_2 - c_2^e > c_2$. Then maximum benefits are

(1) V_{max} $= (b_1 - c_1^e - c_1) S_1 + b_2 - c_2^e - c_2(1 - S_1)$
      $= w S_1 + c_2 S_1 + Z - c_2$
where $w = (b_1 - c_1^e - c_1)$ and $Z = (b_2 - c_2^e)$.

Now suppose $b_2 - c_2^e = Z < c_2$. Then

(2) V_{max} $= (b_1 - c_1^e - c_1) S_1 + (b_2 - c_2^e) S_1$
      $= w S_1 + Z S_1$.

Here the assumption of risk-neutrality enters, telling us to consider only the expected value of benefits, and not, for example, the expected utility of this value. Expected value, at the start of the first period, i.e., before making any investment, is

(3) $E\left[(w + \min(c_2, Z))\,S_1 + \max(Z - c_2, 0)\right]$

where E[ ] is the expected value of the expression in brackets, min ( ) is the minimum of the two terms in parentheses, and max ( ) is the maximum of the two terms in parentheses. Of course if $S_1 = 0$ then (3) reduces to

(3′) $E\left[\max(Z - c_2, 0)\right]$

The difference between (3) and (3′) is

(4) $E\left[(w + \min(c_2, Z))\,S_1\right]$

so if $E\left[w + \min(c_2, Z)\right] > 0$, then benefits are maximized by $S_1 > 0$.

Now suppose that Z and w are known, and equal to $E[z]$ and $E[w]$ respectively. The investment criterion (4) is then $E[w] + \min(c_2, E[Z])$, i.e., either $E[w] + c_2$ or $E[w] + E[Z]$. Suppose the former. Then we wish to prove that $E[w + \min(c_2, Z)] < E[w] + c_2$, i.e., the information about the values of second period benefits and costs contained in $E[Z]$ increases the gain from investment in the first period. Note, first, that $\min(c_2, Z) \leq c_2$, so $P[\min(c_2, Z) < c_2] > 0$, where P[ ] is the probability of the expression in brackets.
Then

(5) $E\left[\min(c_2, Z)\right] < c_2$ and $E\left[w + \min(c_2, Z)\right] < E[w] + c_2.$

An exactly analogous line of reasoning holds for the case in which $E[Z] < c_2$. What this exercise suggests is that there are some values for second period benefits and costs for which an investment should not be undertaken under uncertainty, but should be undertaken were they known to be the true values. The difference between the expressions on the right-hand side and the left-hand side of the inequality (5) can be considered something like an option value loss from (uncertain) irreversible destruction of the environment in the first period. An interpretation of this result might be, if there is uncertainty about the returns to an environmentally destructive activity, social policy should err on the side of refraining from the activity since the destruction is not reversible. With an ability to learn from experience, the opportunity losses that this might entail can be cut at a later date, but the reverse is not possible.

Shifting Vantage Points
and Efficiency over Time

As SUGGESTED IN OUR INTRODUCTION, it is possible that the re-evaluation of a polluting or otherwise environmentally destructive activity may reveal its undesirability only after it is too late to do anything about it—which is another reason for proceeding conservatively with such an activity even if, at the time, it seems desirable. The problem here can be regarded as essentially a special case, though an empirically important one, of Strotz's (1956) problem of inconsistent planning. Suppose an economic agent, say a consumer, is choosing a plan by which to allocate his consumption over time of a fixed stock of goods in order to maximize its utility to his present welfare. If he reconsiders the plan at a later date, will he abide by it or disobey it? The answer, as demonstrated by Strotz, is that in general he will want to disobey—i.e., revise his "optimal" plan—even assuming no change in expectations in the interim.

The explanation for this apparently perverse outcome is simply that the *relative* (discount factor) weights attached to consumption in each period change as the time perspective of the evaluation changes. In other words, the value of a given level of consumption in year $t + k + 1$ relative to that in year $t+k$ will not be the same in year $t+1$ as it was in year t—unless the discount rate has not changed, in which case the relative weights constitute a geometric sequence with the property $W_{t+1}/W_t = W_t/W_{t-1}$, where W_t = the discount factor weight in year t, and so on.

Assuming that the constancy of the discount rate cannot be taken for granted, why is the implied inconsistency in planning from one period to the next a matter of concern? The reason, as demonstrated elsewhere (Krutilla and Fisher, 1975), is that the resulting allocation of goods over time is likely, in important cases, to be inefficient in a sense we shall describe very briefly here. Let each period represent a generation. Now, if the planner for each generation weights consumption by other generations with a discount factor based not only on a (let us assume) constant rate of time preference, but also on a constant adjustment reflecting preference for the consumption of one's own generation relative to all others, we obtain the following intuitively plausible results. In generation 1's optimal plan, generation 2 consumes less, relative to generation 3, than in generation 2's plan. In generation 2's plan, however, generation 1 consumes less, relative to generation 2, than in generation 1's plan. Accordingly, there exists between generations 1 and 2 the opportunity for a mutually beneficial exchange, in which generation 2 consumes less of its inherited stock in return for a larger inheritance from generation 1.

The problem, however, is that the indicated exchange is not possible, assuming there is no way for the generations to get together and negotiate a binding contract. One interpretation of the inability to realize the "gains from trade" is that the present-value maximization criterion is inefficient, or non-Pareto-optimal. Note the similarity here to the way Marglin determines a social discount rate, described in the preceding section. In Marglin's analysis, members of the *same* generation get together to provide, via a lower social discount rate, the public good of increased consumption opportunities for future generations. Since there is no barrier to negotiation, presumably the resulting allocation is efficient. In what we might call the Strotz problem, however, if there is no way in which individuals in *different* generations can get together to arrange a situation in which all would be better off, inefficiency results.

We shall argue that inefficiency in this rather special sense—though for somewhat different reasons—may also arise quite naturally where irreversible decisions concerning the control of water pollution are taken at different points in time. That is, it is possible that decision makers in a later period would prefer that a less environmentally destructive alternative had been chosen in some earlier period, and would if they could, find it worthwhile to bribe those who initially preferred the more polluting alternative to refrain from the undertaking. The problem, of course, is that the indicated exchange cannot be realized.

The situation is readily illustrated with the aid of the simple certainty-case, two-period models of pages 278-79. Recall that one possible solution (of the first model there) is $(b_1 - c_1^e) > 0$, $\alpha_2(c_2^e - b_2) > 0$, with $(b_1 - c_1^e) < \alpha_2(c_2^e - b_2)$, in which case present value is maximized by not investing in the first period, even though this involves an opportunity loss of first period returns $(b_1 - c_1^e)$. Now suppose $(b_1 - c_1^e) > 0$, $\alpha_2(c_2^e - b_2) > 0$, but $(b_1 - c_1^e) > \alpha_2(c_2^e - b_2)$. Then present value is maximized by investing, with $V_{max} = (b_1 - c_1^e) + \alpha_2(b_2 - c_2^e) > 0$. However, as the vantage point of decision makers slides through time to the second period, the net value of the investment becomes negative: $V_2 = b_2 - c_2^e < 0$, where V_2 = present value of the polluting activity from the vantage point of second period decision makers. Moreover, if $(c_2^e - b_2) > (b_1 - c_1^e) + \alpha_2(b_2 - c_2^e)$—or, in other words, net losses from the activity as evaluated in the second period exceed net gains as evaluated in the first—then, in principle, the second-period net losers could bribe the first-period net gainers to refrain, were negotiation possible. Although this result has much the same flavor as the inconsistency-inefficiency result in the Strotz problem described above, note that it is explained differently. In the former case, the inconsistency is due to shifting relative discount weights, whereas in this latter case it arises because values are shifting: the values of those in the earlier period favor the polluting activity, which is not the case when the activity is evaluated at a later period.

Concluding
Remarks

SOME MAY BELIEVE THAT ALL ECONOMIC ACTIVITY IS IRREVERSIBLE, in the sense that time does not move backwards, and others (or perhaps even the same individuals!) that nothing is truly irreversible, since sufficient application of technique and conventional resource inputs can reproduce any desired physical environment. We have argued that meaningful and useful distinctions can be made between reversible and irreversible activities, between replaceable and irreplaceable resources. In making this argument we recognize that, as with other useful abstractions of economic theory such as "perfect competition" or "public goods," this distinction is often a matter of degree. One form of our argument would then be that, in cases of the degradation of bodies of water due to pollution of the sort described earlier in this paper, irreversibility may be an important consideration for those responsible for regulating and controlling waste discharges and related activities.

Once this premise is granted, the question is: How does irreversibility constrain environmental resource use? An important implication of our analysis, perhaps not surprisingly, is that the optimal commitment of resources to activities that are (irreversibly) destructive of the environment is smaller than commitments to activities whose consequences are reversible. This conclusion is strengthened if there is uncertainty as to the magnitude of the consequences, and inconsistency in their evaluation over time.

With respect to these latter considerations it should be acknowledged that the present discussion is not intended to be exhaustive. In closing we suggest at least two promising directions for future study. As Ralph d'Arge has noted, an interesting and probably important type of uncertainty in pollution problems has not been considered: namely, uncertainty over whether a particular course of action is, in fact, irreversible. In the present discussion uncertainty exists only with respect to the magnitude of effects that are assumed known to be irreversible (or reversible, as the case may be). Clearly a more general treatment would permit us to explore the implications of an assumption that, for example, continued discharge of waste materials into a given body of water for a period of x years at the level and composition that prevails today would result in an increase in the probability of irreversible degradation from, say, 10 to 90 percent.

A second area that seems worthy of further study is the relationships of the usual criteria for welfare improvements to shifts in time perspective. The discussion (in the previous section) of inconsistency and inefficiency in present value maximization represents one attempt to explore this territory. These particular issues are treated in more detail elsewhere (Krutilla and Fisher, 1975), but there remain questions of equity, generally in the distribution of goods across generations and over long periods of time,[6] and of possible special considerations with respect to both equity and efficiency in the use of the life support system of air, water, and other environmental resources over long periods of time.[7]

DISCUSSION

Discussant: *Ralph C. d'Arge*

While the analysis of Fisher and Krutilla has been pathbreaking in analyzing irreversibility and option value problems in economics, I do not believe that they have yet achieved a synthesis general enough for policy making.

What also bothers me is the role of the social rate of discount which establishes what happens in the kind of problem dealt with here. For example, it is clear that a social rate of discount could be high enough to justify the net benefit stream emanating from a project, even though preservation would yield substantial future benefits. Likewise, on the basis of negative benefits less costs, a low social rate of discount can prevent the project from being undertaken, if preservation benefits are permitted to grow over time without limit. Thus, one of the essential problems I see with the paper is that there is little generality in the results, even though Fisher and Krutilla have made more than substantial contributions to our understanding of the economics of preservation. Most of the results lack what some analysts call *robustness*.

REFERENCES

1. Arrow, K. J. *Social Choice and Individual Values* (New York, 1951).
2. Arrow, K. J., and Fisher, A. C. "Environmental Preservation, Uncertainty, and Irreversibility," *Quarterly Journal of Economics* (May 1974).
3. Arrow, K. J., and Lind, R. C. "Uncertainty and the Evaluation of Public Investment Decisions," *American Economic Review* 40 (June 1970).
4. Baumol, W. J. "On the Social Rate of Discount," *American Economic Review* 58 (September 1968).
5. Cicchetti, C. J., and Freeman, A. M. "Option Demand and Consumer Surplus: Further Comments," *Quarterly Journal of Economics* 85 (October 1971).

[6] For a discussion of equity and the choice of welfare criteria, see Rawls (1971).
[7] For further discussion of some of these considerations, see Leopold (1949) and, most recently, Page (1973).

6. Cummings, R., and Norton, V. "The Economics of Environmental Preservation: A Note," *American Economic Review* (December 1974).
7. Eckstein, O. *Water Resource Development: The Economics of Project Evaluation* (Cambridge: Harvard University Press, 1958).
8. Feldstein, M. S. "The Social Time Preference Discount Rate in Cost-Benefit Analysis," *Economic Journal* (June 1964).
9. Hirshleifer, J. "Comments," in *Public Finances: Needs, Sources and Utilization* (Princeton: NBER-Princeton University Press, 1961).
10. Hirshleifer, J. "Investment Decision Under Uncertainty: Choice-Theoretic Approaches," *Quarterly Journal of Economics* 79 (November 1965).
11. Hirshleifer, J. "Investment Decision Under Uncertainty: Applications of the State-Preference Approach," *Quarterly Journal of Economics* 80 (May 1966).
12. Krutilla, J. V., and Eckstein, O. *Multiple Purpose River Development* (Baltimore: Johns Hopkins University Press, 1958).
13. Krutilla, J. V., and Fisher, A. C. *The Economics of Natural Environment: Studies in the Valuation of Commodity and Amenity Resources* (Baltimore: Johns Hopkins University Press, 1975).
14. Leopold, A. *A Sand County Almanac* (London: Oxford University Press, 1949).
15. Marglin, S. "The Social Rate of Discount and the Optimal Rate of Investment," *Quarterly Journal of Economics* 77 (February 1963).
16. Marglin, S. "The Opportunity Costs of Public Investment," *Quarterly Journal of Economics* 77 (May 1963).
17. Mishan, E. J. "A Proposed Normalisation Procedure for Public Investment Criteria," *Economic Journal* 77 (December 1967).
18. Page, R. T. *Economics of Involuntary Transfers* (New York: Springer Verlag, 1973).
19. Rawls, J. *A Theory of Justice* (Cambridge: Harvard University Press, 1971).
20. Samuelson, P. A. "The Pure Theory of Public Expenditure," *Review of Economics and Statistics* 36 (November 1954).
21. Sandmo, A. "Discount Rates for Public Investment Under Uncertainty," *International Economic Review* (June 1972).
22. Stankey, G. H. "A Strategy for the Definition and Management of Wilderness Quality," in J. V. Krutilla, ed., *Natural Environments: Studies in Theoretical and Applied Analysis* (Baltimore: Johns Hopkins University Press, 1972).
23. Steiner, P. "Choosing Among Alternative Public Investments in the Water Resource Field," *American Economic Review* (December 1959).
24. Strotz, R. "Myopia and Inconsistency in Dynamic Utility Maximization," *Review of Economic Studies* 23 (1955-1956).
25. Weisbrod, B. "Collective-Consumption Services of Individual-Consumption Goods," *Quarterly Journal of Economics* 78 (August 1964).

Part Four

POLICY ISSUES
AND
INSTITUTIONS

Voting, Cost-Benefit Analysis, and Water Pollution Policy

PAUL R. PORTNEY

The political system is one important mechanism for the allocation of resources (dominant in a socialist system) of major importance in virtually all modern systems. The participants in the political system, the voters, are after all the same individuals as the participants in the market system, and it is at least an interesting hypothesis that their choices as voters are governed by the same preference systems as those which determine market choice.[1]

The other papers in this volume, with the possible exception of Stockfisch's on the distribution of property rights, concern themselves with the *economic* analysis of various aspects of water pollution policy. More specifically, they concentrate on the application of a particular method of resource allocation (i.e., cost-benefit analysis) to problems of water resource management.

As Arrow and Scitovsky observe above, however, resources may be allocated by a political as well as a market mechanism and there is much evidence of the growing importance of nonmarket allocations. In 1972, for example, combined expenditures of federal, state, and local governments amounted to $371.6 billion dollars, more than 32 percent of GNP. [2] Accordingly, it is the purpose of this paper to discuss the role of *political* mechanisms in the allocation of resources to water quality pro-

---

[1]Kenneth Arrow and Tibor Scitovsky, "Political Aspects of Welfare Economics," in *Readings in Welfare Economics,* edited by Arrow and Scitovsky (Homewood, Ill.: Irwin, 1969), p. 113.

[2]*Survey of Current Business,* 53, no. 5 (May 1973): 9, 12.

grams. Special attention will be directed to the uses of referendum voting by individuals and the uses of legislative voting (and vote-trading) by elected representatives of the people. We will also make brief mention of the Lindahl method for determining public expenditures. This method involves no voting, as such, but is clearly a political mechanism relevant to resource allocation.

The first part of this paper examines the nature of water quality programs and discusses the reasons why the market cannot be relied upon to provide a satisfactory allocation of resources. In the second section, we discuss in some detail the way in which individual voting might help us choose the optimal level of output of an essentially indivisible "social good" like water quality. The voting "solution" we analyze there is the one due primarily to Bowen but later modified by others. In the third section we turn away from theory and toward the practical applications of voting to questions of resource allocation, paying special attention to the use of referenda on matters of environmental importance. In the fourth section, we discuss how a different form of voting, legislative voting, may be expected to alter patterns of resource allocation, especially in the area of water quality. The final section is devoted to recommendations.

Private Goods, Public Goods, and Water Pollution Policy

WE MIGHT WELL BEGIN OUR ANALYSIS by asking why we need to consider voting or other nonmarket mechanisms in a discussion of water pollution policy. Other goods are provided to consumers without recourse to voting, and policy makers whose main concern is not economic or social choice theory might be justified in thinking that market forces would direct resources to water quality programs as well.

There are, however, features of water resource management which make market allocation most difficult. This difficulty is due to the "public good" nature of water quality. While the distinctions between public and private goods are drawn most carefully elsewhere (see Samuelson [25, 26], Buchanan [6], and Bowen [5]), it might prove helpful to discuss briefly the public nature of water quality improvements.

In considering the levels of water quality which the many members of a community might desire, it quickly becomes apparent that those members have no recourse to a private market from which they might purchase the exact level they desire. Similarly, it would be most impractical for any community to provide a wide range of water qualities from which its citizens might choose—even if the municipality were large enough to justify more than one treatment facility, it would be exceed-

Figure 1

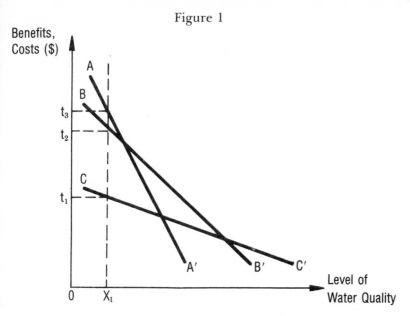

ingly difficult to determine which users received which level of water quality and to get the proper level to each user.

Because of this indivisibility, a single level of water quality is somehow selected to be provided to all the members of the community. In theory, once this quantity is selected (by whatever means the community chooses), efficiency can be guaranteed if the benefit schedules of each citizen are known. In such a case, as illustrated for a three citizen community in Figure 1, tax shares are assigned to each citizen in such a way that the marginal benefits to him of the level of water quality provided are exactly equal to the marginal cost which he must bear.

In Figure 1 quality level $0X_1$ has been chosen. Then tax $0t_3$ is assigned individual A with marginal benefit schedule AA'; $0t_2$ is assigned individual B with marginal benefit schedule BB'; and $0t_1$ is assigned to individual C with benefit schedule CC'. This process is seen to be similar to the operation of a competitive market where individuals adjust quantities in response to fixed prices, except that we have here fixed the level of water quality which each individual is to consume and have varied the price faced by each consumer in accordance with his benefit schedule.

The reason such an apparently simple mechanism is *not* used to determine the output of public goods and tax shares which individuals bear is that information about demand or benefit schedules is not easily obtainable. For reasons discussed thoroughly elsewhere (Samuelson [25, 26] and Musgrave [20], for example), it would be in an individual's interest not to honestly reveal his true benefit schedule.[3] Furthermore, even if such information were discernible or honestly revealed, the fact that municipalities supply many public goods to many users makes an effective bookkeeping system whereby each user is taxed for each public good in accordance with his benefits received a virtual impossibility.

Since individuals cannot purchase and consume their desired level of water quality by market action, and since it seems impossible to arbitrarily select some level of water quality and assign tax shares in accordance with individuals' benefits received, we are faced with the problem of selecting a single water quality level in the face of differing opinions as to what it should be. It is to voting solutions to this problem that we now turn.

Individual Voting

IF A JURISDICTION WERE ABOUT TO EMBARK on a major water cleanup program, and if perfectly accurate information about the potential benefits and costs of the cleanup program were available to the members of the community, the jurisdiction would be able to choose an optimal level of water quality. This is depicted as $\overline{X}$ in Figure 2. That is, at water quality level $\overline{X}$ total marginal benefits (TMB) for the jurisdiction would be equal to total marginal costs (TMC)—assumed to be constant in this example. Therefore, $\overline{X}$ is the quality level at which total *net* benefits are maximized.

In the absence of this "perfect" information, is there a way that we can use individual voting by the members of the community to arrive at water quality level $\overline{X}$? Bowen [5] suggests that we can, by allowing each member of the community to vote for the level of water quality he desires most.[4] Individuals (who, Bowen obviously assumes, have access to considerable information) would presumably select the level at which their own marginal benefits are equal to the marginal cost they would expect to incur for that quality of water. Their marginal cost, of course, depends on their share of the total marginal cost of additional units.

[3]Since writing the original draft of this paper, I have come across a paper by Tideman [27] which proposes a solution to the problem of revealing preferences through a questionnaire approach. This approach, while of theoretical value, appears to be of limited practical value, requiring, as it does, responses to somewhat technical questions concerning marginal benefits and the elasticity of the marginal benefit schedule. It does appear, however, to eliminate in most instances the incentive for an individual to conceal his preferences.

[4]We are modifying Bowen's method to fit the water quality problem. He used education as his example although he clearly intended to include all "social goods" which could not be marketed individually.

Figure 2

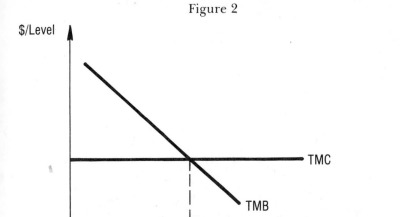

$/Level

TMC

TMB

0   X̄   Quality Level

According to Bowen, that level of water quality which received the most votes (we shall refer to this as the "modal level") would in fact be X̄, the level which maximizes the total net benefits, as long as several key assumptions are satisfied. These assumptions are: (1) that all individuals vote and vote honestly; (2) that the cost of providing any quantity of the social good are known; (3) that these costs are divided evenly among the members of the community; and (4) that for any given quantity of the social good the citizens' marginal evaluations are normally distributed about the modal evaluation. This last assumption is illustrated in Figures 3A and 3B.

Figure 3A

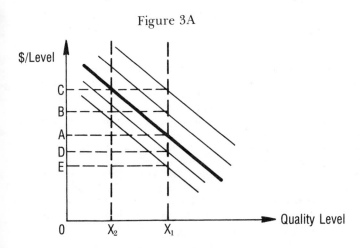

$/Level

C
B
A
D
E

0   X₂   X₁   Quality Level

298

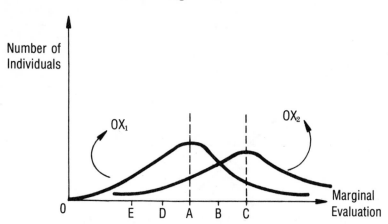

In Figure 3A, individual demand (or marginal evaluation) schedules of the social good are plotted.[5] Bowen's fourth assumption is that, for any output of the social good ($0X_1$ for instance), individuals' marginal evaluations of that level of output will be normally distributed about some modal evaluation, 0A in Figures 3A and 3B. Thus, when we consider a different level of output ($0X_2$ for example), we are faced with another distribution of marginal evaluations about a new mode (0C this time). The curve which connects these modal evaluations is the heavy line in Figure 3A.

If this method is to provide a practical solution to the problem of selecting an optimal level of water quality when benefit estimation is difficult or impossible, these four assumptions must be realistic. Let us consider them *seriatim*.

First, we know that voting turnouts are less than perfect and, in fact, the turnout is generally lowest in elections concerning nonpolitical public expenditure questions.[6] Abstention may not be a serious problem,

[5]The derivation of the demand curve for any good requires that the good be supplied at a constant price for any and all quantities; i.e., for the buyer, marginal cost equals average cost. While this assumption is a reasonable one for most private goods, Buchanan [6], pp. 39-40, points out that quantity discounts may be much more common in the provision of public goods, thus making the derivation of a demand curve impossible. In Figure 3A, we assume that a horizontal supply curve exists.

[6]In Arlington County, Virginia, for example, a recent bond issue for wastewater treatment was voted upon by less than 38 percent of registered voters even though the referendum appeared on the same ballot with the 1972 presidential election. In Montgomery County, Maryland, known for its high voting participation rates, only 57 percent of all registered voters cast ballots on a proposition which involved land acquisition for sewers in November, 1972, even though more than 81 percent of them voted for a presidential candidate on the same ballot.

however, if individuals who prefer a relatively low level of water quality are as likely to abstain as those preferring an intermediate or high level of water quality. If this is the case, the level of water quality which receives the greatest number of votes might still be the optimal level. If, for some reason, however, the likelihood of abstention is skewed in one direction or the other, the voting solution cannot help us in the selection of a water quality level.

Evidence of such a skewness exists, unfortunately. Both Lane [15] and Dahl [8] have established a strong correlation between political participation (voting in particular) and income level, indicating that the poor are much less likely to vote than those from higher income groups. In addition, Wilson and Banfield [31] have analyzed voting data from municipal elections in a number of large cities, which indicate that voters from low income precincts (especially where the percentage of owner-occupied housing is low) are much more likely to approve public expenditure issues than other income groups. Taken together, these studies suggest that individuals most likely to abstain from voting may be those who would most likely approve high outputs of social goods. However, with respect to water quality, those favoring higher levels, may, in fact, be the relatively high income groups.[7] In any event, that level of water quality which receives the most votes in a Bowen-type expression of preference may not be the socially optimal level, as Bowen claims.

The second assumption—that we know the cost of providing various levels of water quality—can also present problems, as the papers dealing with pollution costs at this conference point out (see the papers of Kneese and of Hanke and Gutmanis, in particular). While we are clearly making progress on the technological component of the costs of water quality, accurate estimates of social costs are still not available, and this makes difficult any accurate determination of the total cost of water cleanup programs.

Third, the cost of any quantity of the social good (the level of water quality, here) must be divided equally among the citizens. This is unlikely to be the case, however, if waste treatment facilities for improving municipal water quality levels, or major water pollution cleanup programs, are financed by increases in property taxes. This follows because the ownership of property differs widely between individuals (renters

[7]Myrick Freeman has pointed out to me that the social good "water quality" may, in fact, differ from other social goods in such a way that the wealthy might tend to favor higher levels of it than the poor. If water quality includes the cleanup of a river or lake, in addition to improvements in municipal tap-water quality, the wealthy may indeed desire higher quality levels than others because they can afford boats which provide them access to lakes and rivers. At any rate, it is most unlikely that the first effect (the poor preferring higher levels of public goods than the wealthy) would interact with the effect just discussed in such a way that the modal level revealed in an election turned out to be the true mode.

are considered to have no property), although property tax rates are the same for all residential, commercial, and industrial property; hence, the marginal cost of water quality programs and other public goods will vary substantially between individuals. To the extent that the property tax is not passed on to renters, in fact, increases in water quality financed by property tax increases become available to them at no extra cost. At any rate, because we use the financing of social goods to achieve both allocative and redistributive goals, it is seldom that the costs of such goods are distributed equally among citizens. There is no reason why equal cost sharing could not be the rule, however, with redistribution occurring through other more explicit transfers and taxes.

Bowen's fourth assumption—that individuals' marginal evaluations of any given level of the social good are normally distributed about some modal evaluation—is also unlikely to be satisfied.[8] This is so because individuals' marginal evaluation schedules will depend on the distribution of income, a distribution which is skewed rather than symmetrical.

In considering potable water, for example, it is likely that very wealthy individuals would tend to have low marginal evaluations of any given water quality level because they are much more able to afford substitutes for publically provided drinking water—e.g., bottled water, wine, milk. Very poor individuals, however, cannot afford alternatives to municipally provided drinking water and would, therefore, attach much higher marginal evaluations to any given level of water quality. Given these income effects on marginal evaluations, we would only expect the distribution of these marginal evaluations to be normal about the modal evaluation if the distribution of income were normal or if, by chance, marginal evaluations somehow differed systematically between income levels in a way that produced the required normal distribution. Since the latter eventuality borders on the absurd and since the size distribution of income is skewed in such a way that the modal personal income *is* less than the median income,[9] we must conclude that Bowen's fourth assumption is also unrealistic.

We have devoted much attention to Bowen's voting solution because of its originality and because the attempts to modernize it and put it into practice in the thirty years since it was written have stuck so closely to Bowen's original assumptions. Indeed, Borcherding and Deacon [4]

---

[8] I am indebted to Henry Peskin for pointing this out to me after I had overlooked it in the earlier drafts.

[9] For evidence of this skewness, see Joseph Pechman and Benjamin Okner, "Individual Income Erosion by Classes," in *The Economics of Federal Subsidy Programs, a Compendium of Papers Submitted to the Joint Economic Committee*, Part I, *General Study Papers*, 92 Congress, 2 sess. (1972), pp. 13-40; also, see the Current Population Survey of the Census Bureau, Series P-60.

assume equal cost sharing and merely make explicit Bowen's implicit assumption about the selection of that level of output favored by the median voter. Furthermore, Bergstrom and Goodman [2, p. 281] make the additional assumption that "in each municipality, the median of the quantities demanded is the quantity demanded by the citizen with the median income for that community." This enables Bergstrom and Goodman to predict the changes in the quantities of public goods demanded due to changes in income and other economic variables.

Bowen was pessimistic about the practical applicability of his voting solution because he felt that production under conditions of increasing cost would cause a divergence between marginal and average costs. We, too, remain pessimistic about the usefulness of such methods but for different reasons.

While we feel that water quality cleanup programs, especially those involving municipal water quality, can eventually be operated under constant or decreasing costs over a substantial range, other requirements seem unlikely to be fulfilled. As we pointed out earlier, not everyone votes, and abstention appears not to be a random process.

The dependence of individuals' marginal evaluations on a skewed income distribution is also likely to vitiate Bowen's conclusions about the applicability of the voting solution. Equally as important are the difficulties with the measurement of costs and the obstacles to equal cost sharing. While there is nothing inherent in the economics of taxation which would prevent equal sharing, we would agree with Daly and Giertz [9] who suggest that much less redistribution might take place if taxation were to be done on a "cash" basis, handled apart from the allocation of resources to social goods and services.[10] Therefore, while equal cost sharing might help us locate the optimal production of social goods through voting, it might also imply a reduction in the amount of income redistributed to lower income groups.

Although we will not examine it closely, we should point out that similar problems exist with the so-called "voluntary exchange" approach to the determination of public expenditures. This method, due to Wicksell [30] and Lindahl [16], involves the simultaneous determination of the total amount of public expenditure as well as the allocation of tax burdens among individuals. As Musgrave [20], Johansen [14], and Olson and Zeckhauser [21] have shown, the gains which households can make by concealing their true preferences for the social good make the volun-

[10]McGuire and Aaron [18] make essentially the same point. They say, "For most public expenditure decisions, the planner must regard supporting tax shares as fixed. In such cases, the creation of public goods or services is often the only way to effect income transfers from rich to poor, though from an overall allocative perspective it may be inefficient" (p. 33).

tary exchange method impracticable.[11] In addition, McGuire and Aaron [18] have shown that the voluntary exchange method would be of little practical use even if revelation of preferences were not a problem because of the unrealistic assumption which would have to be made about the pretax distribution of incomes.

Individual Voting in Practice

WE HAVE CONCLUDED THAT MODELS of individual voting like Bowen's or voluntary exchange models like Lindahl's are unlikely to be of much help to us in determining the optimal amount of investment to undertake in water quality programs. Nevertheless, the recent and increasing use of individual voting through state and local referenda on matters of great environmental consequence seems to indicate that individual voting has some appeal. This appears especially true when considering matters which relate to water quality. In California, Proposition 20 on the November, 1972, ballot proposed the creation of a Coastal Zone Conservation Commission which would have strict control over the development of land within 1000 yards of the Pacific coast. The creation of the commission was approved, reflecting the belief, perhaps, that private calculations concerning the development of California's Pacific coast ig-

---

[11]Note that such a problem could arise in the context of the Bowen model. Assume that an individual had demand schedule DD' in the figure below

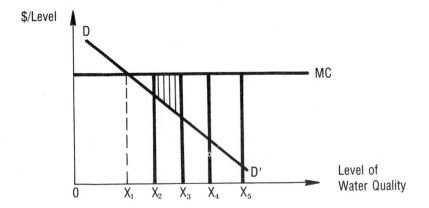

and suppose he knew that the distribution of votes for the 5 possible quality levels was 1, 4, 5, 4, 1, respectively. Although he would actually prefer quality level $0X_1$, he might rationally vote for $0X_2$ instead, if he thought he could change the mode and hence the actual level of output from $0X_3$ to $0X_2$. By doing so, he would reduce his excess of costs above benefits by the area of the shaded trapezoid.

nore important social costs which must be taken into account. Similarly, in New York, voters approved a $1.2 billion bond issuance, $650 million of which was to assist communities in the building of new sewer facilities and a smaller portion of which was to be used for wetland acquisition and preservation.

What accounts for this renewed interest in referendum voting and of what use is it to us in the allocation of resources? We suggest that in cases like those cited above, referendum voting is being used *not* as a complement to cost-benefit analysis in the way Bowen and others have suggested, but rather as a substitute for it.

There are two reasons why individual voting may be increasing in popularity as an allocative device at the expense of cost-benefit analysis. First, it may be the case that cost-benefit analyses, and particularly those involving projects with major environment-altering effects, have been improperly done. While other papers in this volume discuss this possibility at much greater length,[12] we should mention that cost-benefit studies have not included, or at least have not taken adequate account of, all relevant benefits and costs. This is possibly due to the existence of irreversibilities, uncertainty about future benefit and cost streams, and the difficulty in valuing intangibles. If this is true, as evidence presented elsewhere seems to suggest, it would partially explain the increasing use of referenda.

Elsewhere [23] I have examined a second reason for the increasing use of referendum voting for resource allocation. Briefly stated, the reason is this: even if cost-benefit analysis does take perfect account of *all* potential benefits and costs, it may fall into disfavor and come to be replaced by some other, perhaps political, allocative mechanism because of its distributional anonymity.

That cost-benefit analysis suffers because it identifies and compares total and marginal benefits with costs without identifying the recipients of each has been pointed out by Maass [17] and Weisbrod [28]. One of the reasons why this lack of explicit distributional consideration is disturbing is that we do use the provision of social goods—including water quality programs—to redistribute income. Failure to identify cost bearers and benefit recipients makes this process of redistribution through allocation less precise than it might otherwise be.

Another reason the distributional anonymity of cost-benefit analysis is vexing is that it makes possible the disgruntlement of a majority of voters. Notice that it is possible for a number of projects to be approved on efficiency grounds which provide great benefits to a single geographic area or to the members of a particular income class but which, nevertheless, leave a majority of citizen-voters worse off. This is not a

---

[12]Especially those of Fisher and Krutilla and of Bishop and Cicchetti.

problem if projects are designed and proposed in such a way that large
numbers of citizens eventually receive benefits in excess of costs—a kind
of logrolling process would emerge in which each citizen would be willing to accept small net losses on many projects in order to receive
substantial net gains on several others. However, the fact that projects
which have passed cost-benefit tests have begun to be defeated in public
referenda (the Texas Trinity River Project is an example), might be
taken as an indication that net benefits are not being spread around
sufficiently and that voters are reacting to this phenomenon. We must
expect this reaction to continue as long as those receiving the majority of
benefits from projects are not those bearing the majority of the costs.

From that standpoint, individual voting is a rational response to the
frustration of the majority, if that frustration is, in fact, the reason for
increasing use of referenda. For if projects are to be approved by a
simple majority of voters, as is the case on most referenda, then each
project must provide net benefits to at least a majority of those voting.[13]
If we assume, for example, that the costs of a major water pollution
cleanup program will be borne by many citizens in some jurisdiction
—perhaps because it is to be financed through property tax increases
—then at least half of those citizens must also expect benefits from the
program in excess of their cost if the program is to be approved by a
majority when put to a vote. This points, then, in the direction of water
quality programs which will benefit large numbers of individuals, and
away from those which provide substantial benefits to relatively few
individuals but which are financed by charges upon the many. The use
of individual voting under simple majority rule when used to direct the
allocation of economic resources may be expected to reduce the disparity which often exists between the number of individuals benefiting from
any project and the number of individuals who bear the costs.[14]

While allocation via majority rule may indeed prevent a majority of
voters from suffering net losses on any project, it is not without draw-
backs of its own.[15] Its most obvious shortcoming is that it can prevent
highly efficient projects from being undertaken and can permit the
undertaking of grossly inefficient ones. In the former case, projects with
substantial benefits directed at only a very small number of voters but
financed by charges borne by many can be expected to be defeated even
if the cost-benefit ratio is very low. Conversely, inefficient projects which

---

[13]We make the standard and somewhat unrealistic assumption that voters are motivated by strict self-interest (i.e., no utility interdependencies exist) and the more realistic
assumption that vote-trades are *not* generally made on referenda ballots because of the
large numbers of voters and the difficulty of enforcing the trades.

[14]For a more extensive discussion of a method to make benefit recipients the same
group as those who bear the costs of projects, see Olson [22].

[15]These are discussed at much greater length in Buchanan and Tullock [7].

may confer benefits barely in excess of costs to a majority of voters, but which also result in a minority bearing costs greatly in excess of the benefits they receive, may be approved. If taxes could be assigned discriminatorily in fact, a "tyrannous majority" of the sort Madison feared could continually approve projects which primarily benefited them by shifting the costs to the minority.

Allocation by majority rule may also prove problematical when considering new projects to which no previous funds have been committed. Since referenda generally provide us with nothing more than *yes* or *no* answers, it may be difficult to initially determine the proper scale of such projects—this is in contrast to allocation accomplished by cost-benefit analysis which would indicate some specific scale of operation. This problem should prove less serious when considering increments or decrements to existing levels of funding, however, since it is likely that changes which are sufficient to attract majority support would be forthcoming in a form of iterative process. In other words, if a school millage increase were to be initially defeated, the school board could repropose a slightly smaller increase which would stand a better chance of passage. Through a process like this, majority rule allocation could arrive at a determinate project scale.

The possible defeat of highly efficient projects or the passage of highly inefficient ones under an allocation by majority rule arrangement makes an allocative mechanism which combines efficiency and equity objectives appear most desirable. This leads us to an examination of legislative vote-trading and water quality programs, an examination which will comprise the next section of this paper.

Legislative Voting and Water Quality

ANOTHER WAY THAT RESOURCES COULD BE ALLOCATED to water quality programs and other social goods characterized by some degree of indivisibility is by local, state, and national legislatures. In such cases voting would still be used, but it would be done by elected representatives of the citizens and under circumstances considerably different from those which surround individual voting.

The most obvious and important difference between individual and legislative voting on resource allocation is that the latter generally takes place between a small enough number of representatives that vote-trades can be arranged and effective coalitions built which will affect the outcome of the voting. This is unlikely to be the case when individuals vote by the thousands in referenda on such allocative questions. Also, when vote-trades are arranged by legislators, role-call voting makes it possible for them to observe whether or not bargains are kept—this is not possible with individual voting, which takes place by secret ballot.

That legislative voting makes such trading a practical possibility is important because vote-trading has the potential for overcoming one of our objections to allocation by individual voting under majority rule. Specifically, legislative voting makes possible a set of trades which can ensure the passage of a highly efficient project which might, for example, clean up a polluted lake or river for recreational use even though the benefits may accrue nearly exclusively to local residents while the costs are spread over a much larger set of voters. Under individual voting, as we have pointed out, such a project might easily be rejected. If, however, the elected representative of that locality where the preponderance of benefits are expected to be felt promises his vote to other representatives on future projects designed to benefit their constituents, he may be able to "collect" enough votes from other representatives to ensure the passage of the project. If each representative is willing to support a number of very efficient projects which, nonetheless, might mean small net losses to his constituents, he may be able to garner the votes necessary to pass one or two projects which benefit his constituents greatly but which would have been defeated by a majority of representatives in the absence of such arrangements. Thus representatives may come to agreement on one issue (e.g., which harbors are to be dredged as part of an omnibus harbors bill) or on completely independent issues (e.g., "you support the dredging of my harbor and I will vote for your crime bill").[16]

There is no guarantee that such logrolling or vote-trading would occur only on efficient projects, however. To ensure that both efficiency and equity considerations are taken into account, resources could be allocated by a combination of cost-benefit analysis and representative majority rule. A criterion similar to this is supposed to be at work in Congress now. In committees like Public Works, for example, projects are considered which have been subjected to cost-benefit analysis by the Army Corps of Engineers to eliminate inefficient ones. Then, the efficient projects are discussed on the floor of Congress (if they are reported out of committee) where agreements are made between representatives. If the project is approved, it has theoretically passed a cost-benefit test and has also been approved by a majority of representatives.

In practice, however, this system has not worked well. First, from the cost-benefit standpoint, future benefits have been discounted at extremely low rates and certain costs have been ignored, thereby overstat-

---

[16]It may be convenient to call such arrangements between legislators on the formation of any one issue *logrolling* and reserve the term *vote-trading* for arrangements between representatives on independent issues.

ing the attractiveness of many projects and allowing inefficient projects to be considered. Also, committee chairmen have often been able to kill legislation in committee which might have attracted majority support had it reached the floor. In this way the political aspect of the twin criteria is frustrated.

One response to these problems has been to take water resource management decisions out of the political arena and make them the concern of specially created boards or commissions. The Potomac River Basin Commission, the San Francisco Bay Commission, and California's new Coastal Zone Conservation Commission are notable examples. Furthermore, these boards or commissions are often advocated in academic analyses of water management policy. For example, in their discussion of the optimal management of a hypothetical watercourse area, Whinston and Ferrar [29] write:

> In an attempt to insulate the control of the region's water resource from the pressures of party politics, the governing body [federal, state, or local government] sets up a Water Control Board (W.C.B.). This W.C.B. will have the legislative and judicial responsibility of specifying a standard for water quality and enforcing this standard in the region's watercourses.

Elsewhere, in discussing environmental policy reform, Roberts [24] has suggested:

> For each river basin an authority would be created by the federal government as a special purpose public agency to plan for and manage water quality for the basin as a whole.

To the extent that the boards or commissions are staffed by appointed rather than elected officials, this is an unfortunate development. There is no reason why decisions about water quality and wetland preservation or reclamation need to be made apart from decisions about education, garbage collection, or fire protection. The strength of representative government lies in its ability to reconcile conflicts between individuals with highly disparate tastes over a wide range of issues. No compromises can be reached, no winning coalitions formed, and no protection to minorities offered when questions of water quality management are resolved independently of other issues by such boards and commissions comprised of appointed officials.[17] If, on the other hand, water quality management decisions are handled at the appropriate legislative level with a wide range of other issues, those who strongly favor vigorous water pollution cleanup programs will have the opportunity to build winning margins by promising, through their representative, support on other issues of less pressing concern to them. Furthermore, candidates for office can engage in "implicit" logrolling by run-

---

[17]Both Haefele [12] and Freeman and Haveman [11] have made this point elsewhere.

ning on platforms which cater to intense minority interests. As Haefele [12] has demonstrated, a properly functioning two party system which incorporates implicit logrolling can be expected to arrive at the same set of outcomes as would obtain if the citizens of the jurisdiction met in a general assembly and voted on the issues according to their individual preferences.

Recommendations for Water Quality Management Policy

WHERE DOES ALL THIS LEAVE THE POLICY MAKER charged with designing new institutions or making use of existing ones to properly manage water quality? Several points deserve reemphasis. First, we have little reason to believe that individual voting can lead us to the optimal level of water quality. Voting patterns and participation are too irregular, cost shares too unevenly distributed, and preferences generally too unsymmetrical across voters to allow us to claim any real-world relevance for the individual voting models we have discussed. This is not to say that we should give up on such research, however. Perhaps jurisdictions can be redrawn to produce the symmetrical distribution of preferences which individual voting models require. Furthermore, as we pointed out, there is no law against distributing the costs of social goods equally and handling any desired income redistribution through explicit taxes and transfers. Also to be encouraged are studies which would identify the individual and market demand curves for public goods—accurate estimations would greatly facilitate accurate cost-benefit analyses.

Second, the increasing use of individual voting on state or local referenda, particularly with regard to matters of water and air pollution and land use, should be at least partially interpreted as dissatisfaction with the outcomes arrived at by cost-benefit analysis. This dissatisfaction may be due either to inaccurate cost-benefit analyses which have resulted in underestimates of costs (or overestimates of benefits) or to perverse distributional effects resulting from the use of distributionally anonymous cost-benefit analyses for the allocation of resources. This last point may be especially important if members of the "middle class" feel that the provision of social goods has benefited higher or lower income groups at their expense.

Finally, we should reemphasize our conclusion that questions of water quality management should be resolved in federal, state, or local law-making bodies while they are considering other issues which concern them. In some instances where jurisdictional overlap is a problem, water quality problems may require the redrawing of municipal or perhaps even state boundaries or, at the very least, the creation of regional governments or decison making bodies which encompass entire watersheds. Also, officials responsible for water quality decisions should be responsible for decisions on other social goods and political questions, and these officials must be accountable to those people who are affected by their decisions. It is important from an efficiency standpoint that cost-benefit analysis be continually modified in accord with recent theoretical and empirical developments. It is equally important that the legislative trading mechansim envisioned in the not-so-recent *Federalist Papers* be used when important equity questions about water quality are to be resolved.

DISCUSSION

DISCUSSANT: *Henry S. Rowen*

I suggest that much more analysis and design—and I emphasize design—is needed on the voting process questions in this paper.

There are four problem areas which I believe should be thought about in connection with the design of voting processes. One is the *problem of information*—how the voters, legislators, and other participants generally get information, and what can be done to affect this by way of building institutions to generate and develop information on long-term as well as short-term consequences of programs.

The second is the *problem of bias*. It is well known that any given institutional structure runs the danger of bias and must include safeguards against it. Accordingly, the question of bias should be addressed analytically.

The third is the *problem of rule*—the way things are done, the legal procedures. Regulatory procedures have a tremendous impact on outcome. The examination of alternative rules is, therefore, extremely important.

And fourth, there is the *problem of incentives*. What incentives are there within the existing organizations, agencies, legislative committees, and firms involved? And what individual incentives are there —such as incentives to vote?

Although these are not altogether new problems, they are important and require a good deal more systematic study.

DISCUSSANT: *A. Myrick Freeman III*

One of the issues that emerges in the paper is whether voting is a complement to or a substitute for cost-benefit analysis. I am not sure it is the correct question to ask, but nevertheless it leads into some interesting areas. First, it implies that cost-benefit analysis is possible on the issue being posed for the voters, and, second, that a decision that is reached when cost-benefit analysis and voting are used together, is in some sense "better" than if cost-benefit analysis alone dictated the decision.

The issue is: What do we mean by "better"? When we say "better," we are in effect saying there are criteria beyond those that maximize benefits over costs. This may very well be true, and in that case using cost-benefit analysis as a component of the information that is given to voters could lead the voters to reject or modify a proposition, or add other elements into the weighting factors that are used to arrive at a final decision. Second, could cost-benefit analysis be a substitute for voting or vice versa? That is, would voters give identical or equivalent answers to the results obtained from cost-benefit analyses? I do not think that this is likely, for many of the reasons that are suggested in the paper.

Third, it is possible that the question of substitutability or complementarity is irrelevant, in that the two—voting and cost-benefit analysis—really apply to different domains of questions. I think the paper suggests that possibility.

REFERENCES

1. Arrow, K., and Scitovsky, T. "Political Aspects of Welfare Economics." In *Readings in Welfare Economics,* edited by Arrow and Scitovsky. Homewood, Ill.: Irwin, 1969.
2. Bergstrom, T., and Goodman, R. "Private Demands for Public Goods." *American Economic Review* 63 (June 1973): 280-96.
3. Birdsall, W. "A Study of the Demand for Public Goods." In *Essays in Fiscal Federalism,* edited by R. A. Musgrave, pp. 235-95. Washington, D.C.: Brookings Institution, 1965.
4. Borcherding, T., and Deacon, R.T. "The Demand for the Services of the Non-Federal Governments." *American Economic Review* 62 (December 1972): 891-901.
5. Bowen, H. "The Interpretation of Voting in the Allocation of Economic Resources." *Quarterly Journal of Economics* 58 (November 1943): 27-48.
6. Buchanan, J. *The Demand and Supply of Public Goods.* Chicago: Rand-McNally, 1968.
7. _____, and Tullock G. *The Calculus of Consent.* Ann Arbor: University of Michigan Press, 1967.
8. Dahl, R. A. *Who Governs? Democracy and Power in an American City.* New Haven: Yale University Press, 1961.
9. Daly, G., and Giertz, F. "Welfare Economics and Welfare Reform." *American Economic Review* 62 (March 1972): 131-38.
10. Davis, O., and Barr, J. "An Elementary Political and Economic Theory of the Expenditures of Local Governments." *Southern Economic Journal* 33 (October 1966): 149-65.
11. Freeman, A.M., and Haveman, R. "Water Pollution Control, River Basin Authorities, and Economic Incentives: Some Current Policy Issues." *Public Policy* 19 (Winter 1971): 53-74.

12. Haefele, E. "Environmental Quality as a Problem of Social Choice." In *Environmental Quality Analysis,* edited by Bower and Kneese, pp. 281-332. Baltimore: Johns Hopkins University Press, 1972.

13. _____ "A Utility Theory of Representative Government." *American Economic Review* 61 (June 1971): 350-67.

14. Johansen, Leif. *Public Economics.* Amsterdam: North Holland, 1965.

15. Lane, R.E. "Political Involvement Through Voting." In *Voting, Interest Groups and Political Parties,* edited by B. Seasholes. Glenview, Ill., 1966.

16. Lindahl, E. "Just Taxation—A Positive Solution." In *Classics in the Theory of Public Finance,* edited by Musgrave and Peacock, pp. 168-76. New York: St. Martin's, 1967.

17. Maass, A. "Benefit-Cost Analysis: Its Relevance to Public Investment Decisions." *Quarterly Journal of Economics* 80 (May, 1966): 208-26.

18. McGuire, M., and Aaron, H. "Efficiency and Equity in the Optimal Supply of a Public Good." *Review of Economics and Statistics* 51 (February 1969): 31-39.

19. Mishan, E. J. *Cost-Benefit Analysis.* New York: Praeger, 1971.

20. Musgrave, R. *The Theory of Public Finance.* New York: McGraw-Hill, 1959.

21. Olson, M., and Zeckhauser, R. "An Economic Theory of Alliances." *Review of Economics and Statistics* 48 (August 1966): 266-79.

22. Olson, M. "The Principle of Fiscal Equivalence." *American Economic Review* 59 (May 1969): 479-87.

23. Portney, P. "Cost-Benefit Analysis and Majority Rule." An unpublished paper presented at the Annual Meeting of the Public Choice Society, College Park, Maryland (March 1973).

24. Roberts, M. "Organizing Water Pollution Control." *Public Policy* 19 (Winter 1971): 75-142.

25. Samuelson, P.A. "The Pure Theory of Public Expenditure." *Review of Economics and Statistics* 36 (1954): 387-89.

26. _____ "A Diagrammatic Exposition of a Theory of Public Expenditure." *Review of Economics and Statistics* 37 (November 1955): 350-56.

27. Tideman, T. "The Efficient Provision of Public Goods." In *Public Prices for Public Products,* edited by Selma Mushkin, pp. 111-24. Washington, D.C.: The Urban Institute, 1972.

28. Weisbrod, B. "Income Redistribution Effects and Benefit-Cost Analysis." In *Problems in Public Expenditure Analysis,* edited by Samuel Chase, pp. 177-209. Washington, D.C.: Brookings Institution, 1968.

29. Whinston, A., and Ferrar, T. "Taxation and Water Pollution Control." *Natural Resources Journal* 12 (July 1972): 307-17.

30. Wicksell K. "A New Principle of Just Taxation." Reprinted in *Classics in the Theory of Public Finance,* edited by Musgrave and Peacock, pp. 72-118. New York: St. Martins' Press, 1967.

31. Wilson, J., and Banfield, E. "Voting Behavior on Municipal Public Expenditures." In *The Public Economy of Urban Communities,* edited by Julius Margolis, pp. 74-91. Baltimore: John Hopkins University Press, 1965.

Cost-Benefit Analysis, Property-Right Specification, and "Distribution"

J. A. STOCKFISCH

Introduction

THE CONCEPT OF PROPERTY is central to the subject of resource allocation and management. The analytical apparatus of cost-benefit analysis deals with the allocation of resources within the public sector and also between the private and public sectors. But the key words *benefit*, and especially *cost*, possess certain elusive qualities, particularly in the context of an economy as a whole (or what economists term a "general equilibrium" framework).[1] In the same vein, the word *property* presents difficulties because, first, it is too general and, second, it has given rise to most of the complex social arrangements that are the essence of a society and its political economy.[2] When the focus on property is shifted to *property rights*, attention can then be directed to how "things" are used by people and what human ends they might serve. The way property rights are specified with reference to actual things or "instruments" bears on the way the things are used, and so affects the "benefits" that things provide. Since "costs" are foregone benefits, property-right specification becomes relevant to the subjects treated by cost-benefit analysis.

Almost any activity which entails resource use by the state, and which is therefore an object of cost-benefit analysis, will certainly alter property-right assignments between private individuals and groups. Both the value-adding (or income-generating) capacities of privately-owned productive agents and the qualities of the end products produced

---

[1]"When stated with full qualifications, the doctrine of opportunity cost degenerates into the conditions of a general equilibrium"—Paul A. Samuelson, "Welfare Economics and International Trade," *American Economic Review* (June 1938): 263.

[2]For a small but provocative sample of selected passages from the writings of diverse thinkers from Blackstone through Bentham and Ely to Lippmann, see John E. Cribbet, et al., *Cases and Materials on Property,* 3d edition, (Mineola, N.Y.: Foundation Press, 1972), pp. 1357-75.

by those agents will, or can, be affected. Such effects are especially evident in the case of indivisible facilities and projects like bridges, irrigation, flood control, harbors, airports, and so on, that provide instrumental services to private producers, and which were the initial focus of formal cost-benefit analysis. The presence of such facilities and the services they provide have often had happy and dramatic effects on the productivity and hence the value of nearby land, thereby enhancing the holdings of specific parties.

Pervasive but more subtle effects of government use of resources center on the ways in which police and legal effort is allocated to the enforcement of the various property rights relevant to different activities and outputs. Thus, if a city allocates ample law enforcement resources to recovering stolen automobiles and apprehending car thieves, more city residents may feel it worthwhile to own a car. If, at the same time, meager resources are made available to enforce, say, antipollution laws, the community will tend to have more automobiles and suffer from more pollution than it would if enforcement resources were allocated with an opposite emphasis. With limited law enforcement budgets there is opportunity for police officials to weight their enforcement efforts in one direction or another in ways which affect the value of property rights and, hence, the allocation of society's resources in the broadest sense. Of course, police bureaucrats are generally responsive to signals emanating from civil legislators and administrators who control and monitor policy agency budgets, as well as signals from influential and politically powerful citizens. But whatever the rationale by which law enforcement resources are allocated, it will affect the relative quantities of outputs, the kinds of influence exerted on enforcement decisions, and the measures recorded in the marketplace.

Government resource-using activities necessitate creating and sustaining bureaus and agencies. The authorizing legislation and administrative procedures of government bureaus tend to become "property rights" for the specialists who operate these bureaus. Since the content of property rights governs (though it may or may not define) incentives,[3] the behavior of government bureaus with regard to resource use can differ from that of private profit-seeking organizations—often with peculiar results.[4] In a sense, cost-benefit analysis seeks to apply some sort of measuring rod or calculus which will permit comparisons between

---

[3]For further emphasis on this point, see Eric G. Furubotn and Svetozar Pejovich, "Property Rights and Economic Theory: A Survey of Recent Literature," *Journal of Economic Literature* 10 (December 1972): 1137-62, especially 1138-40.

[4]For a development of this point, see William A. Niskanen, *Bureaucracy and Representative Government* (Chicago: Aldine, 1972).

private and government resource-using activities, given the fact that decision makers in the private and public spheres are nevertheless constrained to operate under markedly different incentive systems. Presumably, with such a measuring device, it is hoped that public administration decisions will improve.

Formal cost-benefit analysis derives its theoretical underpinnings from neoclassical welfare economics[5] which, although an impressive body of thought, pays limited attention to certain connotations of the term *property*. In most of the theoretical discussion, it is generally assumed that property rights are clear and unambiguous; hence, there is no possibility that the future services of things and productive agencies will be subjected to uncertainties that would be brought into the open by damage claims, law suits, legislative changes, and administrative rulings on the use of things.[6] This assumption that property rights are unambiguous is similar to other assumptions employed in economic theory, particularly the assumption of "correct foresight" and the assumption that the operation of markets entails no transactions costs.[7] These assumptions have been useful, and even necessary, in formulating rigorous theory, but they pose difficulties when the theory is applied to certain practical policy matters. They have also been responsible —especially the property assumptions noted above—for difficulty in theoretical discourse itself. The interactions between these two sets of difficulties has, in turn, led to the development of such economic adumbrations as "externalities," "common property," and even "public

[5] For an explicit development of this point, see E. J. Mishan, *Cost-Benefit Analysis: An Introduction* (New York: Praeger, 1971), pp. 307-15; also relevant is Mishan's defense of the consumer's surplus concept, which he more narrowly specifies as a compensating variation to take account of the point that a new project is usually the object of a cost-benefit analysis (pp. 31-47 and 325-30).

[6] The property concept which emerges in modern welfare economics and, more generally, in both the classical and neoclassical economic paradigms (including the Austrian and Walrasian streams of thought) resembles the one which had emerged in England by the eighteenth century with individual ownership of things in fee simple, and with labor transactions unencumbered by collective agreements or government-enforced rules. The focus was on the physical thing and, especially, its protection—a condition congenial to capital accumulation.

[7] This assertion may be unfair. The fact, however, is that most works on pure theory are simply silent on the concept of property and related social institutions. Of course, markets must exist if theoretical endeavor is to probe the process of exchanging goods. However, even the latter point is obscured (or questioned) when it is asserted that the algorithm of a giant computer in a centrally-managed society (or on Mars) that is employed for resource allocation would replicate the equations that describe the "hidden hand" workings of a "capitalist" economy (provided all preference and production sets are convex in both instances).

goods."[8] Each of these clearly has a high "public sector" and public policy content, much of which, in turn, comes to focus on the concept of property rights. This paper tries to identify and clarify some of these linkages.

On the Meaning of "Property"

THE WORD "PROPERTY" MEANS MANY THINGS, entails many complex relationships, is the object of interest to students in diverse specialties, and is hence ambiguous. Yet there is a fairly standard definition of the word that is worthwhile to ponder. It is, "in law, a scheme of relationships recognized or established by government, between individuals with respect to an object." Further, the expression

> "My property" probably means at a minimum that the government will help me exclude others from the use or enjoyment of an object without my consent, which I may withhold except at a price.[9]

The second statement about the meaning of "my property," in relation to "exclusion" and "price," succinctly describes what most laymen and economists have in mind when they use the term. But the first statement, which defines property as "a scheme of relationships . . . between individuals with respect to an object," contains the seeds for insights that bear upon the concept of property as a social institution and is worthy of emphasis and development.

The idea that property is a scheme of relationships between individuals with respect to objects clearly states that objects, or things (*res*, in Latin), are *not* property. It is the relationships, which specify how persons may use a particular thing, that is the key concept. The idea of a relationship further implies two or more parties and, of course, a social setting. Hence, for a Robinson Crusoe the concept of property is meaningless until a Friday enters the picture. This scheme of relationships between individuals toward things consists of a list of specifications on how things may and may not be used in the rendering of services. The particular specification may be a right, or a constraint, or an

---

[8]My use of the term *adumbrations* is not meant to be pejorative. However, the concepts mentioned do pose difficulties, in part because they have become vehicles for much ad hoc theorizing. For rather strong arguments in this connection, see, especially, Steven N. S. Cheung, "The Structure of a Contract and the Theory of a Non-Exclusive Resource," *The Journal of Law and Economics* 13 (April 1970): 49-70; and Harold Demsetz, "The Exchange and Enforcement of Property Rights," *The Journal of Law and Economics* 7 (October 1964): 11-26.

[9]"Property," *Encyclopedia Britannica*, 1970 ed., vol. 18, p. 631.

obligation—generally, all three. If I have a right to use a thing, then you have an obligation to respect that right. In using a particular thing, I may simultaneously be constrained not to use it in certain ways—so that you will be able to enjoy one or some of your rights. All of this, of course, has a tautological flavor—like that of accounting, with its associated concepts of assets, liabilities, claims, and net worth.

Yet, these property-related tautologies are worth pondering so as to avoid confusion from careless use of words. If the essence of property is relationships, then it is straightaway a public concept. More to the point, any *thing* to which the term *property* applies has a public quality or is even a public good.[10] Another way of putting this idea is that any *thing*, depending on how its owner uses it, is capable of generating an "externality." If an externality has a high nuisance content or is regarded as socially pernicious, then those who use the thing may be obligated to constrain their behavior in deference to the rights of others or to the survival of the group. In some instances, it might even be decided that certain classes of things should be owned and managed collectively. In virtually all instances, the use of things for the purpose of violence is supposed to be the monopoly of a group, and that monopoly is the essence of a "state." This particular monopoly is necessary both to protect the group's endowment of things from appropriation by outside groups who may think they possess a "comparative advantage" in the conduct of organized violence;[11] and—most important and paradoxically—to specify and enforce internally the rules that are the *sine qua non* of property itself. In brief, the word property therefore entails quite a mouthful.

If the word property connotes the set of rules governing the use of scarce things—or more accurately, a bundle of rights and associated constraints—then what are the meanings of such expressions as "a given distribution of property," or "a given distribution of wealth," or even a "given distribution of property and wealth"? Economists often use such expressions when they specify initial conditions to develop models of a private economy undertaking its spontaneous production and exchange of goods and services. Each of these expressions can be vague and can present difficulties.

---

[10]This assertion smacks of paradox; however, it can also derive a rationale from the idea that property, as a "bundle of rights," is a creature of the state. Yet the idea of publicness should not be pushed too hard. Hannah Arendt, in exploring the meaning of property among the Greeks and Romans, emphasizes its significance with respect to privacy, and, especially, its function in sustaining a sharp distinction between the private and public realms. See Hannah Arendt, *The Human Condition* (Chicago: University of Chicago Press, 1958), pp. 58-71.

[11]Hence the concept of "territoriality" is the second essential quality of the state.

"A given distribution of property" could be unambiguous if we had: (1) a given set of explicit and clear rules governing the use of things; (2) an established and evenhanded allocation of resources in the enforcement of those rules; (3) a given endowment of the physical stocks, specified in whatever detail necessary with respect to the diverse things that comprise the aggregate physical endowment; and (4) the distribution of legal claims against the endowment between households and other legal entities. Included in this concept of "endowment" are people and their individual capabilities to perform various services. Thus one individual may "own" his capacity to operate a machine tool; another, to perform violin solos; and so on. There are also natural endowments, or the "gifts of nature," in the classical economists' phrase. Finally, there is an accumulation of physical assets like capital goods and inventories, of diverse types and vintages.

Given the rules and the enforcement apparatus (including the allocation of law enforcement activity) that govern how things may be used, the process of production, which transforms the services of agents into outputs demanded by spenders, endows the services with values through exchange in the marketplace. The value of any particular service flow is the income attributed or imputed to the agent rendering those services. If one capitalizes these individual income streams attributed to agents, and adds them, we have a stock value or present-worth measure which is "wealth."[12] Normally, however, this wealth measure is not applied in the marketplace to human beings or, more specifically, to the production services they are capable of rendering, although conceptually it is possible to do so. The reason, of course, is that there is a limited opportunity to buy and sell human productive agents in a society that enforces adverse value judgments about the institution of slavery.[13]

Given the "distribution of property"—when defined to include a given specification and enforcement of property rights—then the combined workings of people's preferences and technology determine an economy's general equilibrium, i.e., the aggregate level of economic

[12]The word *wealth* may seem to imply a capitalization rate, which brings up many of the long-standing issues that are the body of capital theory. However, the capitalization rate consistent with opportunity cost doctrine is determined—like any other price—by the combined workings of production and preference functions, when one of the demanded items is new capital goods. Then the ratio of the money cost of producing new capital goods, relative to their prospective earnings, is the capitalization rate, and the cash flows of existing assets are appropriately discounted so as to yield that rate. The result is the wealth measure. However, production of a comparable income stream from any existing asset can also be undertaken on the margin; and the opportunity cost of so doing can provide the appropriate "stock" measure. Hence a capitalization rate is not really necessary to derive a wealth measure.

[13]Some labor earnings *are* capitalized, however, as illustrated by athletes' bonuses and authors' royalty advances.

activity,[14] the relative amounts of diverse final outputs, and the relative prices of both outputs and resource services. The earnings of any household (or individual) are, then, a function of the amount and composition of its physical endowment, and the particular prices (or earnings rates) obtainable from the sale of the services of the "things" it owns. An important characteristic of modern societies—or of modern technology—is that resources are highly specialized and are, hence, "idiosyncratic." Moreover, production processes are highly idiosyncratic in that they often combine diverse service inputs in ways peculiar to themselves.[15] Hence, such expressions as *capital-intensive, labor-intensive,* or *land-intensive* are frequently used to describe differences in production processes.[16]

In such a general equilibrium framework, the wealth of any individual or household is, then, a function of its possession of (or its ownership claims against) physical endowments, and the manner by which the interaction of relative demands for all outputs and the technologies describing each production process determines the values imputed to the services of the physical things each household owns. Thus a man's capacity to render services as an oboe player may be great; but the man may still be poor or obtain meager earnings in a society which overwhelmingly prefers the outputs of electric guitars. This linkage between the demand for (and value of) goods, and the demand for (and value of) productive services is the essence of the general equilibrium that characterizes an economy. Thus the distribution of income and wealth are dependent variables that result from preferences and wants, on the one hand, and technology on the other. Changes in wants and technology will also change the distribution of earnings, or, more accurately will change the earnings of specialized and idiosyncratic physical resources.

Such expressions as a *given distribution of wealth* or *income* do not seem to have meaning unless they refer to one of the consequences of a prior set of physical endowments (e.g., the distribution between households), a given specification of property rights, and the legal and police mechanisms which are charged with enforcing those rights. If a *given distribution of property* means these three sets of conditions, then one can

[14]By *aggregate level of economic activity*, we mean the neutral concept of value-adding activities the outputs of which are exchanged in markets, and are summed. This amount is roughly measured by the concept of gross national product, which does not, never has, and never will measure "welfare."

[15]The term *idiosyncratic* is borrowed from Joan Robinson, "Rising Supply Price," reprinted in *Readings in Price Theory*, edited by G. J. Stigler and K. E. Boulding (Homewood, Ill.: Richard D. Irwin, 1962), pp. 233-41.

[16]Such expressions as *labor-intensive* or *capital-intensive* are tricky to use in a general equilibrium context. When some production functions exhibit a high elasticity of substitution, changes in relative factor prices can evoke a "factor reversal" whereby a given process which was initially labor-intensive becomes capital-intensive, or vice versa.

develop a general-equilibrium pricing and output model. In this context, such assumptions as a given distribution of wealth are out of place or even misleading unless one intends to invoke some compensation scheme, given changes in any of the systems' parameters. Notice, however, that the assumption of a given distribution of property—when it entails the triple assumptions regarding the physical endowment, the specification of property rights, and the enforcing mechanisms—also seems to assume some sort of political apparatus or similar means of specifying and enforcing the "property rights" themselves.

In the same vein, such an expression as the *distribution of property rights* also seems unclear. For example, it could mean a given set of laws (including court and administrative interpretations of those laws as well as the allocation of resources to enforce them). Given this condition, plus the distribution of the physical endowments and the parameters encompassed by the concepts of preference and production functions, there will be a "given distribution" of income and wealth.

Pro tanto, any change in the property laws or the resources allocated to their enforcement will cause a change in relative resource earnings, output prices, and the welfare of different households. But notice, we then describe how the structure and workings of the legal system, including the allocation of law-enforcement activity, affect the distribution of earnings between things, the qualities of the services that things render in providing utility to different individuals, and the welfare of individuals as buyers of outputs in the context of a given physical endowment. In principle, any particular legal system can be neutral in its distributive impact. For example, if each household owned an aliquot share of each of the diverse physical resources in the economy, a change in the law affecting adversely the earnings of one class of resource would favorably affect earnings of another resource, and any household would find its relative earnings unaffected. If, in addition, each household consumes each final output in identical proportions, then any associated change in relative output prices would leave the consumers in the same relative condition.

This pair of assumptions is exceedingly strong. But it illustrates why changes in the laws or in the allocation of resources for enforcement acquire critical significance in almost any conceivable modern society. Each household does not own an aliquot share of the diverse physical agencies. Nor do households or individuals exhibit identical preferences, partly because of unequal physical endowments. *Privacy* derives its meaning from the notion that individual tastes are to be respected or at least be accorded some weight. And from this latter condition, the concept of "private property" derives a partial justification as a social institution.[17]

[17]See Arendt, op. cit.

Given the unequal physical endowments of different households, specialization in the ownership of different physical agents, and divergent preferences, any changes in the rules governing the use of physical agents will necessarily increase the earnings of some things and lower the earnings of others. Simultaneously, the relative prices of outputs change, and these will change the relative amounts of total utility or satisfaction that a given dollar amount will afford different individuals.[18]

Each of these mutually exclusive outcomes could be Pareto-optimal in the sense that all the relevant marginal equalities are satisfied. But in each outcome, some individuals will be better off than others, and vice versa.

This example of two mutually exclusive general equilibriums, both of which may be Pareto-optimal, might be questioned by the argument that one of the situations cannot be optimal, because if "tastes" and technology were both unchanged, then the change in the rules must either improve or worsen aggregate welfare. Such an argument is, however, an assertion in the absence of any independent way of measuring aggregate welfare. Nevertheless, it can be maintained that a change in rules reflects a change in wants registered through a new political consensus on, say, income distribution, or the "environment."

The resulting change in rules simply reflects the casting of political votes, either at the ballot box or in the legislative process. And in its way, the political process that respecifies property rights, and allocates law enforcement resources to the competing claims by which rights are made operative, reflects choices which are just as relevant as voting with dollars in the marketplace.

One can also say that a *given distribution of property rights* means a change in some rule or law affecting resource usage, but that the change is accompanied by measures (perhaps implemented by a change in some other law, or by means of a taxing-transfer program) which compensate those who would otherwise have been made worse off.[19] This would be moving from a non-Pareto-optimal condition to one that more closely

[18]This is another way of saying that the distribution of welfare will be changed, which is admittedly a slippery if not nonoperational concept.

[19]Compensation through this nexus, however, is necessarily approximate, since laws affecting property (including *ad rem* taxes) must necessarily be directed toward the "thing" rather than toward the individual. (Such is the operational meaning of "government based on the rule of law rather than the rule of men.") Given that physical resources are often highly idiosyncratic, it is difficult to envisage, for example, how land owners could have been individually compensated for the repeal of the English Corn Laws by changes in other laws that would have enhanced land earnings in other ways. Such a change would be likely to overcompensate some owners and undercompensate others, with the total amount either exceeding or falling short of the initial "injury." Thus ownership patterns and the structure of the political decision making process acquire key significance in the rate at which the "rules" change and social reforms occur.

approximates the ideal. But even this condition, whereby no one is any worse off, is not sufficient to be called a "given distribution of property rights." There is still the question of how the residual (i.e., some aggregation of "consumers' surpluses" in excess of the initial compensations) is itself distributed.

Although much of this section is devoted to raising questions, two positive assertions can be derived from it. First, the idea of property rights necessarily applies to things or activities, and as such it is neutral with respect to the earnings of individuals or groups of individuals. Changes in property rights do, however, affect the relative earnings of specialized physical resources. Because the costs of goods and services demanded by consumers and spenders (including investors and government) are nothing but the hiring prices of resource services, the relative prices of final outputs are changed. It is in this manner that something called "distribution" is changed, whether it is defined to include only the claim that an individual is able to exercise against the economy's aggregate of value-adding activities, or the more elusive output of utility (or welfare), either singly or in combination. For these reasons, such expressions as a *given distribution of wealth, property,* or *property rights* should be used very gingerly or specified very carefully. Tentatively, it is our preference to couch economic models to be used for policy analysis in terms of a given physical endowment and its initial composition and ownership distribution, and a carefully specified set of property rights.

If the idea of a "given set of property rights" is adhered to in economic discourse, a second assertion can be advanced, namely that the concept of "social cost" (or opportunity cost), which is a mainstay of cost-benefit analysis, possesses certain fragile if not politically-charged qualities. Costs are merely foregone benefits. But the benefits most items provide consumers depend on the individual's ability to keep others from using them. This ability, in turn, depends on police power and the related decisions which influence the allocation of enforcement effort. Since this effort ultimately resides in and draws upon the coercive power that is the monopoly of the state, the benefits provided by different outputs (and hence the social cost of a given output) will, perforce, be a consequence of the political system. The concepts of opportunity cost (and social cost) do not seem free of political content or of the consequences of choices registered and made operational through the political process.[20]

[20]An important question in this context is: How should law-enforcement effort be allocated to the protection or policing of the different kinds of property rights? One answer is that it should be related to the social values of diverse outputs, as is somewhat vaguely suggested by Demsetz (op. cit., pp.,17-19). But if market value-adding measures are taken as the measure of social value (even if we waive the problem posed by monopoly),

Externalities, the "Theory" of the "Common-Property" Resource, and the Relevance of Exclusion

IN THIS SECTION THE FOCUS IS NARROWED to the second part of the definition of property quoted at the beginning of the previous section —specifically, the idea of exclusion. According to that definition, " 'My property' probably means at a minimum that the government will help me exclude others from the use or enjoyment of an object without my consent, which I may withhold except at a price."

Although exclusion is the key to individual privacy and its deeper social significance, and although the quoted passage strongly smacks of the social institution of private property (and hence its ideological overtones), the quality of exclusion is critical in another very fundamental and somewhat different context: that of physical production and the absolute quantity of outputs obtainable from the use of physical agents. This interaction has come to be associated with the concept of "production externalities" in economic literature, particularly external economies and diseconomies. Indeed, so close is this association that the rationale for observed supply price and output behavior is often embedded in the concept of the production function as homogeneous to a degree greater than (external economy) or less than (external diseconomy) one. Thus it is often argued that the nature of a production externality is such that production functions, either between the firms comprising an industry, or between different industries, are interdependent. This way of looking at the social activities of production has produced a theory of external effects and associated trappings like a "common property resource," to say nothing of a body of political doctrine (we hesitate to say political theory) on how to deal with its beneficial and, especially, its toublesome aspects.

---

and if these values depend on how enforcement resources are allocated to different activities, the problem of allocating such resources remains. Demsetz suggests (op. cit., p. 19) that a mix of property rights (and presumably the associated allocation of police effort) which would increase the value of property rights should be the appropriate criterion "if we seek to maximize wealth." A resulting increment of wealth is a "surplus of value sufficient to compensate those harmed by the change" (i.e., "welfare," although, critical as he is of Pigou and the new welfare economics, Demsetz does not use the word).

It would seem that, given the inability to measure the aggregate of welfare, Demsetz is obliged to adhere to this wealth measure. Yet, it is unclear, in terms of capital theory, just how a reallocation of police activity from the lesser protection of one set of outputs to the greater protection of another set will affect (a) the aggregate of value-adding activity, or (b) the capitalization rate. Hence, we are at a loss to apply the wealth-maximizing measure as a rationale for allocating resources to the policing of property rights which are assigned to different outputs.

An alternative interpretation of this view might be that police power should be allocated in proportion to the current value of property or value-adding activity. The broader ethical and political implications of this possible position warrant scrutiny.

The key idea can be illustrated in a number of ways, as with a housewife cooking in her kitchen. Any production process entails two elements. The first is the list of ingredients and service inputs—like one hour's use of an oven, and so much time spent sifting, beating, and so on. The recipe is the production function. But production also necessitates that inputs be combined in a particular sequence, and that certain "external conditions" be maintained. These are the rules, the procedures, and so on that characterize the sequence of the activities and the conditions under which the prescribed results will most likely be forthcoming. To adhere to the appropriate procedures requires organization as well as concentration. However, if in the course of producing the output, something causes a break in the routine (like a child opening the oven door as the cake is rising), the output is apt to suffer. Thus many housewives prefer privacy (the exclusion of others) in the kitchen when preparing the meal. If it is lacking, the service capacity of the inputs or the resources are not utilized to full potential.

This interdependence between production and external conditions also applies to the "common resource." Commercial ocean fishing is an oft-cited and well-researched example.[21] Fish reproduction and growth are such that in a given fishing ground there is some optimum rate of harvest, which depends on a complex interaction between different kinds of fish feeding on each other, marine growth, and so on. Given that independent fishermen are earnings maximizers, they each will try to catch as many fish as they can. But if there is no control on the aggregate annual catch, and if some fish are taken at too high a rate, the ecology can be upset and after a period the maximum yield is either greatly reduced or eliminated. It is the same with an oil pool if each person owning ground surface rights strives to maximize his share of the yield of the pool: at the much greater cost of drilling more holes than are necessary, underground gas pressure is more quickly dissipated, and more of the oil must be extracted by costly stripper operations than would be the case if the pool were operated as a unit. The overcrowding of a bridge causes traffic congestion and lowers the bridge's "production." Waste dumped into streams and lakes can reduce their capacity to handle waste efficiently and simultaneously make the water unfit for other uses.

Although each of these (and similar resources) have their peculiar technical aspects, the general consequence of their "common property" characteristic is that they are overused, overcrowded, or not managed at all. The rent or value of output potentially imputable to the resource is

---

[21]See, e.g., Ralph Turvey, "Optimization and Suboptimization in Fishery Regulation," *American Economic Review* 54 (March 1964): 64-76, and Cheung, op. cit.—which contain references to the literature.

dissipated, in part or entirely.[22] The social prescription to cope with the problem is simple: find a way to simulate what the cook does when she invokes an element of exclusion with regard to her operations. With the fishery example, this is hard to do because it is difficult to identify the resource itself. With the bridge example, it can lead to excessive prices (or tolls), or—when the operation is owned and managed by a government instrument—facilities might be oversized. Finally, it should be noted that the use of an exclusion mechanism which functions like private property rights creates some income distribution problems that can be troublesome. The following tries to illustrate some of these points.

A Property Rights Approach to Cost-Benefit Analysis

WHAT ARE SOME OF THE IMPLICATIONS of the role of property rights in cost-benefit analysis as it is applied to the present concern with the environment? Specifically, in a setting where it is increasingly fashionable to assert that a different assignment of property rights may be called for (in the interest of enabling people to enjoy a right to a cleaner environment) it might even be contended that since conventional economic measures of costs and benefits reflect the existing structure of property law they may not be relevant. Hence it may be necessary to search for a set of appropriate shadow costs and benefits. Or it may even be asserted that allocative efficiency, which has been the principal focus of cost-benefit analysis, may not be particularly germaine at all.

Although such thinking can mirror a valid concern with a real problem, in our view it is off the mark; moreover, it masks some critical difficulties in any attempt to improve things. Rather, we feel it can be argued that a property-rights approach to the subject may itself constitute the model of how to undertake a cost-benefit analysis. The model may also be used to illustrate some critical points that tend to be overlooked in the public sector problems to which cost-benefit analysis is normally applied.

Let us assume a moderate-sized community that surrounds a lake. The lake affords a capacity to handle some industrial effluents, and it also has beaches and a fish population that provide recreation. Presently, it is regarded as a free good, or common property, by all parties. As a result of growth and shifts in demand, the capacity to accommodate industrial use and recreation has become overtaxed. Water used for

[22]See, e.g., D. A. Worcester, Jr., "Pecuniary and Technological Externality, Factor Rents and Social Cost," *American Economic Review* 59 (December 1969): 873-85; and J. R. Gould, "Externalities, Factor Proportions and the Level of Exploitation of Free Access Resources," *Economica* (November 1972): 383-40.

industrial purposes must be specially treated by some of its users (interdependent production functions). Beaches are crowded, and therefore less pleasant (interdependent utility functions). Finally, the level of industrial use impacts adversely on recreational use (interdependent production and utility functions). Thus all the externalities or interdependencies are present.

Let us now assume that the lake is sold to a private group that is permitted to treat it like any other private property. To maximize the value of that asset it would, of course, operate in such a way as to maximize its dollar earnings. The owners' property right enables them to exclude others from use of the asset's services, except at a price. This could entail the installation of meters, the building of fences, and so on. A market would be created. Some knowledge of the lake's capacity to handle effluents would have to be acquired; and this knowledge would be applied to the matter of how various capital investments might be made to augment the lake's capacity to handle effluents and provide recreational outputs. Simultaneously, selling the lake's outputs provides information about consumers' relative valuations of those diverse outputs, including signals to the owners on how much effluent is tolerable, how congested the different beaches should be, and so on.

When the operation shakes down and the owners are maximizing the lake's net earnings, those earnings measure the sum of the marginal valuations which consumers (and industrial users) place on the different outputs, less the annual (or similar period) cost of improvements, meters, fences, attendants, and so on. Any increase in net output may also lead to lower prices of certain items, which would generate some amount of consumers' surplus. But if the owners are unable to practice price discrimination, the creation of some consumers' surplus will not affect their decision making and management, nor will it enter into their calculations.

In such a context, the costs of hiring labor and investing capital to operate the project are those confronted in the existing market and include the prevailing opportunity return on investment before taxes. Is there anything "wrong" with this set of prices as the appropriate measure of the project's cost and, more fundamentally, as a constraint on the mix of inputs its managers employ and the size at which the project is run? One answer to this question is that, because of market and other distortions, existing relative prices are not optimal. Although one of the consequences of such distortions is that they cause the prices of certain outputs and factor earnings to be higher than they would otherwise be, those same imperfections also cause the prices of other outputs and factor earnings to be lower. Only if the project in question employs a greatly disproportionate share of either higher cost or lower cost inputs might there be concern about this point. And even if there is concern, the problem is to determine what the relevant input prices should be. Private operators, of course, have neither the knowledge about such prices nor the incentive to use them even if they did have the knowledge.[23]

Do the net benefits generated by the privately-owned lake project, and recorded in the market value of its saleable outputs, adequately reflect the appropriate social benefits? The answer, though complicated, is a provisional *yes*. The basis for this answer is that some increments of diverse real outputs are a result of the owner's behavior. The values at which they are sold say something about consumers' relative demands. The fact that consumers buy these outputs voluntarily, and simultaneously make voluntary adjustments with regard to some of their other purchases, suggests that they are to some extent better off. This is generally about as far as economics (and formal cost-benefit analysis) can go.

It may be argued, however, that the management of the lake project should act differently than do other types of private operations because there should be a larger supply of some of its outputs. For example, it might be that under private ownership the amount of beach recreation facilities services is X whereas the optimal amount, X*, would be greater. The reason it is X instead of X* is that, given consumer preferences, the distribution of cash holdings, and relative output prices, there simply is no *effective* market demand for the additional facilities. At this point it is possible to invoke an externality argument to the effect that a larger beach facility keeps children off the street, and so on. This assertion is a subject for cost-benefit analysis in its own right. Upon settling the question as to how large the beach facility should be, an implicit decision will have been made to provide an in-kind subsidy; and there is a further necessity to ration it in some way. Such a program can be compatible with private ownership of the lake as long as the owners obtain payment for the cost of providing the additional facilities. (Indeed, many in-kind income programs often elicit strong political support precisely because they enhance the earnings of specialized producers.)

If we can isolate the issues centering on in-kind income, it is difficult to see how one, in conducting a cost-benefit analysis on an environmental issue, could improve upon a property-rights approach to the subject. First, the latter strongly suggests that, as a method of constraining behavior, effluent charges for the use of the environment have considerably more merit than administrative rulings which are often heavy-

[23]This discussion surrounding the adjustment of prices because of market distortions may be somewhat academic. McKean has argued that only marginal improvements could be expected if prices were adjusted for market imperfections. See Roland N. McKean, "The Use of Shadow Prices," in Samuel B. Chase, Jr., *Problems in Public Expenditure Analysis* (Washington, D.C.: Brookings Institution, 1968), pp. 33-65.

handed. Second, that private owners could make such programs operational, suggests that compensation for enlarging the program to provide
added social benefits would not be difficult to arrange within our political framework. Whether compensations should be made, and to whom,
by way of assigning a property right or rearranging existing ones, is a
matter that warrants intensive examination in its own right. However, it
is a subject that cannot be fruitfully addressed in isolation from the
workings of our political process.

Although we suggest that a property-rights approach to cost-benefit
analysis may be useful, the approach also has limits, an awareness of
which would be helpful to those who demand and supply cost-benefit
analyses. These limits can be illustrated by our hypothetical lake example. Such a conversion to private property is conceivable with an asset
like a lake, or possibly even a river. But with other assets such as portions
of the air mantle, there are difficulties. In the latter case, some sort of
government intervention is necessary to effect the changes. Yet when a
private property apparatus can be employed, two important ingredients
are present which are usually absent in public-sector resource-using activities.

First, the market mechanism provides much useful information,
particularly about valuations. That information, in turn, often permits
the owner of a resource to manage his resource experimentally or by
trial and error methods. For example, in the case above, the owners of
our lake do not really have to know much about the lake's underlying
ecology. All they really must do, as earnings maximizers, is to employ the
services of the lake and to manage the lake in a way that maximizes the
value of their property. Of course, this can be achieved more quickly if
they have prior knowledge (to say nothing of a good, coherent theory)
about their resource. But the point that knowledge is acquired indirectly
or by a rough empiricism may have favorable cost implications that are
inadequately appreciated.

A second insight suggested by our lake example has to do with
incentives. If we make a simple distinction between a private property
apparatus on the one hand, and a public bureaucracy charged with
managing some physical resources on the other, the private property
owner has a strong incentive which the public bureaucrat generally does
not: i.e., to economize on the absolute quantities of physical resources
used. Indeed, managers of public resource-using activities usually have
an opposite incentive. The bureaucrat's power (and ability to do good or
to serve the public interest) tends to be equated with the size of his
budgets; and those budgets become his "property rights," so to speak.
Just as private earnings-maximizers might use some of their earnings for
influencing the political process and protecting their holdings, bureaucrats may try to influence politicians and the latter's constituents so as to

affect the budgetary process. Conversely, politicians—through the minute attention they give to detailed line item budgets—strongly influence
the incentives of bureaucrats and, specifically, the way the latter spend
some of their budget. (In all of this, it is not entirely clear who "corrupts"
whom.)

One of the dimly-understood consequences of this bureaucratic behavior is that a peculiar and distinctive "production function" can
emerge. Whether the expression *production function* is right in this context is a matter of definition. But whatever it might be called, it is not
something that closely (or even remotely) resembles the concept
economists have in mind when they treat a private-sector activity and
derive their marginal social cost functions. This divergence can be particularly acute if the production process and the knowledge about it have
been the object of research and development effort which, in turn, has
been directed and managed by the specialist bureaucrats.[24]

For these reasons, it is suggested that the really formidable problem
in cost-benefit analysis is to discover the relevant production functions.
This task is far more complicated and difficult than it may seem. The
main reason it is difficult is that the specialist, budget-maximizing
bureaucrats are the primary source of knowledge with respect to their
technical areas. Their organizations are often the main, if not sole,
source of statistical information that might be available to apply to cost-
benefit or operational research models. One way to try to deal with this
information problem is to conduct experiments. Yet the bureau often
controls much if not all of the resources needed for an experiment, as
well as the expertise needed to conduct it.

Coping with these problems is usually considered outside the scope
of that aspect of applied welfare economics which is formal cost-benefit
analysis. However, the problem is primarily one of incentives, not engineering technicalities. And in its broadest sense, property rights describe and determine the incentives that impact on the use of scarce
resources.

Conclusions

THE INTERRELATIONSHIP BETWEEN THE PRICES OF GOODS and the prices of
productive resources is well known. There is also a powerful interdependence between the structure and workings of the legal system, on
the one hand, and the behavior of the economy on the other. Because

---

[24]For an illustation of how these forces have affected military management, see J. A.
Stockfisch, *Plowshares Into Swords: Managing the American Defense Establishment* (New York:
Mason and Lipscomb, 1973).

the structure of property rights both constrains and motivates people with respect to the use of scarce things and services, it therefore impacts on the utilities provided by various outputs and hence on relative demands and resource allocation. This same force affects relative resource earnings. The operation of the legal system, by the degrees of exclusion it permits, influences the quantity of services produced by the physical resource. The ability to exclude, combined with the opportunity to exchange in markets, also has a profound effect on incentives to manage resources efficiently.

Changes in law and the content of property rights cannot help but affect the distribution of resource earnings, wealth, and welfare. Hence both relative prices and costs observed in the economy are functions of the legal and political systems.

The apparent technical and consumption interdependencies (or externalities) alter the quantity of useful services available from a physical resource, and the consumer satisfaction provided by a product or service. These are also direct consequences of people's behavior and of the incentives that constrain or fail to constrain their behavior. The composite effect of this behavior may also be termed the *environment*.

Attempts to improve the quality of the environment can generally take two forms, both of which fall into the public realm. One is to constrain people's behavior by means of law. The other is to undertake public resource-using projects or activities that attempt directly to cope with a problem. Antilitter laws are examples of the first form; devoting resources to cleaning up litter is an example of the second. Since enforcing the antilitter laws requires resources, the optimum mix of these two policy devices is conceptually, at least, a straightforward analytical problem.

Although changes in the law and the outputs of government projects affect income distribution and the relative prices of resources and other inputs—and through this nexus can change government program costs—the impact on costs will be in the future. This aspect of general interdependence, however, should not be used to reject existing prices (including the private, before-tax cost of capital) as appropriate measures of program or project costs. They do measure the worth of alternatives that must be given up now (or in the immediate future) so the government project can be carried out.

The estimation of benefits derived from projects designed to improve the environment constitutes the major difficulty that confronts cost-benefit endeavors. The problem is overwhelmingly technical, although it is mixed with severe bureaucratic and political elements. The sad truth is, our knowledge about the anatomy of the basic subject is meager, especially with respect to specific water and air sheds. These uncertainties may also be couched in terms of opportunity costs: the degree to which other outputs must be sacrificed to get some given level of environmental improvement is largely uncertain. Similarly, the distributional impact of attempts to make sweeping changes can be even more uncertain.

DISCUSSION

DISCUSSANT: *Anthony C. Fisher*

Since the title of this paper is "Cost-Benefit Analysis, Property-Right Specification, and 'Distribution,' " I expected a discussion of determining the costs and benefits of water pollution abatement projects, especially (though not exclusively) where common-property resources are involved. That is, which of the alternative measures of consumer surplus—or, what is the same thing, which specification of property rights—is appropriate in this situation? And to pose a prior question, do they differ in allocative as well as distributive implications?

The individuals who currently benefit from an unspoiled environment are consuming the many services that the environment provides. Modification of that environment would diminish the flow of environmental services which these individuals consume. Is the appropriate measure of benefits their income-constrained willingness to pay for the services of the environment, or is it the amount by which they would have to be compensated to give up the services they currently enjoy?

Or take the case of kidnapping. As John Krutilla pointed out in another context, the victims of kidnapping are generally wealthy families, not because the value that they set on their children is necessarily higher than anyone else's, but simply because of their ability to pay. In that case, an income-constrained measure of consumer surplus would be relevant in deciding the *"allocation"* of the child as between the kidnapper and the parents of the kidnappee, as well as the distribution of income between them.

Project-by-Project Analysis vs. Comprehensive Planning

FRED H. ABEL

Cost-Benefit Analyses
as Criteria

THIS PAPER WILL EXAMINE some of the problems involved in choosing the scope of an environmental cost-benefit analysis, both with respect to the geographic area included in the analysis and in the number of pollutants considered.

Although it is generally understood that cost-benefit analysis is useful for decision making, pollution control laws, except for special situations in water control, do not permit, much less require, cost-benefit criteria. In air pollution control, for example, achieving ambient conditions with zero health effects is assumed to justify any cost, and control decisions are based upon that assumption. The same is true for the 1977 projected requirements of the amended water bill, where the benefits of control by the "best practicable technology" (BPT) are assumed to exceed costs. If all sources use BPT and ambient quality standards are still not met, then something beyond BPT will be required (unless it can be shown that the benefits do not exceed the costs). By 1983, sources must use the best *available* technology (BAT) if benefits exceed costs.

In the 1972 amendments to the Water Quality Act, some sections require that social and economic costs and benefits be determined for all actions taken under the act. Permits issued, as well as any regional or state plans, qualify as such actions. In principle, a cost-benefit analysis should be performed for each permit application, or any other contemplated action, to insure that the level of effluent from each plant will not exceed the level at which the marginal social costs of reducing emissions equal the marginal social benefits of the reduction.

However, if a project is so defined that the scope of the cost-benefit analysis covers the control of individual emission sources, then the following problems result. The first is, simply, the number of sources involved. For a typical region or state there are several hundred, if not thousands, of such sources. To describe in detail the costs and benefits of controlling every one of these would be a monumental task.

Second, it would be nearly impossible to determine the marginal benefits (assuming we could quantify benefits) of control in any but the simplest cases. In the more usual, complex case, where there are multiple polluters, the marginal benefit of controlling any one source depends on the level of control at all other sources, including those which are natural. The marginal benefit function for any single source in a complex situation can be obtained only by making arbitrary assumptions about the level of control at other sources and equally arbitrary assumptions about the share of benefits deriving from control of all the sources (a joint product) that should be assigned to any one source.

Third, it will also be difficult to determine the economic impact of projects. This may be true even of the direct costs of the project if the project resorts to change of process rather than add-on technology. The indirect economic effects, particularly if firms go out of business or reduce employment, are especially difficult to determine.

Fourth, the analysis may be defined to include only one or a very few pollutants, but the control method may regulate a number of other pollutants as well. In that case, the share of the total control costs that apply to each of the targeted pollutants must be determined.

Project vs. Area Analysis

SOME OF THE ABOVE PROBLEMS MAY BE SOLVED if the analysis is at a higher level of aggregation, as in area or regional analysis, where all sources and all pollutants in a geographic area would be considered.

Concern about the utility of cost-benefit analysis for individual sources as opposed to areas or regions has often been discussed in connection with management of water resources. A. V. Kneese and E. T. Haefele [3] have discussed why the area, or regional, approach is essential. In another study, Kneese [2] selected a river basin as the area for analysis. He did not, however, discuss the problems which arise when the analytic unit, the basin, does not correspond to (1) statistical units for which data are usually available, (2) the political units responsible for implementation, or (3) economic units, including Standard Metropolitan Statistical Areas (SMSAs), etc., which minimize the spillover, or external, economic effects.

Area analysis does simplify some problems that exist with analysis by source. In particular, a comprehensive method can be used to assign costs and benefits that are essentially joint products of individual sources. The cost-benefit trade-off here is at the area level. A function that relates ambient quality to level of damages is derived for the area as a whole even though damage functions cannot be obtained for specific sources. With cost of control functions for each source and with knowledge about secondary economic effects on the area, a least-cost strategy can be determined for any desired level of ambient quality. With these functions the cost-benefit trade-off level of ambient quality for the area is determined. The least-cost control function also determines the level of emissions from each firm that gives the desired level of water quality at the least cost. This procedure does not ensure that the level of emissions for each firm is necessarily the level where the firm's control costs equal the damages caused by the firm's emissions. However, this procedure avoids the problem of assigning damages, a joint product, to individual firms and brings most of the external economic impacts of the firm's response into the analysis.

Not all impacts of the control actions in an area will be contained within the area. Some will spill over the area boundaries, and, at the same time, effects of activities in other areas will spill over into the study area. In water development project analysis, it is sometimes asserted that considering these spillover effects constitutes a "regional" analysis of the project [4]. In regional development analysis the spillover effect is considered by analyzing the flow of goods and services into and out of the area. For environmental quality evaluation an equivalent procedure would be to consider the flow into and out of the area of (1) residuals, (2) costs, and (3) damages or benefits.

In general, past water project analyses have not adequately considered these spillover effects. The best example of this is the counting of the total value of additional recreation as benefits to a project. The reduction of recreation in other areas because of the project is seldom taken into account. From a regional, state, or national point of view, only the net benefits should be counted. With water quality improvement this becomes a key point, since many activities affected by water quality improvement are very mobile. Thus, spillover effects are potentially very large. It may be that, to reduce damage in any given area, part of the optimum strategy, politically as well as economically, is to convince neighboring areas to reduce the flow of residuals that are coming across the borders.

Size and Type of Area

WE HAVE DISCUSSED THE DIFFICULTY OF DETERMINING the costs and benefits of effluent control by individual source and have argued that an area approach is more practical. But what are the criteria for delineating an "area"? Ideally, the kind and size of area used for analysis should be small enough to be manageable, but large enough to encompass the entire area for which significant economic, political, and social interactions exist. Boundaries should correspond with political boundaries since plans and permits must be implemented by political institutions. Some of the jurisdiction-implementation problems are discussed by Kneese [3]. Areas also must be geographically exhaustive and must not overlap. Such areas are analogous to the functional economic areas delineated by Brian Berry. His delineation has been modified and extended to be geographically exhaustive and the resulting areas are called Basic Economic Research Areas (BERAs) [1]. The BERAs are somewhat idealized units for analysis and planning. They often encompass part of more than one state because of the way economically and socially interdependent areas have evolved.

The BERAs have been delineated using the concept of an urban node linked to a surrounding dependent hinterland. For parts of the urbanized eastern United States, the delineations are somewhat arbitrary since there is no hinterland separating urban nodes. Thus some BERAs have more than one urban center. In many parts of the western United States, rural areas are not closely dependent upon any urban center. This has resulted in some BERAs with very small "urban" centers (population 6,000). Nevertheless, these BERAs are a meaningful basis for both planning and analysis. However, because political and planning organizations are not currently organized on the basis of the BERAs some other multicounty units may have to be used initially.

One alternative is the substate districts delineated in the Office of Management and Budget Circular A-95. Many of these districts have been delineated on the basis of urban centers or functional economic areas. Also, in many states these A-95 districts have established planning boards and governing boards of local officials which can implement the plans. In those cases where a major economic region, like an SMSA with contiguous rural counties, involves more than one A-95 district, more than one state, or both, an overall governing unit similar to a Council of Governments (COG) should be (and often is) organized to plan and implement environmental improvements. Unfortunately, such planning and governing boards typically have limited power and influence for implementation.

One concern, of course, is that these planning areas are not congruent with geographic entities such as river basins or airsheds. This causes some problems for analysis and planning. It would be ideal if the geographical entity and the political entity were the same. They are not and never will be, because there are different natural geographical entities for different planning needs. The airshed does not correspond with the water basin, and neither corresponds with areas of high economic and social interdependence.

Water basin and airshed plans must be on the basis of geographical units. This requires converting data from political units for analysis and converting results back to political units for implementation. It is easier to plan on the basis of social-political areas because it is easier to track environmental residuals across political boundaries than to track economic effects or political (implementation) actions across natural boundaries. Such social-political areas could correspond to the regional waste treatment management areas called for in Section 208 of the Federal Water Pollution Control Act Amendments of 1972.

For a single planning area like a BERA, an A-95, or a COG, a single planning and implementation organization should be responsible for all environmental quality improvement, economic development, and land-use plans. If different organizations are responsible for planning different parts of an interdependent environmental quality improvement program, the result will be confusion and long delays in implementation. Implementation plans for water quality should be consistent with air quality implementation plans, land-use plans, etc.

Comprehensive Environmental Planning

AS USED HERE, the term "comprehensive" simply means an enumeration of all costs and all benefits in the area resulting from the implementation of pollution control plans. It should include costs and benefits in the area from environmental programs outside the area and vice versa. It should also present secondary benefits, direct costs, and the distribution of the costs and benefits by geographic areas and population groups.

Comprehensiveness is also concerned with the range of pollutants included in the analysis. Is the analysis concerned with only one pollutant or a set of pollutants? Since water quality and effluent standards are often established for specific pollutants, a cost-benefit analysis for each pollutant would seem appropriate. However, this is not always adequate for correct control decisions, or even possible. When there are synergistic effects with other pollutants, damage estimates for a single pollutant have to be conditional upon the level of other pollutants. Or, when control technology treats several pollutants simultaneously, the joint cost has to be arbitrarily allocated to each pollutant. When either of these conditions exists, analysis of a single pollutant will not necessarily lead to decisions which maximize social welfare. In this context, comprehensive analysis means including in the analysis all significant "interdependent" pollutants.

If there are multiple objectives, the program impacts, both good and bad, should be included in comprehensive planning. These objectives include national economic development, regional economic development, improved environmental quality, and conservation of natural resources. When multiple-objective analysis is done, a single cost-benefit ratio may not be the most meaningful or useful result. This is because any proposed program may affect the objectives in different ways, enabling some to be accomplished and interferring with the accomplishment of others. On the other hand, a single ratio can be obtained only if the analyst weighs the costs and benefits of the different objectives. This is one of the key functions of decision makers: to weigh the many impacts of any action and on that basis make a decision. To be useful, an analysis should therefore produce a display of the costs, benefits, and impacts on each major objective. The results of the analysis should be aggregated and summarized to some extent, since excessive detail is cumbersome.

Summary

WE HAVE ENTERED AN ERA OF CONCERN about the quality of the environment and are taking some actions to improve it. These actions include the development of area implementation plans, although there are still unanswered questions about the size and nature of the planning area and the scope of environmental program. It is recommended that the geographical basis for planning and implementation should be an area in which significant social, economic, and political interdependencies form a relatively self-contained unit, and that the environmental plan for the unit should include all major pollutant categories, i.e., air, water, solid waste, and hazardous materials.

There are also major questions on how to conduct cost-benefit analysis of environmental actions. Should it be by individual source or on an area basis; and on a single-pollutant, media, or multimedia and multipollutant basis? It is recommended that the analysis should cover an area but whether it should be on a single-pollutant, media, or multimedia and multipollutant basis depends on synergism among pollutants and the degree to which the control technology can be refined.

Other questions concern the level of detail for comprehensive analysis and the ability to aggregate for state or national analysis. *At minimum*, the analysis should take account of the flow of residuals *into* the area, the generation and assimilation of residuals *within* the area, and the flow of residuals *from* the area, and should also quantify the effects of these residuals and the costs of control functions for all sources.

REFERENCES

1. Abel, F. H., and Nelson K. M. "Basic Economic Research Areas: A Delineation and Prospects for Use." Unpublished report, Economic Development Division, Economic Research Service, United States Department of Agriculture (1971).
2. Kneese, A. V. "Socio-Economic Aspects of Water Quality Management." *Journal of Water Pollution Control Federation* 36, no. 2 (February 1964).
3. Kneese, A. V., and Haefele, E. T. "Environmental Quality and the Optimal Jurisdiction." Joint Institute on Comparative Urban and Grants Economics: University of Windsor, Canada, November 1972.
4. McKinsick, R. B., and Snyder, J. H. "A Regional Approach to Project Evaluation." *Water Resources Bulletin,* of the American Water Resources Association 8, no. 3 (1972).

Cost-Benefit Analyses of Cost-Benefit Analysis

THOMAS D. CROCKER

Introduction

> There is no imperfection in a market possessing incomplete knowledge if it would not be remunerative to acquire (produce) complete knowledge; information costs are the costs of transportation from ignorance to omniscience, and seldom can a trader afford to take the entire trip.[1]

THIS STATEMENT OF GEORGE STIGLER'S succinctly expresses the problem of information production and choice. Present in the statement are the notions that: (1) information is a scarce good the acquisition of which, as with any other good, constitutes a problem of economic choice; and (2) therefore one can obtain some understanding of this problem by analyzing informational questions within the formal framework of economic theory.

Cost-benefit analysis is one of several means whereby decision makers can acquire information about the probable pecuniary, social, and ecological effects of a water pollution control investment. Those who employ the analysis try to ascertain the quantity of some numéraire (most frequently current dollars) which the gainers and losers from some proposed public investment will consider equivalent in value to their respective gains and losses. The algebraic sum of these dollars is then used in deciding whether or not to invest, and to what extent. If the net sum is positive, the investment is commonly considered worthwhile; if negative, it is viewed as undesirable. My purpose here is to apply cost-benefit analysis to the question whether or not to undertake a cost-benefit analysis of a water pollution control investment. That is, I wish to discuss some of the considerations involved in doing cost-benefit analyses of cost-benefit analysis.

---

[1]Stigler, G. J., "Imperfections in the Capital Markets," *Journal of Political Economy* 75 (June 1967): 291.

A Comment on the Distinction between Information and Decisions

IT IS FREQUENTLY ASSERTED THAT a natural distinction exists between the information accumulated by cost-benefit analysis (or any other technique for assessing the impact of an investment on an economic system) and the decision which the information is expected to elicit. Haldi, for example, states:

> I believe it is fair to say that in a number of Government agencies, perhaps most, financial records and statistical data are *not* now kept in ways that enable them to be brought together to assess the need for their programs, the results of their programs or full program costs.[2]

Similarly, Herfindahl says that one of the purposes of providing natural resource information is to "provide factual information which can be used to improve the way in which a natural resource is being used, that is to improve its current use."[3] Is the distinction between information and decisions really this clear?

Consider, for example, the social cost imputed to an industrial chemical released into a watercourse. The data on social costs which an information system supplies to a decision making agency are useful for some decision making purposes. But underlying each imputed cost are assumptions about the decision behavior of the chemical producer and his customers—decisions that determine the production process and consider the marketability of the chemical being released. To resolve one set of problems—those for which cost-benefit analysis is typically considered useful—it is necessary to assume the resolution of many others. Sometimes these assumptions can lead to serious and costly errors. The dilemma is universal and is central to the problem of information production and choice. It is nearly always necessary to reduce the complexity of real systems to dimensions that are computationally feasible and analytically tractable. However, this creates its own decision problems since in designing an information system one must determine which decision elements shall remain explicit and which shall be considered invariant for the problem in question. The choice of a particular information system thus implies the use of a particular class of decision models since certain types of information are relevant to some models and not relevant to others. Conversely, the choice of a decision model implies the use of a particular class of information systems yielding the parameters of that decision model.

[2]Haldi, J., "Program Monitoring, Evaluation and Control," in H. H. Himricks and G. M. Taylor, *Program Budgeting and Benefit-Cost Analysis* (Pacific Palisades, Calif.: Goodyear, 1969), p. 344.

[3]Herfindahl, O. C., *Natural Resource Information for Economic Development* (Baltimore: Johns Hopkins University Press, 1969), p. 21.

A distinction between information and decisions is thus not valid. The connection between the two requires instead a deliberate choice that may be made more or less stringent. Given that choice is involved, one would like to make it in the most economic manner. The choice is between enlarging the range of alternatives—which increase the demands placed upon the decision maker's ability to calculate and to understand—and a contraction of this range, which simplifies the decision maker's task. An intuitive grasp of this point is readily conveyed.

An information system useful for assessing water quality can be defined as the relation between a set of observable attributes of water quality states and the descriptive measures (signals) of these states. If a decision maker's choices among alternative water pollution control programs depend in part on the attributes of the existing state, then it can be shown that the range of alternatives available to him will be systematically narrowed as the information system loses its ability to discriminate between levels of the same attribute.[4] In effect, the actual state of affairs becomes relatively uncertain and the decision maker becomes relatively less able, for a given information processing capacity, to predict the exact results of alternative programs. Unlike programs may appear to be similar in terms of their measured results and may thus be treated as identical. Or, equivalently, if differences in programs are economically distinguishable only insofar as they generate measurably different results, the number of alternative programs can be no greater than the number of measurable consequences. It is up to the decision maker to decide the extent to which it is worthwhile to make explicit the consequences of his actions in choosing among alternative programs.

A Framework for Decisions about Information

PRESENTED IN THIS SECTION IS A FRAMEWORK, or model, for the evaluation of information. The model, adapted from the theory of teams as developed by Marschak and Radner,[5] is general rather than particular. That is, the model is designed to be applicable to a wide variety of informational systems and real world circumstances including cost-benefit analysis and water pollution control activities. In its generality, however, it abstracts from the details that distinguish cost-benefit analysis from other information systems, and water pollution control activities from other activities. It is properly viewed more as a model

---

[4]For proof under an assumption of independence of benefits and costs, see J. Marschak and R. Radner, *Economic Theory of Teams* (New Haven: Yale University Press, 1972), chap. 2.

[5]Ibid. The Marschak-Radner notation is retained.

from which useful tools can be forged than as an instrument capable of direct empirical application. Nevertheless, at a minimum, it can make those who must decide about information structures aware of the nature of the relations among the relevant variables, including those variables that are important for normative purposes. In later sections, we will provide some considerations relevant to cost-benefit analysis and water pollution control.

To begin, assume there is an information assessor whose function is to decide whaι information to supply a decision maker. The decision maker and the assessor may or may not be the same person. Although in this simple framework no explicit recognition is accorded the assessor's current stock of information, wisdom, and experience, it should nevertheless be noted that his perception of the decision problem will be influenced by this experience and information. Given his perception of alternatives, conditional returns, events, and probabilities, the assessor's decision problem is composed of several elements.

As perceived by the information assessor, the decision maker must make a set of resource commitments. These commitments are termed *decision acts*. Let a denote the act the decision maker chooses and A the set of mutually exclusive acts the assessor perceives might be selected by the decision maker. Act a need not be contained within A; that is, the assessor's perceptions of the alternative acts available to the decision maker may prove to be completely in error.

If the decision maker is concerned with the scale on which to build a sewage treatment plant, an example of an act is the decision to build a plant designed to provide primary treatment for the equivalent of 50,000 persons. In this case, the set of alternative acts, A, available to the decision maker would have been the various scales of plant that could have been built, given the legal, administrative, and budgetary constraints he faced. Note that the intent of act a and the result of the act need not be identical. Coincidence between intent and results would occur only when implementation was perfect.

Usually only a few of the many factors that can affect the result of a decision act can be controlled by the decision maker. This is the major cause of discrepancies between the planned and realized results of a decision act. The results that actually occur during a specified time interval are indicated by r, while R denotes the set of all possible results that could occur. Of course, it must be that r is contained within R. These results, r, may not be independent of past outcomes or results. In fact, most information assessors will behave as if there is some formal relation between results prior to the period in which the current decision is being undertaken and the results expected from this decision. Let the results or states occurring prior to the current act be denoted by x and the set of all possible prior results by X. Embedded in x and X may be the results of decision acts identical to the current act.

To assess or evaluate alternative informational structures, the assessor must have some preferential ordering of the set of all possible results, R. The measure of that preference, denoted by u, is dependent on realized results and is represented by

$$u = \omega(r) \qquad (1)$$

where u is the payoff stated in terms of some numéraire good and $\omega(r)$ is the payoff function. If an expected value formulation is assumed, (1) implies that if the assessor has two conditional probability distributions defined over the elements of R for each act, a, that may be specified by the decision maker, then the assessor will always prefer the situation producing the conditional distribution with the highest expected payoff. $\phi(r|a)$ will denote the subjective or the objective probability that result r will occur, given that the decision maker has chosen a particular act a. The expected payoff from a particular act is therefore readily expressed as:

$$E(u|a) = \sum_{r \in R} \omega(r)\phi(r|a) \qquad (2)$$

where E is the expectations operator. This expression simply states that the payoff to be expected from an act is equal to the arithmetic sum over all possible results, R, of the expected payoff of each result, r, multiplied by the probability of that result occurring whenever the act on the left-hand-side of the expression is undertaken.

Trial lawyers sometimes earn their keep by being aware that different individuals can be present at the same event, yet completely disagree on what actually transpired. Natural scientists have also been known to express differences on the observed facts of a situation as well as its cause and meaning. These notes of discord are interpreted here as being caused by different observers viewing different subsets $x_1, x_2, \ldots , x_n$ of the set of all possible prior results, X. That is, different observers partition the set of all possible states in diverse ways. When interpreted in this manner, incorrect information cannot exist, since all information must refer to some aspect of the set of all possible prior results. However, information may indeed be utterly irrelevant to the payoff function.[6]

Consider a particular partitioning of the set of states, X. A given partition means that certain signals, y, are assigned to particular subsets

[6] Ibid., p. 47.

of X. The signals are the output of information gathering efforts. These efforts and the signals they produce define an information structure, μ. That is,

$$y = \mu(x). \tag{3}$$

The information assessor knows only those prior results that are accurately depicted to him by the signals. An information structure, μ, is defined as "perfectly precise" if and only if the same signals are generated whenever the same results have occurred. Any structure can also be characterized as a composition of an aggregation function, $\Theta(x)$ and a measurement function $\gamma(x)$.* Thus by writing

$$\mu(x) = \sigma[\Theta(x), \gamma(x)], \tag{4}$$

the information production process of, for example, cost-benefit analysis can be viewed in terms of an initial assignment of monetary value to prior results (the determination of "shadow" transactions prices) and the aggregation of shadow prices into net gains and losses for various individuals, groups, and locations.

Decision acts by the decision maker are sensitive to the signals he receives. If the decision maker and the information assessor are the same person, this implies that

$$a = \alpha(y,\sigma), \tag{5}$$

where α is termed the decision function or rule. Given (5), the information assessor must develop a conditional probability distribution defined over the expected results for each possible signal. Let $\phi(r\,|y,\ \sigma,\ a)$ be the conditional probability that result r will occur, given that signal y is generated by informational structure σ and act a is chosen. Similarly, $\phi(y|\sigma)$ denotes the conditional probability that the signal generated by the information structure σ will in fact be y. For decision making purposes, the information assessor must somehow therefore arrive at three fundamental (subjective or objective) conditional probability distributions: $\phi(r\,|a)$; $\phi(y|\sigma)$; and $\phi(r\,|y,\ \sigma,\ a)$.[7]

*In (4) the function σ is used to indicate the composition of two separate functions θ and γ into the information structure μ. Henceforth the author often uses σ to denote the information structure. [Editors' Note.]

[7]These three distributions can be calculated as follows. Assume the information assessor has available three probability distributions: (1) a distribution taken over past results, $\phi(x)$; (2) a conditional distribution taken over the expected results given prior results and the relevant act, $\phi(r|x, a)$; and (3) a conditional distribution take over the signals to be

If the information assessor is simultaneously the decision maker, he will select an act, *a*, that maximizes the expected payoff, given the signals and information structure. That is, he attempts to establish

$$E^*(u\,|\,y,\,\sigma) \equiv \max_{a\in A}\left[\ \sum_{r\in R} u(r)\phi(r\,|\,y,\,\sigma,\,a)\ \right], \tag{6}$$

where the * attached to the expectation operator indicates a maximum.

If the decision maker and the information assessor are not the same person, the former will base his decisions on his own decision model. His choice problem may then be stated as

$$E^*(\hat{u}\,|\,y,\,\sigma) \equiv \max_{a\in A}\left[\ \sum_{r\in R} \hat{u}(r)\phi(r\,|\,y,\,\sigma,\,a)\ \right], \tag{7}$$

where the circumflex refers to the decision maker. Whenever the decision maker and the information assessor are different individuals, the

---

received given prior results and the information structure to be used, $\phi(y\,|\,x,\,\sigma)$. The fundamental distributions referred to in the text can then be calculated as

(a) $\phi(r\,|\,a) = \displaystyle\sum_{x\in X} \phi(r\,|\,x,a)\,\phi(x)$

(b) $\phi(y\,|\,\sigma) = \displaystyle\sum_{x\in X} \phi(y\,|\,x,\sigma)\,\phi(x)$

(c) $\phi(r\,|\,y,\sigma,a) = \displaystyle\sum_{x\in X} \phi(r\,|\,x,a)\,\phi(x\,|\,y,\sigma)$

$$= \sum_{x\in X} \phi(r\,|\,x,a)\ \frac{\phi(y\,|\,x,\sigma)\,\phi(x)}{\displaystyle\sum_{x\in X}\phi(y\,|\,x,\sigma)\,\phi(x)}$$

The calculation of (c), which requires the use of Bayes' Theorem, is rather involved. See Marschak and Radner, ibid. pp. 59-66 for the use of Bayes' Theorem in similar calculations. Together, these three expressions, (a), (b), and (c), lend consistency to the various probability distributions that underlie (10). That is, the calculations that the expressions represent enable each distribution to be based on identical prior results. If the probability distributions are not consistent, the solution of (10) for various information structures will be useless since the calculated payoffs will not be comparable.

assessor must be aware of, and therefore make predictions about (7). Thus

$$E^*(\hat{u}\,|\,y,\,\sigma) \equiv \max_{a\in A}\Big[\,\sum_{r\in R}\hat{\hat{\omega}}(r)\,(r\,|\,y,\,\sigma,\,a)\,\Big], \tag{8}$$

where the double circumflex indicates the assessor's perception of the model the decision maker will employ.

As noted earlier, the problem of the information assessor is to select an information structure. The establishment and operation of an information structure requires the expenditure of resources and therefore some recognition of the costs of the structure. Let $\omega'(y,\,\sigma)$ be the cost or negative payoff from a particular structure σ when it produces signal y. Net payoff, u', can then be expressed as

$$u' = \omega(r) - \omega'(y,\,\sigma), \tag{9}$$

given that the gross payoff and information structure costs are independent of each other. Then, given $\alpha(y,\,\sigma)$ as determined by (6) or (8), the expected net payoff for a given information structure as perceived by the information assessor is

$$E(u'\,|\,\sigma) = \sum_{y\in Y}\Big[\,\sum_{r\in R}\omega(r)\,\phi\,(r\,|\,y,\,\alpha(y,\,\sigma)) - \omega'(y,\,\sigma)\Big]\phi\,(y\,|\,\sigma). \tag{10}$$

Using (10), it is possible in a cost-benefit context to approach two related problems. First, for any list of alternative information structures the assessor is able to develop, the structure that maximizes (10) is the optimal structure since it maximizes the difference between information costs and information benefits.[8] Second, if the information assessor must deal with marginal changes in information, then $E(u'\,|\,\sigma_2) - E(u'\,|\,\sigma_1)$ is the expected change in net payoff if he adopts information structure σ_2 rather than σ_1.

In summary, the entire information evaluation process can be viewed more or less accurately as a sequential process in which, for any particular decision problem, information structure alternatives are likely to produce different information signals. These signals, in turn, induce

[8]Note that the analytical quandry, referred to earlier is relevant here. That is, the point at which to terminate the list is itself a problem in the cost-benefit analysis of information structures. It is obviously easy to conceive of an infinite regression of such decision problems. For more on this issue, see R. Radner, "Competitive Equilibrium under Uncertainty," *Econometrica* 36 (January 1968): 31-58.

different selected acts that will bring about varied results, each of which can lead to a different payoff for the information assessor. In order to work through this sequence, the assessor must connect or relate the elements of the sequence in the following way: the signal generation probability, $\phi(y|\sigma)$; the decision maker's decision rule, $\alpha(y, \sigma)$; the relation between the acts selected and the results that will occur, $\phi(r|y, \sigma, a)$; the gross payoff he will derive from the expected results, $\omega(r)$; and, finally, the cost, $\omega'(y, \sigma)$ of establishing and operating the alternative information structures from which he can choose.

Some Factors That Influence the Costs and Benefits of Cost-Benefit Analysis for the Cost-Benefit Analyst

THE MODEL IN THE PREVIOUS SECTION is helpful in identifying the general elements of the information evaluation decision problem. Nevertheless, unless greater specificity is introduced, it is incapable of indicating whether an information assessor should regard a structure as good or bad. In this section, we provide some qualitative criteria for this decision problem and point to some factors that might influence the relative costs and benefits of alternative information structures. We shall assume that the payoff functions (i.e., preferences) of the information assessor and the decision maker are very similar. The material in this section will be presented in the context of the generalized model of the previous section, but no effort will be made to establish the exact analytical derivation of the criteria and the factors which influence costs and benefits.

Consider the following example. Assume that the information assessor has to evaluate a water pollution control problem in which there are several perpetrators and sufferers. He is supposed to employ cost-benefit analysis to evaluate any number of control alternatives (low-flow augmentation, changing the perpetrator's processes, etc.) that are capable of changing the present level of pollution and are within the authority of his agency to prescribe or enforce. This evaluation process might employ a variety of the types of inputs commonly used in research projects—e.g., personnel with various types of training and accumulated intellectual capital, monitoring equipment, computer time, office, storage, and travel facilities, library access, and a host of other inputs.

In formulating his decision problem, the information assessor will rarely be able to account for all the variables that influence all the relevant acts he might consider. He will recognize that some acts are implied by variables whose influence he does not take into account. For example, if an effluent variable is assumed to be a constant proportion of a certain production input variable, it may not be necessary to measure both variables. As noted earlier, the fewer the variables the assessor considers, the less his expenditure on research and the less he knows about the possible results of alternative actions. This implies that some economically preferred acts may be neglected. Ultimately, the only basis the assessor can employ to decide whether to include or exclude a particular variable is whether or not the partition of prior results that occurs with inclusion of the variable would cause at least some of the signals to be different—and whether or not any difference in those signals would have an influence upon the responses of the decision maker.

Of course, the key acts of interest to the decision maker are those whose results can be influenced by his manipulation of key variables. The assessor must therefore attempt to predict the relation between the signals the decision maker receives and the acts he chooses—or, in our notation, $a = \alpha(y, \sigma)$.

In water pollution control, three types of activities (coarsening of partitions) are typically undertaken by information assessors: separate pollutants are transformed into a composite measure such as BOD or aggregated into a mass such as tons of waste load; dissimilar inputs are treated as similar; and enterprises (perpetrators and sufferers) are grouped by the types of salable outputs they produce. Each of these three classes of transformation and aggregation activities can, for any given assessor, cause plausibly efficient acts to be disregarded in the evaluation process.

For example, assume that the assessor evaluates the results of certain polluting acts in terms which represent aggregations of various pollutants. If the assessor makes the transformation and evaluates the results of the acts in terms of the transformed variables, he might then be faced with the task of evaluating the results of acts for specific enterprises, locations, and pollutants. This cannot be done without some comprehension of the specific pollutants different enterprises will release at various locations. However, in performing the original transformation and carrying out his initial evaluation in terms of the aggregations, the assessor has given up a good deal of information. In particular, knowledge about the relative quantities of the various pollutants that enter into the aggregations is destroyed. If, for example, the aggregates are simply weighted sums of the original pollutants, there are an infinite number of pollutants, locations, and enterprises consistent with the magnitude of any given aggregate.[9] The assessor is unlikely under these

---

[9] Assume, for example, that a disaggregated "control cost" function for water polluters is to be estimated. Let this function be given by

$$\text{(a)} \quad c_{ij} = a_i + b_i p_{ij} + u_{ij} \qquad \begin{array}{l} i = 1, \ldots, n \\ j = 1, \ldots, r \\ n \geq r \end{array}$$

where i refers to a particular pollutant, j to a particular location, a and b are coefficients to be estimated, and u is an error term having the customary properties. For simplicity, it is assumed that waste loads, p, is the only argument of control cost, c. The latter, of course, is expected to differ from one firm to another.

Average enterprise control costs and waste loads are clearly

(b) $c_j = \dfrac{\sum\limits_i c_{ij}}{n}$

(c) $P_j = \dfrac{\sum\limits_i p_{ij}}{n}$

With aggregation, the intercept and the error term are

(d) $a = \dfrac{\sum\limits_i a_i}{n}$

(e) $u = \dfrac{\sum\limits_i u_{ij}}{n}$

The aggregate relation is therefore

(f) $C_j = a + bP_j + u_j$

where b, the coefficient of P_j, is apparently

(g) $b = \dfrac{\sum\limits_i b_i}{n}$

In other words, the average control cost depends on the waste loads released by the n polluters. This perhaps seems reasonable enough since (f) continues to be linear and includes an error term the expected value of which is zero for all c_{ij}. Disregarding a and u_j, note, however, that both b and P_j are aggregated. Thus

(h) $bP_j = \left(\dfrac{\sum\limits_i b_i}{n}\right)\left(\dfrac{\sum\limits_i P_i}{n}\right) = \dfrac{\sum\limits_i b_i P_{ij}}{n}$

Therefore

(i) $b = \dfrac{\sum\limits_i b_i\, P_{ij}}{nP_j}$

Nothing goes awry if the cost functions are identical. However, if one permits these functions to differ somewhat among various collections of polluters—by placing all paper mills in one group and all sewage treatment plants in another group, for example—it is apparent from (h) that the value of the waste load parameter, b, will be a *weighted* mean of the same parameters for the individual polluters. In particular, those enterprises having high control costs will have a disproportionately strong influence upon a group's contribution to the value of the waste load parameter in (f). Similarly, those polluters having low control costs will have a disproportionately weak influence. The conclusion is the rather dismaying one that the *measure* of control costs, employing some group or aggregation of polluters as the fundamental unit of observation, can differ from one group to another. There could conceivably be as many unique measures employed as there are groups.

circumstances to identify the optimal combination of pollutants, locations, and enterprises unless the weights used for aggregating happily
coincide with the weights the assessor would actually adopt before "coarsening the partition." Given the low probability of this happy occurrence, assessors should recognize that the likelihood of neglecting to
evaluate potentially efficient acts probably increases with increases in the
extent to which unlikes are treated as likes or likes are treated as unlikes.

An awareness of the importance of interdependence to the above
point perhaps enhances its clarity. Interdependence implies the existence of substitutive or complementary relations. For example, consider
a lake containing landlocked salmon and togue. Assume the assessor's
payoff function has an argument measured in terms of the pounds of
game fish in the lake. Further assume game fish consist solely of salmon
and togue. If the assessor already knows there is an optimum weight of
togue, (and if he considers togue a close substitute for salmon) he need
not devote much attention to the weight of salmon. When the assessor
learns there are no togue, he is not well informed about the weight of
game fish, as long as he does not know about the weight of salmon.
Unless the assessor has reason to view salmon and togue as having little
or nothing in common, the two can readily be transformed into a measure of the weight of game fish. In short, choose the components of
aggregated and transformed variables (indices) such that the indices in
terms of the assessor's payoff function end up being strongly independent of each other.

Assessors must also recognize that if they evaluate the results of
alternative acts in terms of aggregates, the perpetrators can subvert the
assessor's predicted results if they know what the assessors are doing.
Assume, for example, that evaluations are stated as the costs and benefits of reducing BOD, and that it is feasible for a plant to modify its
operation so as to substitute other pollutants for BOD. If the cost of
controlling BOD is greater than the cost of modifying the operation and
emitting the other, perhaps more harmful pollutants, the perpetrator
will have every incentive to shift to these latter pollutants. This suggests
that assessors should be cautious in using aggregate variables for purposes of evaluation whenever the production processes of perpetrators
permit the production or the disposal of waste products in a variety of
forms and ways.

Another difficulty with evaluations employing aggregated variables
involves the assumption of constancy of quantities of pollutants, inputs,
and salable outputs across alternative acts. Where relative waste loads
and the output quantities of the relevant enterprises are expected to
remain fairly constant and where production processes remain more or
less unchanged, aggregate measures may be adequate for comparatively
crude cost-benefit analyses. That is, these aggregate measures may be

useful as first-line indicators of the desirability of investigating certain alternative acts in greater detail. The crude analysis can be employed to dismiss acts the results of which clearly generate negative net payoffs. If relative output quantities and production processes do not remain fairly stable across enterprises, however, the gross measures do not even provide a consistent basis for indicating the desirability of more detailed investigation.

The preceding points refer to what are commonly termed specification errors. That is, although the cost-benefit analysis may describe some partition of prior and expected results, it fails to consider adequately the relations fundamental to the choices the decision maker faces. Another example of this which is frequently found in cost-benefit analysis involves the simplifying assumption that sufferers and perpetrators are unable to influence the prices at which they sell their outputs or buy their inputs. In many cases, signals may be quite insensitive to this assumption; however, they are likely to be quite sensitive for an analysis of pollution control in, say, the electric utility industry or a factory in a remote one-employer town. An assessor who carries out a cost-benefit analysis based upon a model embodying this assumption can be greatly misled and can therefore convey misleading information to the decision maker.

The assessor may also mislead himself and the decision maker because of measurement error—the deviation of the result of any single measurement effort from the mean of the results of repeated applications of measurement effort under the least constrained conditions. This sort of error can be reduced by devoting more resources to constructing measurement devices and techniques, by allowing more time for measurements to be made, and by better training of measurement personnel.[10] But resources are expensive. Whether the expenditure is worth it depends on the assessor's valuation of the research payoff.

The cost-benefit analyst whose research resources are economically scarce should recognize that there may frequently be a trade-off between specification error and measurement error. That is, the assessor may opt for the theoretical delights of ever increasing generality in the specification of the models supporting his empirical analysis. His ultimate objective would be the ability to predict the results of every alternative act he or the decision maker deems worthy of attention without having to alter any of the relations expressed in the model. The generality and realism of the ideal model would be so great that there would

[10]The use of the term "error" implies the existence of a "true" measure or, at least, a most preferred measure. The idea of a true or ideal measure has caused great philosophical difficulty in science. See, for example, K. R. Popper, *The Logic of Scientific Discovery* (New York: Basic Books, 1969), chap. 1.

never be any doubt in the assessor's mind as to whether an observed change in a signal was random and thus transitory in nature or whether it was due to changes in the values of fundamental model parameters. However, the greater the progress of the assessor toward this intellectually captivating state, the greater are likely to be the number of variables for which he must make observations, collect and organize data, and establish parameter values. Furthermore, the complexity of relations among these model variables may be so great that estimating techniques are either extremely costly or perhaps even nonexistent. In effect, the elaboration and detail of the model may be so great relative to the availability of research resources that only superficial attempts can be made to ascertain the true value of any one parameter. The problem in this case is not with a model that involves dangerous simplification of reality but with a model which, given the resources available, is alarmingly complex. The model is insufficiently artificial. Just as one fails to capture the truth when he fails to comprehend the complete structure of a system, he also fails when he is unable to measure with some fair accuracy the parameters of any given comprehension of the structure.

Unfortunately, this writer is unable to offer any precise procedures for those faced with decisions about choosing combinations of specification error and measurement error when they undertake cost-benefit analyses. There are, nevertheless, some general principles that can be established from the literature of experimental science on the propagation of measurement errors in complex models.[11] Let u be a derived property related to the directly measured properties, $x_1, \ldots, x_n$, by $u = f(x_1, \ldots, x_n)$. Given that the x's are not independent of each other—they might, for example, be the parameters of a model for simultaneously managing air, water, and land uses in a region—the error in u due to the accumulation of errors in the separate estimates of the x's is given by

$$l_u^2 = \Sigma \; f_{x_i}^2 \; l_{x_i}^2 + \underset{i \neq j}{\Sigma \Sigma} \; f_{x_i} \; f_{x_j} \; l_{x_i} \; l_{x_j} \; r_{ij}$$

where l_u is the error in the estimate of u; f_{x_i} is the partial derivative of f with respect to x_i; l_{x_i} is the measurement error in x_i; and r_{ij} is the correlation between x_i and x_j. The presence of the correlation coefficient in the above expression makes apparent at least one thing to avoid in the construction and use of complex models in cost-benefit analysis: do not employ variables in the same model that are highly correlated with one another. It seems probable that the greater the number of attributes of

[11]See, for example, L. G. Parratt, *Probability and Experimental Errors in Science* (New York: John Wiley, 1961), pp. 109-18; and W. Alonso, "Predicting Best with Imperfect Data," *Journal of the American Institute of Planners* 34 (July 1968): 248-54. The material that follows is mostly taken from Alonso.

reality introduced into a model in the form of directly measured properties, the more likely are some pairs of these properties to be highly correlated. Relatively simple models, by definition, require fewer directly measured properties for their solutions.

Further inspection of the expression readily suggests two more rules. First, the presence of the partial derivatives, the f_{x_i} and f_{x_j}, indicates that measurement resources are more likely to be allocated efficiently if they are assigned to those directly measurable properties thought to have a significant influence upon the derived property. Since the variables that have a significant influence upon a derived property will frequently be the same in both complex and simple models, the use of the simple model is perhaps to be preferred if avoidance of substantial error in the estimate of the derived property is of high priority. For example, if one is interested in deriving the willingness to pay for a water-based recreational experience, it is well known that the permanent income of the individuals in the sample will be a very important factor. This will be true whether one takes account only of income and age or whether one includes in a demand structure every conceivable influence upon the willingness to pay.

Second, given the presence in the expression of the measurement errors associated with the directly measured properties, it pays to devote resources to reducing the larger of these measurement errors. Since in simple models there are fewer estimates of directly measured properties to be obtained, it follows that, to a greater extent than in a complex model, a given stock of measurement resources can be used to reduce the error associated with any one property. Thus, given the cumulative nature of measurement error in models where measured properties are tied together in long chains of reasoning, this rule along with the previous two implies that simple models are advantageous in applied work. The advantages exist apart from the fact that simple models are easy to use and, in spite of possibly gross specification errors in the signals they provide, that they will usually give quick answers to questions.

Some Factors That Influence the Costs and Benefits of Cost-Benefit Analysis for the Decision Maker

A WATER QUALITY CONTROL AGENCY consists of decision makers and staff personnel who, for any given stretch of a particular watercourse, are responsible for seeing that a predetermined level of water quality is maintained. As noted earlier, the effectiveness of an assessor, whose agency function is to evaluate alternative information structures, is determined by the correctness of his understanding of the (subjective or objective) model the decision maker employs to make choices. Decision makers who sponsor cost-benefit analyses expect findings and implied

recommendations that they can support. On occasion, information assessors and decision makers may have incompatible perceptions of the
utility of different acts and the probabilities of various results. Because
of these differences in perception, the decision maker must "communicate" his model to the assessor and cause the assessor to give heed to it.
The decision maker thus has the problem of evaluating alternative ways
of motivating the assessor so as to attempt to maximize the decision
maker's own effectiveness. When the decision maker delegates choice of
the information structure to an assessor, he must make prior and retrospective evaluations of the assessor's performance.

Since the assessor can be depended on to pick the information structure that accords with his own best interest, the decision maker's problem is to select an appropriate incentive system. This selection can be
viewed as a two-part process. First, the decision maker must specify a
performance measure and then he must specify a reward structure.
Here a simple, formal framework—essentially the one introduced earlier, but with some embellishments—will again be presented to establish
the essential elements of the decision problem. We will then discuss some
practical considerations.

Let $\beta(\cdot)$ denote a particular performance measurement method
dependent on the information structure, σ, chosen by the assessor; the
realized results, r, of the decision; the decision act, a, itself; and the
payoff, u_D, to the decision maker. For a cost-benefit assessor, a retrospective measure of performance might be the percentage discrepancy between the assessor's estimate of a cost-benefit ratio for a particular project
and the actual, realized costs and benefits of that project. Of course, the
measure of performance would probably not be this simple but would be
tempered by the decision maker to allow for changes in exogenous factors beyond any assessor's ability to predict, e.g., a change in agency
budget. In any event, denote such a performance measure by p. Thus

$$p = \beta(\sigma, r, a, u_D). \tag{11}$$

Since the measurement problems the decision maker faces in his evaluation of assessor performance are quite similar to those the assessor faces
in his evaluation of the costs and benefits of alternative investment programs, it will be assumed the decision maker's measurement error is
subsumed in r. Now let $\Omega(\cdot)$ denote a specific reward structure dependent on the performance measure, p, and the results, r. The performance measure itself may not be independent of current or prior states.
For example, one would expect the reward an assessor received to be
dependent on his performance record and the current demand for his
services by parties other than the decision maker. In addition, one would

expect the choice of an information structure to be a function, ∂, of the incentive system represented by (β, Ω). That is,

$$\sigma = \partial(\beta, \Omega). \tag{12}$$

The results of any decision act by the decision maker are now a function, Ψ, of prior states, the act itself, the assessor's incentive system, and the information structure selected by the assessor. That is,

$$r = \Psi(x, a, \beta, \Omega, \sigma). \tag{13}$$

The monetary equivalent of the assessor's reward, m_A, is thus given by

$$m_A = \Omega(p, r) = \Omega[\beta(\sigma, r, a, u_D), r]. \tag{14}$$

Of course, the monetary equivalent of the payoff to the decision maker and his constituency, m_D, is reduced by at least the amount, m_A, paid to the assessor. Therefore $m_D = m - m_A$, where m is the gross monetary equivalent.

For a specific (β, Ω) pair, the assessor's decision problem is then to employ an information structure, σ^*, for which

$$E(u_A|\beta, \Omega, \sigma^*) \equiv \max_\sigma \sum_y \left\{ \max_{a\epsilon A} \sum_{r\epsilon R} \Omega\,[\beta(\cdot), r]\ \phi_A(r|y,\sigma) \right\}$$
$$\times\phi_A(y|\sigma). \tag{15}$$

The optimal information structure the assessor chooses generates a set of signals for the decision maker, and the decision maker then rewards the assessor in accordance with the expression in the braces.

To summarize, the assessor selects σ, given $\beta(\cdot)$ and $\Omega(\cdot)$ on the basis of his payoff and his perception of the decision maker's model. The decision maker generally will not be able to influence the form of u_A except insofar as the assessor identifies himself with the decision maker. However, if, prior to the assessor's evaluation, the decision maker explains his decision model to the assessor and lets him know what is expected of him, $\phi_A(r|y,\sigma)$ is certainly likely to be influenced.

Any incentive system the decision maker provides for his assessor will contain a performance measure and a reward structure. We shall not, however, discuss the performance measure in depth since, in choosing it, the decision maker faces the identical set of general problems the assessor faced in the choice of an information structure, and the same type of specification and measurement simplifications the assessor used for information evaluation may also be used—at a cost—in assessing the assessor's performance. Instead, we will focus on the reward structure.

As (15) implies, and as we have noted, the assessor's effectiveness in maximizing the decision maker's payoff depends in part on how well he understands the latter's decision model, which, in turn, depends on how well the decision maker has explained it. But his explanation may be far from clear because he cannot predict the accuracy of his model with sufficient precision before he examines the information the assessor provides him. This means that frequent revisions of the model may be required as the results of various decisions come in, especially in the initial period and especially if actual results are markedly different from those anticipated.[12] Revisions of the model provide other advantages beyond increasing its accuracy. They provide a tangible, ongoing control over the assessor's performance, independent of subjectivity in the decision maker's appraisals and of possible self-serving interpretations advanced by the assessor. Moreover, the assessor's knowledge—at the initial stages of the investigation—that his performance will be subject to repeated evaluation before the investigation is completed will restrain his possible impulses toward opportunistic behavior. On the more positive side, with a decision model that is not prescribed once and for all at the beginning of the investigation but, rather, remains open to modification and revision, the assessor realizes that the model can be improved by what he does and says—and this adds a new dimension to his performance incentive and to the rewards which that entails.

The likelihood of periodic review and revision of the model is enhanced somewhat if the assessor must use resources drawn from within the decision maker's organization. These resources are then likely to be involved in the analysis on a continuing or, at least, intermittent basis. Given that there is a real possibility of substantial discrepancies in assessor and decision maker payoff functions, having personnel who are not dependent on the assessor for their advancement and livelihood participate in the cost-benefit analysis can serve as a check on the assessor's choice of information structures. An organizational structure of this sort can be particularly useful if ex ante evaluation of the assessor's performance is difficult and costly.

Devoting to cost-benefit analysis the efforts of organizational personnel answerable to someone other than the assessor provides for a continuing, independent audit of the assessor's performance. Independent audits can also be achieved by using experts from outside the organization. However, the use of referees of this sort generally means that auditing of assessor performance occurs only ex post rather than ex ante or continuously. If assessor performance is discovered to be unsatisfactory only after completion of the cost-benefit analysis, resources will have been expended and perhaps decisions made, yielding little or

---

[12]For some operationally adaptive, sequential decision models in a cost-benefit context, see A. H. Packer, *Models of Economic Systems* (Cambridge, Mass.: MIT Press, 1972).

no return. With continuing audits and adaptations, unproductive lines of endeavor would be observed and terminated earlier. Moreover, independent auditing of the assessor's performance by involving personnel from within the agency in the analysis would probably cost no more (and might well cost less) than ex post auditing which has left the assessor with complete discretion in the selection of evaluation resources for the duration of the analysis. The choice, of course, ultimately depends on the assessor's payoff function and the degree of expertise in cost-benefit analysis present within the agency.

From the decison maker's point of view, the value of using agency personnel for cost-benefit analysis may be great, although the value from a societal perspective is likely to be somewhat less. The reason is, simply, that the agency is given the responsibility not only to evaluate the costs and benefits of alternative decisions but also to make the choice and implement it. Frequently, the agency's mandate limits the range of alternatives it is able to consider. For example, the water pollution control activities of the U.S. Army Corps of Engineers are limited to structural alternatives by both legislative mandate and agency expertise, and therefore one would not expect the Corps to do careful cost-benefit analyses of such nonstructural alterntives as re-oxygenation or the adoption of effluent charges. Given its legislative and historical mandate, the payoff would be negative.

If the cost-benefit assessor's payoff function is independent of the agency responsible for producing water quality, the assessor is better able to consider objectively all technologically feasible alternatives. Preventing cost-benefit assessors from being influenced by the benefits that the agencies derive from their own production decisions reduces the likelihood that cost-benefit analysis will be subverted to serve the particular, and perhaps narrow, interests of the agency.

The Role of Cost-Benefit Analysis in Policy Making

HENRY ROWEN

The Objectives of
Cost-Benefit Analysis

Two QUITE DIFFERENT POINTS OF VIEW towards cost-benefit analysis exist. One, which dominates most of the scholarly literature on the subject, sees it as a method for making choices, a *decision rule* which enables one to select the most *efficient* of several alternative courses of action. The other, less often discussed, sees cost-benefit analysis as a member of a class of analyses which contribute to *decision processes* by assisting in the formulation of objectives and of alternative actions as well as contributing to the process of choice between them. Although these two perspectives are not inconsistent, they are quite different. The first assumes that values and objectives exist in the minds of decision makers, or at least that they can readily be formulated when the need arises. The other makes no such assumption. It is based on the premise that objectives and alternatives must, often with great difficulty, be discovered, designed, and invented. It is the principal argument of this paper that cost-benefit analysis is very much more important in this latter function than in the former and that there are good reasons why this should be so.

The argument in summary follows.[1] First, over a wide range of policy areas and types of decisions, people, including those responsible for making public policy decisions, do not have preferences among broad goals nor, in general, among policy objectives, especially when those objectives overlap differing policy areas. Moreover, the policy issues involved here usually concern public goods—goods not sold on the market. The value placed on these goods by members of the community may be largely unknown to the policy maker. Indeed, this value may also be largely unknown to the members of the community since they have

---

[1] This formulation draws on an essay by James G. March, "The Technology of Foolishness." For a similar view of the nature of preference formulation and the role analysis plays in this process see Albert Wohlstetter, "Analysis and Design of Conflict Systems," in *Analysis for Military Decision,* edited by E. S. Quade (Skokie, Ill.: Rand McNally, 1964). This essay was based on lectures given at The RAND Corporation in 1954-55.

few occasions to reflect on what these goods are worth or find out what they cost. In short, for choices that are normally called "policy decisions," ordered preferences will usually be confined to a small part of the relevant policy universe.

Second, there will usually be important gaps in knowledge about the alternative means of achieving any given objective, and also in knowledge about the several objectives associated with any given course of action.

Third, the issue, or problem, and the context in which policy decisions are made will often be unclear. The occasion for believing that some kind of action is required may be the emergence of a symptom (an unexpected increase in the wholesale price index), an event (a riot in New York City), a new technological possibility (a report that supersonic transportation is technically feasible), or a proposal (for building a dam on the Colorado River). The event which prompts the consideration of action often will not in itself reveal enough information to define the issue to be decided. At best, it will trigger attention and generate a search for information about the problem, alternative courses of action, and the objectives to be sought.

Fourth, policy decisions are generally not made by individuals, nor by people acting in committee. They are usually shaped by the interaction over time and space of individuals with different attitudes, skills, information, and influence. Agreement on the consequences of choices and on values is not needed for action and therefore normally does not exist. All that is essential is agreement on the next step to be taken.

To a decision maker in this condition, a method which purports to provide a rule for making decisions is of limited utility. Whatever objectives are embedded in the analysis, only by chance are they likely to be his (whatever his are), and he himself is not likely to know them initially. If it is held that cost-benefit analysis is useful primarily in those situations where the decision maker does have preferences among objectives and where alternatives can be described and outcomes can be estimated, then the decision making procedure will seem relatively familiar and routine. Under such restricting circumstances, the decision maker is apt to conclude that cost-benefit analysis does not deal with policy problems—that such issues are dominated by bureaucratic and political considerations which are not amenable to the methods of cost-benefit analysis.

In general policy making, a decision maker who is unclear about both objectives and alternatives and who faces a vaguely-defined problem must develop or invent his preferences and alternatives for meeting them. Many methods are used in this process. A decision maker might turn to approaches that have worked in the past for himself or for others in similar situations; or define away the problem by declaring that it falls within existing policy; or use intuition; or call on expert advice; or experiment with data in different ways and try out different objectives and alternatives; or defer decision in hope that the situation will clarify or the problem disappear.

How can analysis, which deals systematically with costs and benefits, help in this evaluation process? It can provide a conceptual structure and a set of techniques for relating means to ends, for arraying the various costs associated with each course of action, and for describing, comparing, and assessing possible outcomes. It may be able to express something about the values people attach to various outputs. It can be used to test the sensitivity of alternatives to different contingencies. If done competently, cost-benefit analysis is very useful in constructing and assembling values, facts, and relationships. It should not be assumed that facts and values are simply "discovered." There are potentially an infinite number of facts and values; they are not equally useful in all given situations, and part of the task of cost-benefit analysis is to find the facts, values, and relationships that appear to be important.

Practical men sometimes complain that cost-benefit analysis is too complicated to be useful, that analysts are more interested in exercising their analytical skills or adding to the sum of human knowledge than in helping to solve policy problems. More fundamental criticisms of policy analysis in general, largely from scholars, are that: (1) analysts concentrate on tangible, quantifiable factors and ignore or depreciate the importance of those which are intangible and unquantifiable; (2) analysts altogether ignore certain "fragile" values, such as the ecological or the aesthetic; (3) they focus on results and, in a search for common measures, ignore both the processes by which preferences and decisions are formed and the significance of qualitative differences in the results; (4) they tend to operate within the interests and values of their clients; (5) they, in an effort to be objective, employ neutral and detached language in dealing with intensely moral issues; (6) they make an artificial separation between facts and values; and (7) analysts emphasize efficiency objectives while distributional objectives are largely overlooked.[2]

The type of analysis recommended by Laurence Tribe, one of the most trenchant critics, would point "in the general direction of a far subtler, more holistic and more complex style of problem solving, probably involving several iterations between problem-formulation and problem-solution and relying at each stage on the careful articulation of a wide range of interrelated values and constraints through the development of several distinct 'perspectives' on a given problem, each couched in terms true to its internal structure rather than translated into

[2]See Laurence Tribe, "Policy Science: Analysis or Ideology?" *Philosophy and Public Affairs* (October 1972).

some 'common currency.' "[3] The best analysis being done may fit this description, but the average analysis does not come near it.

The critics, whether scholarly or "practical," are on strong ground in noting that analysis, including cost-benefit analysis, tends to be of limited use as a decision rule. Conventional cost-benefit analysis is a specialized technique of restricted applicability and comparatively little utility for policy making. On the other hand, in a broader sense, the process of analyzing costs and benefits in a variety of different ways is an approach of great utility. In this form it comes closer to the "subtler, more holistic" style of analysis advocated above. In order to get a clear understanding of the strengths and limitations of analysis it is necessary to be more explicit about the various functions it performs.

Problems in Applying Cost-Benefit Analysis

FIRST, A DESCRIPTION OF SEVERAL TYPES OF ANALYSES. *Cost-benefit analysis* has been used in two senses. Narrowly, it is a method of aggregating all costs and all benefits associated with a given project, program, or decision in monetary terms, converting them to present value, and combining them in a single index, such as the present value of net benefits. In the broader sense, cost-benefit analysis is an activity which investigates the costs and the benefits that are associated with a project, program, or decision. It may also deal with the distribution of costs and benefits and the sensitivity of results to different contingencies. In operation it is coupled—whether closely or loosely—to the policy making process. It does not seek to produce a single measure of merit; in fact, in the spirit of cost-benefit analysis broadly defined, single indices are counterproductive.*

Closely related is cost-effectiveness analysis, which expresses the outputs of activities in nonmonetary, rather than monetary, terms and describes inputs in both monetary and nonmonetary terms. Operations research is similar to cost-effectiveness analysis but it usually is limited to current operational problems. Systems engineering may also be regarded as a kind of cost-effectiveness analysis with an orientation to engineering projects. There are, of course, many other types of analysis that are applied to policy problems, such as bureaucratic analysis, psychological analysis, and implementation analysis.

---

[3] Ibid.

*In the "Introduction and Overview" the editors have used the term "policy evaluation" in referring to cost-benefit analysis in Rowen's "broader sense." [Editor's Note.]

Problems in Analyzing Inputs and Outputs

The concept of input (or cost) and output (or benefit) provides an extremely useful framework for organizing pertinent facts and relationships in dealing with policy problems. Indeed, it is difficult to conceive of a policy choice for which it is not relevant. In the following discussion, we will deal with various problems associated with inputs and outputs, or both. First, the costs. Work on costs associated with various alternatives can sometimes go far to help structure a problem: What tangible costs are associated with actions regarding personnel, material, capital expenditure? What other kinds of costs are involved—spillover costs, goodwill lost, community disruption, hostility of opponents? In what time periods will these costs fall? How certain or uncertain are they? Who will pay them? Even partial answers often enable decision makers to reduce the number of alternatives to be considered further.

However, given the importance of such questions, from what source is the decision maker to get the answers? Although the information produced by accounting and management information systems may be useful, in most cases it is not nearly enough to deal with policy problems. For example, if a highway project cuts a neighborhood in half, what is the social cost of the disruption? What alternative type of compensation—and in what amounts—would leave people no worse off? Polling a community does not insure unbiased results. Observation of demographic changes over time in the community may suggest something about community stability and homogeneity, although inferences on social costs will still be tenuous. But while the information he has to work with is imperfect, if the analyst is careful in defining his objectives, is imaginative in his use of data and techniques, and is sensitive to the limitations imposed on his results, he can probably bound the limits of the social costs in a way that will provide a better basis for decision making than can be produced by unaided intuition.

As the foregoing suggests, estimating the various kinds of costs is arduous, and in the end many questions will probably remain wholly or partially unanswered. Still, a good cost analysis capacity is essential for competent decision making, but it should not be narrowly focused. Many costs, including many of those which are most important, are not computable in financial terms or are not even quantifiable. But they are real costs and may be suspectible to logical analysis.

A second important analytical problem is specifying the *relationship* between inputs of resources and outputs. These relationships are, in economic terminology, production functions, which define the alternative courses of action available. There are two principal sources of production function data: one is the current operation of the decision maker's agency (or of similar agencies); the other is advice from experts who have relevant skills. Both sources have limitations. This is especially true for problems which are novel and which involve more than incremental changes from existing operations. The degree of information available about the production function also varies greatly by policy area. It is relatively high for activities which have a stable technology and measurable outputs (as with water projects or garbage collection) and low for those activities which have a changing technology and hard-to-measure outputs (as with auto air pollution, education, or health care). Although knowledge on production functions may exist somewhere, it may not be accessible to the decision maker with the problem. It costs to search for and use the information. Moreover, the incentive to use the available information may be weak. Bureaucratic preferences may exclude some important alternatives (e.g., water resources agencies prefer alternatives that involve construction rather than alternatives that use the price mechanism). Disciplinary specialization among experts often limits the usefulness of their contribution. The division of responsibility between different agencies limits a given decision maker's area of authority and, as a result, probably also his perspective and range of interests. Many decision makers have found that in their decision making processes they need the work of analysts trained and motivated to search out existing production functions as well as to design and invent new ones.

A third major problem facing decision makers is defining outputs. For example, what should be regarded as the output of a program to reduce auto air pollution: (1) the reduction in grams per mile of specific pollutants per automobile; (2) the reduction in total weight of emissions of these pollutants into the atmosphere from all of the autos in a region; (3) the concentration of various pollutants in the atmosphere over defined regions—perhaps expressed as a frequency distribution over time; or (4) various measures of the effects of the pollutants, such as effects on health or on the clarity of the air? These are increasingly higher levels of objectives, but the higher the level the more tenuous the relationship between inputs and outputs, i.e., between actions and their intended effects. In other words, the production function cannot be defined with precision unless the output measure employed is a lower level one. But outputs so defined may bear only a remote relationship to perceived needs. (What is the relationship between health and the grams of pollutants emitted per mile?) This problem of defining the relevant outputs is inherent in the production of such public goods as defense and police protection, and such quasi-marketed goods as health care and education. What is the relationship between a specific missile program and "security" or between school test scores and "education"? A student who was helping to assure compliance with federal civil rights legislation recently highlighted this problem with the comment, "We get a lot of contract compliance; of course we don't get many jobs for blacks, but we sure do get contract compliance."

One way to deal with this problem is to avoid single measures of output, since there is no *right* measure. Rather, there are several measures, each valid for some purpose and none of them valid for all purposes. For example, trying to collapse output measures into a dollar measure of benefits usually obscures more than it reveals. Determining which measures of output should be used for a given decision consequently requires considerable analysis.

The fourth major problem is deciding what outputs should be produced. Welfare economics often defines such high-level objectives as efficiency, equity, national security, and economic stability. But this list can easily be extended to include justice or cultural diversity or ecological values. Although these objectives are often abstractly expressed, they can directly affect decisions on specific issues in ways that are quite complex. Consider the concept of equity. If a policy decision will affect two individuals differently, how should the difference be regarded? Is equity satisfied if: (1) the absolute position of both parties is improved; (2) the absolute position of both is improved but the position of the one who was better off initially is improved more; (3) both are better off but the one who was initially worse off benefits more; (4) the better off person's position is worsened absolutely and relatively, while the worse off person's position is bettered absolutely and relatively; (5) both are worse off absolutely but the difference between them is narrowed? Further, should equity in the allocation of resources to neighborhoods be measured in outputs (e.g., cleanliness of streets) or inputs (e.g., street cleaning efforts). Are we concerned only that a certain level of some good should be attained, or are we also concerned with the *rate* of change? Should equity issues be dealt with in each policy decision, or should they be dealt with through a separate income distribution policy? (The usual procedure in cost-benefit analyses, of course, is to neglect distributional consequences and limit the analysis to a measure of aggregate efficiency—one reason that policy makers show little enthusiasm for such analyses.)

General Limitations of Analysis

Cost-benefit analysis in its conventional form has very limited utility. There are two principal reasons. First, as we noted earlier, there is the difficulty—in some cases, the impossibility—of attaching money values to such outputs as scientific results, large areas of national security, or basic education. In these areas, valuation problems are so severe that cost-benefit analyses are rarely done. Even in more tractable areas, like housing, transportation, and pollution control, each of which involves important intangibles, it may not be worth the allocation of an analyst's time to produce (often with great ingenuity) a conventional cost-benefit analysis.

The second reason is the limited applicability of the criterion of Pareto-optimality which is at the heart of conventional cost-benefit analysis. In specific cases there usually is no way to identify all of the gainers and losers, and the information costs of attempting such identification are often high. Moreover, the mechanisms for compensating losers are weak or nonexistent. Some values (like justice) are without price. They are considered absolute rights (even though, under certain circumstances, some people are willing to trade them off). This is why condemnation of property even with compensation at market values is widely resented; this exercise of the power of the state conflicts with a widely held right to possession of property. Decision makers therefore give great weight to distributional effects, believing that equity and other objectives of society will be better revealed and satisfied if affected interests act on their own behalf, that is, through a political process. (This may be regarded as a means of achieving some approximation to Pareto-optimality on the grounds that potential gainers and losers will advance their respective interests; this, however, assumes that citizens have sufficient information about potential courses of action, and about the distribution of access to political influence, to understand where their best interests lie.) An analysis which omits distributional effects and discusses only aggregate efficiency deals with a part of the decision maker's problem, and perhaps only a small part. An analysis which does deal directly and competently with distributional effects is much more valuable, although its chief value is apt to be the information contained within the body of analysis, the costs collected, the production function relationships determined, the outputs defined, and the various objectives analyzed, rather than the final "answer." The measure of net benefits is apt to be one of the less interesting quantities included.

Given these difficulties, it is not surprising that cost-effectiveness analysis is much more common than cost-benefit analysis. What is apparently lost in the generality of results is more than compensated for in added information. But many of the problems inherent in cost-benefit analysis when it is used as a decision rule also apply to cost-effectiveness analysis. The alternative which is most cost-effective is not often unambiguously best. With multiple measures of output and of cost, as well as a broad range of contingencies, a dominant alternative is rarely produced. Dominant alternatives are made, not born; they are usually crafted by designers who have a deep understanding of the relevant production functions, have thought hard about objectives and measures of effectiveness, and are able to shape and modify alternatives until one or more emerge as winners. And even if the winner is not uniformly dominant

—since there almost always is *some* contingency for which the preferred alternative is not best—it will be preferred under most criteria and for most contingencies.

Finally, some think it wrong to define cost-benefit analysis as broadly as we have done here, and hold that cost-benefit analysis should be limited to maximizing economic welfare as conventionally defined.* No doubt, it is sometimes useful for a conventional cost-benefit analysis to be undertaken from a detached, more or less neutral, standpoint. There is much to be said for having independent observers, free of the usual bureaucratic biases (but not free of biases of other sorts, for that is impossible) to analyze decisions in aggregate efficiency terms. But there is a price to be paid for this narrowing of focus, since it usually makes the results of much less interest to policy makers. The practitioners of this narrower type of cost-benefit analysis should therefore not expect to contribute much to the policy making process.

Conclusion

THIS PAPER HAS SOUGHT TO DO TWO THINGS. One has been to discuss the limited utility of cost-benefit analysis when it is employed as a decision rule. The other has been to describe the important function of cost-benefit analysis in constructing and assessing values, facts, and alternatives and to highlight the neglected role of design and invention. This discussion has dealt only partially with the criticisms cited and the plea for a broader, subtler style of problem solving. We are all, to a degree, prisoners of our theories and techniques. Much work is required before we can reason adequately in the manner appropriate to our needs. But the scope of policy analysis has expanded greatly in recent decades and there is no evidence that this process is slowing down. So there is little question that we will able to do better, even if much will remain beyond the limits of analysis for the indefinite future.

DISCUSSION

DISCUSSANT: *A. Myrick Freeman III*

How one views the question whether cost-benefit analysis is a decision rule or a part of the decision process, depends on whether you are a policy maker, a staff support person, or a professional who is working for policy makers. It also depends on whether you are out-

---

*There will, of course, be objections to equating the concept of maximizing efficiency—the principal objective of cost-benefit analysis—with the concept of maximizing economic welfare. [Editor's Note.]

side of the system and are trying to understand it. If you are, and if you start with the hypothesis that cost-benefit analysis is a decision rule, you are not going to get very far in understanding the real world. On the other hand, if you are an analyst, as distinguished from policy maker, I think it is valid to view cost-benefit analysis as a decision rule. That is, if you accept the postulates, assumptions, the criteria of cost-benefit analysis—maximizing economic welfare, economic efficiency benefits, and so on—then you must consider cost-benefit analysis as a decision making rule.

I make this statement in an attempt to prevent the contamination of cost-benefit analysis by the introduction of spurious influences that lie outside of the logical postulates on which cost-benefit analysis is based. Eliminating such influences is the job of the decision makers to do in their collectivity. The professional analyst must adhere to his strict set of postulates and interpret them as his decision rules.